MAY 0 1 2013

CALIFORNIA NATURAL HISTORY GUIDES

FIELD GUIDE TO AMPHIBIANS AND REPTILES OF CALIFORNIA

NAPA COUNTY LIBRARY
580 COOMBS STREET
NAPA, CA 94559

California Natural History Guides

Phyllis M. Faber and Bruce M. Pavlik, General Editors

FIELD GUIDE to AMPHIBIANS and REPTILES of CALIFORNIA

REVISED EDITION

Robert C. Stebbins and
Samuel M. McGinnis

UNIVERSITY OF CALIFORNIA PRESS
Berkeley   Los Angeles   London

To Anna-rose Stebbins and Molly McGinnis. Herpetologists' wives are very special people. They continuously put up with dead specimens in the freezer, live salamanders in refrigerator containers, escaped lizards in their kitchens, snakes in living room terrariums, tortoises on the sundeck, sea turtles in the backyard pool, and "vacations" that often consist of camp-outs at hot, desert sites. Within this exclusive grouping, Anna-rose and Molly are two of the very best!

University of California Press, one of the most distinguished university presses in the United States, enriches lives around the world by advancing scholarship in the humanities, social sciences, and natural sciences. Its activities are supported by the UC Press Foundation and by philanthropic contributions from individuals and institutions. For more information, visit www.ucpress.edu.

California Natural History Guides

University of California Press
Berkeley and Los Angeles, California

University of California Press, Ltd.
London, England

© 2012 by The Regents of the University of California

Library of Congress Cataloging-in-Publication Data

Stebbins, Robert C. (Robert Cyril), 1915– .
  Field guide to amphibians and reptiles of California / Robert C. Stebbins and Samuel M. McGinnis.—Rev. ed.
      p. cm.—(California natural history guides)
  Rev. ed. of: Amphibians and reptiles of California / by Robert C. Stebbins. 1972.
  Includes bibliographical references and index.
  ISBN 978-0-520-24466-5 (cloth : alk. paper)—ISBN 978-0-520-27051-0 (pbk. : alk. paper)
  1. Amphibians—California. 2. Reptiles—California. 3. Amphibians—California—Identification. 4. Reptiles—California—Identification. I. McGinnis, Samuel M. II. Stebbins, Robert C. (Robert Cyril), 1915– . Amphibians and reptiles of California. III. Title.
  QL653.C2S67 2011
  597.8′1709794—dc23   2011020261

Manufactured in Malaysia
16 15 14 13 12
10 9 8 7 6 5 4 3 2 1

The paper used in this publication meets the minimum requirements of ANSI/NISO Z39.48-1992 (R 1997) (*Permanence of Paper*).

Cover illustration: Sidewinder (*Crotalus cerastes*). Painting by Robert C. Stebbins.

The publisher gratefully acknowledges the generous contributions
to this book provided by

the Gordon and Betty Moore Fund
in Environmental Studies
and
the General Endowment Fund of the University of California
Press Foundation.

CONTENTS

Preface: Natural History and This Book	ix
Acknowledgments	xiii

INTRODUCTION 1

Amphibians and Reptiles: The Vertebrate Terrestrial Pioneers	**2**
Amphibian and Reptile Distribution throughout California's Habitat Complex	**5**
Ongoing Loss of Habitats and Numbers of California's Amphibians and Reptiles	**23**
Protective Status of Some Amphibians and Reptile Species	**29**
Common and Latin Name Changes	**31**
New Species, Genera, and Families	**31**
Field Identification Keys	**32**

AMPHIBIANS 35

Taxonomy, Anatomy, Physiology, and Behavior of Amphibians	**36**
Salamanders (Order Caudata)	**67**
Key to the Families and Genera of Salamanders	68
Mole Salamanders (Family Ambystomatidae)	69
Giant Salamanders (Family Dicamptodontidae)	84
Torrent or Seep Salamanders (Family Rhyacotritonidae)	89
Newts (Family Salamandridae)	92
Lungless Salamanders (Family Plethodontidae)	104
Frogs and Toads (Order Anura)	**149**
Key to the Families and Genera of Frogs and Toads	150
Tailed Frogs (Family Ascaphidae)	150
Spadefoots (Family Pelobatidae)	154
True Toads (Family Bufonidae)	161
Chorus Frogs (Family Hylidae)	179
True Frogs (Family Ranidae)	186
Tongueless Frogs (Family Pipidae)	212

REPTILES — 217

Taxonomy, Anatomy, Physiology, and Behavior of Reptiles — 218

Lizards and Snakes (Order Squamata) — 233

Lizards (Suborder Sauria) — 235

Key to the Families and Genera of Lizards — 236
Geckos (Families Eublepharidae and Gekkonidae) — 245
Eyelid Geckos (Family Eublepharidae) — 247
Geckos (Family Gekkonidae) — 251
Iguanids (Family Iguanidae) — 253
Collared and Leopard Lizards (Family Crotaphytidae) — 260
Family Phrynosomatidae — 270
Night Lizards (Family Xantusiidae) — 307
Skinks (Family Scincidae) — 313
Teiids (Family Teiidae) — 318
Alligator Lizards and Relatives (Family Anguidae) — 325
North American Legless Lizards (Family Anniellidae) — 331
Venomous Lizards (Family Helodermatidae) — 334

Snakes (Suborder Serpentes [Ophidia]) — 339

Key to the Families and Genera of Snakes — 340
Blind Snakes (Family Leptotyphlopidae) — 350
Boas and Pythons (Family Boidae) — 352
Colubrids (Family Colubridae) — 355
Sea Snakes (Family Hydrophiidae [Elapidae]) — 400
Vipers (Family Viperidae) — 401

Tortoises and Turtles (Order Testudines) — 423

Key to the Species of Tortoises and Turtles — 423
Land Tortoises (Family Testudinidae) — 426
Pond Turtles (Family Emydidae) — 430
Softshell Turtles (Family Trionychidae) — 439
Snapping Turtles (Family Chelydridae) — 442
Sea Turtles (Families Cheloniidae and Dermochelyidae) — 445
Sea Turtles (Family Cheloniidae) — 448
Leatherback Sea Turtle (Family Dermochelyidae) — 454

Observing and Photographing Amphibians and Reptiles — 457
Capturing Amphibians and Reptiles — 469
Amphibian and Reptile Husbandry — 481
Checklist of California Amphibian and Reptile Species — 485
Abbreviations — 493
Glossary — 495
Selected References — 501
Index — 505

PREFACE
Natural History and This Book

This is one of many new books in the University of California Press Natural History Guide series. It was written by two emeritus professors who taught courses annually in vertebrate natural history at our respective universities over a combined period of more than 80 years. One observation that we made during this tenure is that many people are not sure just what natural history really entails. A large part of this confusion may arise because "natural history" is an old-time term, so we felt that perhaps a couple of "old-timers" might be qualified to shed some light on this matter and, in doing so, introduce the theme and composition of this book as well.

The one factor most responsible for confusion here may well be the second word of this discipline: "history." When people think of history, it is nearly always in connection with an account of past human events, occurrences that have taken place and will never be repeated except in books, films, or reenactments. However, this revision and other natural history books like it contain accounts of the lives of animals in their natural habitats that, although based on previous observations, are still happening today and will continue tomorrow. Readers of these books therefore have a chance to witness firsthand enactments, not reenactments, of such natural events. In other words, this is "living" not "dead" animal history.

Unfortunately, most dictionaries do not point out this all-important fact. Webster, for instance, defines natural history as the natural development of something over a period of time. Another definition seen in numerous dictionaries states that natural history is the description and classification of objects in nature. This view most likely stems from the perception that natural history began with the pioneer taxonomic works of the Swedish physician Carl von Linné, known to us by the self-assigned Latin name *Linnaeus*. According to this latter definition, he and a host of other early taxonomists were the first "natural historians." Names like Wallace, Darwin, Audubon, and even Lewis and Clark are forever linked with the pioneer collections of animals and plants from areas formerly unexplored by Europeans and immigrants to the New World. Their work launched the great period of specimen collecting and species designations.

These massive early collections soon gave rise to a lineage of natural history museums, repositories where preserved specimens could be further measured, described, and finally assigned a place in a case or storage box. In one sense this is indeed the type of natural history to which a literal definition of the term refers, because all of the actual specimens are no longer alive. But is this what our book and several others in the Natural History Guide series address? We believe not.

For a very long time there has been another kind of natural history study in which the naming and categorizing of a species is only a first step. It also has not been the unique discipline of Europeans and their descendants but instead was, and in some areas still is, the practical science of all native peoples, past and present. Native Americans, regardless of geographic location or tribal affiliation, had to be very familiar with every aspect of the lives of the animals and plants upon which their very existence depended. For these people, natural history was an everyday practical application of facts, acquired from tribal elders and their own firsthand observations, to the never-ending task of obtaining plant and animal species for food, clothing, and shelter. Each group had its own set of animal and plant names, as valid as the Latin genus and species names that we use, but it was their knowledge of the lives of these organisms that was the key to their existence.

In the academic world a major shift from "museum" to "field" natural history began in the 1930s and 1940s. It was during this period that Aldo Leopold began recording the activities of wild vertebrates on his farm in Sauk County, Wisconsin; these records were to become the foundation for his classic book, *A Sand County Almanac*. At this same time the Dutch behaviorist Nikolaas "Niko" Tinbergen began emphasizing the observations of animals in their natural habitat instead of in laboratory cages, a theme that culminated in his book, *The Study of Instinct*, and inspired the wealth of field behavioral studies with which we are so familiar today.

The old and new approach to natural history is readily seen in two long-standing publications that use this term as their name. The *Journal of Natural History*, founded in 1841, is devoted primarily to the description of new species, animal and plant systematics, and the revisions of genera. In contrast, *Natural History* magazine presents accounts of various aspects of the lives of organisms written in a popular but still scientifically sound style. It is this latter approach that we have followed in writing this book.

One further area of confusion concerning the term natural history lies in the many secondary disciplines that have evolved from this original field of study. The term "ecology," classically defined as the relationship of an organism to its environment, is often the first to come to mind. Ecological studies that closely follow this definition are indeed synonymous with fieldwork in natural history. However, modern ecology texts are filled with formulas and statistical models, hopefully but not always based on firsthand natural history observations, which attempt to explain such phenomena in a highly abstract form.

Ethology, the study of animal behavior, also has its roots in firsthand field observations of animals in their natural habitats but often relies on laboratory or controlled-enclosure studies. Population biology, conservation biology, environmental biology, and physiological ecology are additional parts of the discipline of natural history. However, it is from the initial study of organisms in their natural habitats that all valid questions within these areas arise, and only there can they ultimately be fully answered.

With all of the preceding in mind, we have organized this book along the lines of the authors' respective courses in vertebrate natural history at the University of California, Berkeley, and California State University, Hayward. These offerings contained the three standard zoology course segments: laboratory, lecture, and field trips. The laboratory sessions were to a large extent a mirror image of the "old-style" natural history, where students learned to identify both preserved specimens as well as small live species that could be successfully maintained in terrariums. Lectures, usually accompanied by slide presentations of wild animals in their natural habitats, were devoted to reviewing some of the highlights of the lives of California vertebrate species in preparation for the weekly full-morning and, occasionally, full-weekend field trip. These firsthand field experiences were the most important course segment, for here participants were required to make and record in field notes their own observations of segments of the lives of wild vertebrate species.

As for our book, the family keys, along with the illustrated species descriptions and range maps, represent the course laboratory or classic natural history portion of this guide to the 69 amphibian and 98 reptile species that inhabit California and its coastal waters. The natural history accounts, supplemented by illustrations and photos of species and their habitats, are the book's "lecture sessions." Additional chapters in this area address allied subjects such as amphibian and reptile watching and photography plus the limited capture and husbandry of locally abundant species for classroom and scientific study.

However, it is here that the authors' efforts must end and those of you, the reader, begin as you embark on your own field portion of this "course." We hope that this natural history guide will function as your own personal instructor to assist when needed as you make firsthand observations of the everyday lives of California's great wealth of amphibian and reptile species.

ACKNOWLEDGMENTS

Much of the information presented in this book was acquired through discussions with numerous colleagues and students during the authors' combined total of over 100 years of conducting research in and teaching about herpetology, and to all these talented people we express our heartfelt thanks. In this regard we are especially indebted to David Wake, Director Emeritus of the Museum of Vertebrate Zoology, University of California, Berkeley, for bringing us up to date on the rapidly evolving field of DNA-based species analysis in amphibians. We also thank Jackson D. Shedd, section editor of *Herpetological Review*, for his fine in-depth review of the manuscript. We would also like to thank Robert Hansen, editor of the *Herpetological Review*, for his helpful feedback on the manuscript.

The publication of amphibian and reptile color illustrations would not have been possible without the foresight of Kira Od, who realized the need for compiling a digital record of this one-of-a-kind art and then proceeded to accomplish this enormous task with true artistic precision before the originals were permanently archived. The assistance of Les Chibana in digitally adjusting each illustration at the direction of the artist is also appreciated. We are grateful to Lisa White, Houghton Mifflin field guide editor, for her generous permission for the use of many of the illustrations, and to Kate Hoffman, University of California Press project editor for science, whose periodic updates and suggestions along the lengthy path to publication are much appreciated.

We also thank editors Laura Cerruti, for her enthusiasm and guidance in the initiation of this project, and Kim Robinson, for moving it steadily down the home stretch. The authors assume that production editor Nancy Lombardi of P. M. Gordon Associates ranks their manuscript among her most challenging, and we appreciate her dedication and professional approach to editing our text.

The dedicated effort and Swiss precision that our graphic artist, Madeleine Van Der Heyden, applied to the production of the range maps and temperature profile graphs are outstanding. We also thank Carole Richmond for the use of her fine photo of slender slamander embryos.

This book would have never progressed beyond the scribbled note and poorly typed first draft stage without the editing, word processing, and Photoshop skills of Molly McGinnis. Her support and assistance along with that of Anna-rose Stebbins are greatly appreciated.

INTRODUCTION

Amphibians and Reptiles: The Vertebrate Terrestrial Pioneers

About 400 million years ago at the beginning of the period of Earth's history we call the Devonian, vertebrate life consisted solely of several major fish groups and was therefore confined to water habitats. This was a period of major land uplifting and mountain building, with the result that some areas once covered by seas were transformed into shallow, brackish-water basins. It was on the shores of such habitats that a series of seemingly insignificant events took place that would eventually affect all future vertebrate life on our planet, including the readers of this book: fish began to come out of the water. The species that periodically did this belonged to the fish subclass Sarcopterygii, the lobefin fishes. This group is nearly extinct today, but an extensive fossil record plus a couple of key living species such as the Coelacanth (*Latimeria chalumnae*) and the Australian Lungfish (*Neoceratodus forsteri*) provide insight into what these vertebrate land pioneers were like.

Life in these shallow basins where dissolved oxygen may have been quite low apparently promoted the development of a lunglike structure similar to the present-day fish swim bladder, an air-filled buoyancy organ. Air was supplied to this lung through a passage from the nasal openings, another feature unique to this fish group that prevents extensive water loss to the air from the lung surface. A third feature was the lobed fins themselves. Unlike the relatively flimsy rayed fins of most present-day fishes, the pectoral (front) and pelvic (rear) paired fins had numerous bony elements in the elongated fin base that could provide support and some movement to a fish body on land. However, it is likely that such supportive fins were first used for locomotion on the bottom substrate of shallow, vegetation-choked inshore waters where conventional fish swimming was not effective. Other than these structures, the fossil record shows that most of the features we associate with fish today, such as a caudal (tail) fin, body scales, and a fused head and body skeleton, were already present in these terrestrial pioneers. This strongly suggests that their primary mode of existence was that of a fish and that their initial experiences on land were only brief excursions away from the water world.

The other major stimulus for the exploration of shoreline areas by various lobe-finned fish species was the Devonian climate. It appears that these sojourns proceeded during humid and mild microclimatic periods that would have favored a fish out of water by slowing the rate of body water loss and fin drying. There were also seasonal climatic variations associated with periodic rains and the resulting desalination of the brackish basins and lagoons in which these fish lived, all of which may have accelerated the acquisition of adaptations to land life.

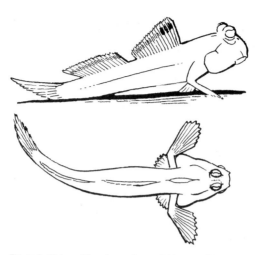

Figure 1. Side and top views of a mudskipper, a fish that uses its strong pectoral fins to "walk" along moist shoreline areas.

As to just why various species of Devonian lobefin fishes began to frequent shore habitats at various times during this geological period, much speculation remains. Some reasons may be found in several living species of gobid fish, especially those known as mudskippers (genus *Periophthalmus*). Although not lobe-finned fish, they have especially stout rayed pectoral fins that they use for rapid land movement as they forage for small invertebrates on tropical mudflats. The shoreline invertebrate feeding niche must have been an especially productive one in the Devonian because, unlike now, there was no competition from other vertebrates. While on land, mudskippers also engage in aggressive displays and other social behavior that may be more effectively carried out on land than in water. The junior author has kept two species of mudskippers in heated, humid land/water terrariums, and each specimen often spent more time on the wet soil surface where they were fed than in the shallow water that was provided. The apparent advantages accruing to these present-day fishes by shoreline foraging and displaying may closely reflect those that the Devonian lobe-finned shore explorers also attained. However, in both of these past and present fish groups the fin structures that eventually enabled them to move on land had developed as part of their particular aquatic lifestyle and only later became advantageous for terrestrial movement (figure 1).

Two other living fish species, the American Eel (*Anguilla anguilla*) and the "walking catfish" (*Clarias batrachus*), are known to occasionally make extensive overland journeys during wet weather, presumably in search of new aquatic habitats. These trips appear to be stimulated by unfavorable conditions in their original habitats that range from overcrowding to seasonal drying. Situations such as these in the Devonian could also have promoted land forays by lobe-finned fishes. This ability and other advantages of a terrestrial existence favored additional adaptations for life on land, and for the next 50 million years or more these new early tetrapod lineages produced a great variety of species, some as large as crocodiles with skulls over one meter (39 in.) in length. Throughout this evolution these first land vertebrates, which eventually gave rise to modern-day amphibians and reptiles, appear to have led an amphibious-type existence, relying on a water habitat for breeding and prevention from desiccation but foraging primarily on land. In this latter activity they were probably like most amphibians today, which rarely roam far from standing water or a saturated soil habitat.

However, natural selection usually favors those variations in anatomy, physiology, and innate behavior that lead to better survival for some individuals within a species complex. For these first land vertebrates, any new feature that promoted better foraging away from heavily populated water sites was most likely of high selective value. Thus, by the end of the Carboniferous Period, more than 50 million years after the first tetrapods "came ashore," the fossil record begins to exhibit a new type of tetrapod amphibian. These had larger, more ventrally positioned legs for more efficient land movement, a rib cage for better lung ventilation, and a thicker skin that reduced body water loss. Perhaps of greatest importance was that these new forms produced an egg that did not have to be laid in a water habitat.

This new lineage, which gave rise to the reptiles, began to evolve herbivorous as well as carnivorous species, and for the first time vertebrates could directly utilize the wealth of terrestrial plant life on Planet Earth. This produced a reptilian herbivore/carnivore food chain that reached its zenith in the Jurassic and Cretaceous periods. Because of the attractiveness of the larger members of this complex to various authors and movie producers, many people today are more familiar with the dinosaurs and their allies than with many of the living mammal and bird groups. These large, spectacular reptiles and their equally impressive amphibian predecessors have long since vanished, but a miniature complement of amphibians and reptiles has survived 100 million years of rigorous competition with evolving bird and mammal species. Drawing on what appears to be a widely varied and fluid gene pool, present-day amphibians and reptiles have been able to adapt to an unusually wide range of both natural and human-wrought environmental conditions. Nowhere is this more evident than in California, with its great extremes in climate, topography, and human habitat intervention.

Amphibian and Reptile Distribution throughout California's Habitat Complex

Amphibians and reptiles are far more dependent on the basic physical properties of their habitats such as temperature, moisture, and substrate characteristics than most bird and mammal species. They are ectotherms, and thus the body temperature of most species is that of the surrounding air, soil, or water except when sunlight is available to basking species. The thin, highly vascularized skin of amphibians allows water molecules to readily pass in and out of their bodies, and thus habitat features such as the presence or absence of standing water, the amount of water vapor in the air, and the degree of soil saturation places further restrictions on this class of vertebrates.

Given these close relationships of amphibians and reptiles with their physical environment, a familiarity with the major geographic subdivisions of a state or country is helpful in understanding its herpetological biogeography. In some regions of North America such as the Great Plains, this subject would require very little text space due to the small amount of topographical and climatic variation throughout such an area. However, geographic variability and extremes in California may well exceed those in any other state or province. On a late spring day in Yosemite National Park you may encounter frogs and toads spawning in meadow snowmelt ponds, while 25 miles to the east lizards are foraging on the desert floor of the Mono Basin at temperatures approaching 38 degrees C (100 degrees F).

The distribution of lizards and salamander species throughout California further illustrates this point. More than 80 percent of the state's 44 lizard species occur in the Mojave and Colorado deserts, but only two of its 42 salamander species are found there. This distribution is sharply reversed in the cool streams and moist forests of northern coastal California, which is home to 50 percent of California salamander species but only five (15 percent) of its lizard species. This trend is mirrored by the distribution of anurans and snakes in these two regions.

However, these general distribution patterns should not be viewed as representing absolute species separation between various regions. Most of California's amphibians and reptiles are "habitat generalists," species that are not dependent on a specific type of food, ambient temperature régime, plant association, or substrate type but instead have a wide range of habitat tolerance. The Pacific Chorus Frog (*Pseudacris regilla*), Western Fence Lizard (*Sceloporus occidentalis*), and Common Kingsnake (*Lampropeltis getula*) are prime examples of habitat generalists and occur in all 12 of the state's geographic subdivisions. Indeed, the majority of California's amphibians and reptiles are to varying degrees habitat generalists and are found in several geographic regions.

In contrast to these generalists are the "habitat specialists," species that are better adapted to a specific habitat than most other similar forms and

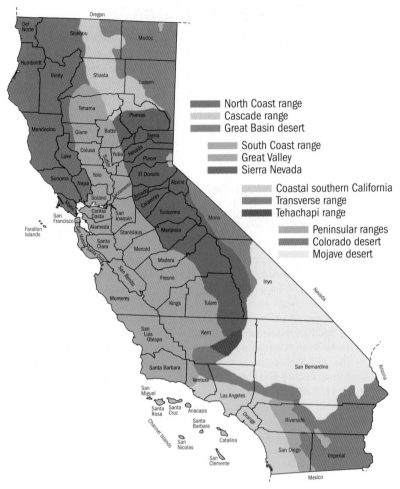

California's 12 geographic subdivisions as defined by Hickman (1993) superimposed on the state's 58 counties.

therein thrive if not disturbed by human activities. Some examples here are California's three web-toed salamander species (*Hydromantes*), which inhabit crevices and caves of the Sierra and Cascade ranges, and the three fringe-toed lizard species (*Uma*), which thrive only in desert sand hummocks, flats, and dunes where wind-blown sand deposits are present. Outside of their unique habitats, such specialists usually compete poorly with other closely related forms.

The preceding map displays California's 12 geographic subdivisions as defined by Hickman (1993), superimposed on the state's 58 counties. The

Plate 1. Herpetology field trip in North Coast Range.

counties are referred to throughout this book when discussing species distribution. Most species accounts also refer to the geographic subdivisions in which a species is most often found, so that you will have a better idea of the general habitat situation in which it may be encountered. In the following brief descriptions of each geographic subdivision or region we present general comments on yearly temperature range, annual rainfall, and the dominant plant communities that occur there. Because California's dynamic topography has produced markedly different thermal, moisture, and vegetation differences within most of these regions, one brief description cannot fit all circumstances. For instance, the most common initial impression of the North Coast Range is that of cool, fog-bathed conifer forests. However, open scrub areas in the interior valleys of this region experience little or no fog or rain during the summer months while often recording mid-day temperatures of 40 degrees C (104 degrees F) or higher.

In the following brief descriptions of these regions, major climatic features and most plant community designations follow those of Hickman (1993) and Mayer and Laudenslayer (1988). The small complement of amphibian and reptile species listed for each region represents those types whose habitat requirements are well met by conditions within these areas. It is also important to note that such listings in no way imply that the species given are found only in that area. Indeed, most of the amphibians and reptiles present in one region are habitat generalists and thus are usually present in adjacent regions as well. Exceptions here are the extreme habitat specialists whose preferred habitat occurs exclusively within one or two of these areas.

Plate 2. Dense conifer forest surrounding Mt. Shasta.

Plate 3. Coastal scrub stand in Tilden Regional Park, Alameda County.

Plate 4. Amphibian breeding pond, coastal San Mateo County.

Plate 5. Urban freshwater marsh, San Mateo County.

Plate 6. Lower Stanislaus River, west side of Great Valley.

Plate 7. Snowmelt pools, Tuolumne Meadows, Yosemite National Park.

Plate 8. Coastal Southern California uplands.

Plate 9. Oak savannah, Tehachapi Range.

CALIFORNIA HABITATS 11

Plate 10. Seasonal amphibian breeding pool, Peninsular Range.

Plate 11. Herpetology class camp at Granite Mountains, Mojave Desert.

Plate 12. Sand dunes, Mojave Desert.

Plate 13. Yucca and rocky upland, Mojave Desert.

Plate 14. Lava outcrop, Mojave Desert.

Plate 15. Darwin Canyon, Mojave Desert.

Plate 16. Encroaching golf course, Colorado Desert.

North Coast Range

Summer temperatures range from mean daily highs of 22 degrees C (72 degrees F) at northern coastal sites to 40 degrees C (104 degrees F) or higher in inland valleys. Subfreezing temperatures and snow are common in winter at the high elevations of this range. Annual rainfall as high as 230 cm (90 in.) occurs in some coastal areas, while summer drought is common inland. The coastal portion of this region is famous for its groves of old-growth redwood (*Sequoia sempervirens*) and also includes coast oak woodland, Douglas-fir (*Pseudotsuga menziesii*), ponderosa pine (*Pinus ponderosa*), Port Orford cedar (*Chamaecyparis lawsoniana*), and tanoak (*Lithocarpus densiflorus*). A major plant community change occurs in the east slopes of the North Coast Range that border the Great Valley. Here, oak woodland, gray pine (*Pinus sabiniana*), and mixed chaparral replaces dense conifer forests in most areas.

The perennially moist woodland areas of the coastal portion of this range are the only sites in California where three chiefly forest-dwelling amphibians, the Del Norte Salamander (*Plethodon elongatus*), Dunn's Salamander (*Plethodon dunni*), and the Wandering Salamander (*Aneides vagrans*) are found. Cool-stream forms like the Coastal Tailed Frog (*Ascaphus truei*) and Southern Torrent Salamander (*Rhyacotriton variegatus*) also thrive here. Only reptile species such as the Western Fence Lizard (*Sceloporus occidentalis*) and Sagebrush Lizard (*Sceloporus graciosus*), which can efficiently thermoregulate under limited-sunlight conditions, or those like the Northern Alligator Lizard (*Elgaria coerulea*), Western Skink (*Plestiodon skiltonianus*), Northern Rubber Boa (*Charina bottae*), and Common Sharp-tailed Snake (*Contia tenuis*), which function efficiently at relatively cool temperatures, are relatively common here. Exceptions to this general trend are found in the eastern scrub areas of this range, where semiarid desert habitat generalist species such as the Western Whiptail (*Aspidoscelis tigris*) occur.

Other broad-spectrum habitat generalists also occur in various areas within this region and a number of other geographic subdivisions. Most common among these are Ensatina (*Ensatina eschscholtzii*), Western Toad (*Bufo boreas*), Southern Alligator Lizard (*Elgaria multicarinata*), Common Kingsnake (*Lampropeltis getula*), Gopher Snake (*Pituophis catenifer*), Racer (*Coluber constrictor*), Western Terrestrial Garter Snake (*Thamnophis elegans*), Ring-necked Snake (*Diadophis punctatus*), and Northern Pacific Rattlesnake (*Crotalus oreganus*). Because of their wide range of occurrence throughout much of California, these will not be listed for most other regions where they occur.

Cascade Range

Temperatures in this region range from –13 degrees C (9 degrees F) in winter at the snow-laden high elevations to midsummer highs approaching 32 degrees C (90 degrees F) in the eastern valleys. The average monthly range is 14–22 degrees C (57–72 degrees F). The maximum annual rainfall recorded for the west side of this range is 177 cm (69 in.), while the maximum for its eastern slopes is 60 cm (24 in.). Winter snowfall at the highest elevations may approach 150 cm (60 in.).

Ponderosa pine and white fir (*Abies concolor*) dominate most higher elevations but give way to aspen and juniper on the eastern slopes. Large bitterbrush (*Purshia* spp.) stands occupy many high valleys on the east side of this range. Two cool climate–adapted amphibians, the Southern Long-toed Salamander (*Ambystoma macrodactylum sigillatum*) and the Cascades Frog (*Rana cascadae*), occur here along with one habitat specialist, the Shasta Salamander (*Hydromantes shastae*). The same spectrum of temperate climate habitat generalist reptiles found in the northern portion of the North Coast Range is also present.

Great Basin Desert

This is the largest desert region in North America, but only two small segments along its western border extend into California. It is a high desert, with elevations ranging between 615 m (2,000 ft) and 1,538 m (5,000 ft). It is therefore not surprising to find annual temperature extremes from winter lows around –10 degrees C (15 degrees F) in the north to summer highs well over 38 degrees C (100 degrees F) along its southern border with the Mojave Desert. Annual precipitation in most other areas of this desert averages 46 cm (18 in.) in the north to 20 cm (8 in.) in the south.

The chief unifying feature of this region is its extensive sagebrush stands, of which big sagebrush (*Artemisia tridentata*) is the most prevalent. These are for the most part open-canopy stands that provide an extensive sun-shade mosaic for heliothermic reptiles such as the Sagebrush Lizard, Common Side-blotched Lizard (*Uta stansburiana*), and Western Fence Lizard. All are preferred prey for two carnivorous species, the Long-nosed Leopard Lizard (*Gambelia wislizenii*) and the Great Basin Collared Lizard (*Crotaphytus bicinctores*), which are also found at numerous sites within this region. The Desert Striped Whipsnake (*Masticophis taeniatus*) occurs almost exclusively within this desert in California but is widespread in adjacent states. A fair number of both temperate and desert habitat generalist snake species also exist here.

As previously mentioned, deserts are not normally hospitable to salamanders, but the Long-toed Salamander occurs at wetland sites in the northern segment of this region, and the Inyo Mountains Slender Salamander (*Batrachoseps campi*) is found in the Inyo Mountains in the southern segment near springs and streams in upland areas north of Owens Lake. Among the anurans, the range of the Great Basin Spadefoot (*Spea intermontana*) extends west into this region, and two habitat generalists, the Pacific Chorus Frog (*Pseudacris regilla*) and the Western Toad, are found at a number of wetland sites.

South Coast Range

This region comprises both the Coast and Inner Coast Ranges, which efficiently separate the cool Pacific coastal area from the hot-summer weather of the Great Valley. Thus, in July the daily high may reach only about 16 degrees C (61 degrees F) in redwood groves in northern San Mateo County, while along the eastern edge of Alameda County it is above 40 degrees C (104 degrees F). The north-to-south annual rainfall gradient is also quite dynamic in this region, ranging from highs approaching 100 cm (40 in.) in San Francisco in wet years to less than 30 cm (12 in.) in Santa Barbara during drought periods.

Oak woodland becomes more common than in the North Coast Range, where moist conifer stands are more prevalent. This region has also been the recipient of more plantings of species of eucalyptus trees than any other. Coastal scrub covers most of the coastal hills, while chamise-dominated chaparral is more abundant in the south. The great variability in plant communities is reflected in the wide spectrum of amphibians and reptiles found here. A relatively short west-to-east transect through the northern part of this region begins in cool coastal areas populated by forest-dwelling lungless salamander species, cool-stream adapted anurans like the California Giant Salamander (*Dicamptodon ensatus*) and Foothill Yellow-legged Frog (*Rana boylii*), and temperate reptiles common to the

North Coast Range. It ends in warm, dry valleys such as Corral Hollow, which supports small populations of arid-adapted forms such as the Western Spadefoot (*Spea hammondii*), Coast Horned Lizard (*Phrynosoma blainvillii*), Western Whiptail, Common Side-blotched Lizard, Western Long-nosed Snake (*Rhinocheilus lecontei*), and Glossy Snake (*Arizona elegans*). This change from a temperate-climate herpetological fauna to a more desert or dry-adapted one can also be observed as you travel from north to south through this region.

Great Valley

The Great Valley has the longest north-south span of all of the California geographic regions and yet is the most thermally consistent. On a given summer day temperatures may reach a high above 40 degrees C (104 degrees F) at both the northern and southern extents of this area. More contrast exists in winter, when the north border of this region may occasionally experience snow and freezing temperatures while the mean monthly winter low in the south rarely falls below 4 degrees C (39 degrees F). Average annual rainfall ranges from more than 76 cm (30 in.) in the north in wet years to less than 15 cm (6 in.) in the south during drought periods. The thermal regime of this region results in a growing season ranging from seven to 11 months and including up to 350 frost-free days in many areas. These features plus the relatively level contour of the land and the summer input of water from rivers that drain the Sierra Nevada has made the Great Valley one of the prime agricultural regions of the world.

This distinction was attained at the cost of a great loss of natural plant communities and wetland habitats. Over 97 percent of the once-dominant valley riparian forest community of this region has been destroyed, along with most of the native grass and forb species. The lingering presence of a fair number of arid-adapted lizards and snakes in isolated sites along this area's southwest border hints of the once-rich reptile complex that existed in the southern half of the pristine Great Valley.

That any native amphibian and reptile species persist in this region is a testament to the great adaptability of many generalist species. The ubiquitous Western Fence Lizard and Pacific Chorus Frog appear to be equally successful in many farmsteads as in more pristine habitats, and the Gopher Snake often finds more rodent prey in alfalfa fields than coastal or foothill grasslands. One habitat specialist, the Giant Garter Snake (*Thamnophis gigas*), has successfully adapted to foraging in the irrigation ditches and canals of the lower Sacramento Valley area but verges on extinction to the south, and an introduced amphibian, the American Bullfrog (*Rana catesbeiana*), has unfortunately found many of the artificial aquatic sites of the Great Valley to its liking, with the result that this region now houses the largest number of these frogs in California.

Sierra Nevada

Winter temperatures range from −10 to 36 degrees C (14 to 97 degrees F), depending on the elevation and latitude of a given area. As in the Cascade Range, summer highs may approach 38 degrees C (100 degrees F) in large, sheltered valleys. Precipitation, mostly in the form of snow, is as high as 330 cm (90 in.) in the north in very wet years and below 76 cm (30 in.) in southern areas during drought periods. Subalpine fir (*Abies lasiocarpa*) and mountain hemlock (*Tsuga mertensiana*) stands near timberline give way to red fir (*Abies magnifica*), white fir, and ponderosa pine at lower elevations. Ghost pine, blue oak (*Quercus douglasii*), and mixed chaparral dominate the western foothills, while stands of pinyon (*Pinus edulis*) and juniper (*Juniperus* spp.) occupy the lower eastern slopes.

Seventy of California's 160 (44 percent) nonmarine reptile and amphibian species occur within or immediately adjacent to the Sierra Nevada geographic range. The understory of the woodlands of the middle and lower western slopes is home to eight species of slender salamanders and two subspecies of Ensatina. Three amphibian habitat specialists, the Mount Lyell Salamander (*Hydromantes platycephalus*), Yosemite Toad (*Bufo canorus*), and the Sierra Nevada Yellow-legged Frog (*Rana sierrae*), occur at higher elevations.

Only highly efficient heliotherms like the Western Fence Lizard and Sagebrush Lizard and species with relatively low preferred body temperatures such as the Northern Alligator Lizard and Northern Rubber Boa can persist in the very "high Sierra" (around 3,200 m or 10,500 ft). Cool-adapted subspecies of the California Mountain Kingsnake (*Lampropeltis zonata*) and Western Terrestrial Garter Snake occur at slightly lower elevations, and the east and west foothills are home to a number of habitat generalist lizard and snake species. In addition, desert-like conditions at low elevations at the southeast border of this region also allow extensions of the range of some typically desert lizard and snakes into this area.

Coastal Southern California

This is the southernmost extension of the Coast Range in California, and like the South Coast Range to the north, it exhibits a true Mediterranean-type climate. Summer daily high temperatures often exceed 38 degrees C (100 degrees F), with mean monthly lows in midwinter to about 6 degrees C (43 degrees F). Rain rarely falls during the summer months, and annual precipitation here approaches that in true desert regions during drought years. The annual precipitation for Los Angeles occasionally falls below 13 cm (5 in.), while yearly rainfall highs may exceed 76 cm (30 in.).

Coastal scrub and coastal oak woodland still persist in those increasingly few areas that have not undergone extensive urban and suburban development. These are mainly steep upland areas and intervening valleys that fall under regional, state, or federal protection. Here two species of slender salamander and three desert-adapted anurans, the California Chorus Frog (*Pseudacris cadaverina*), Arroyo Toad (*Bufo californicus*), and Western Spadefoot may still be found. The ability of these latter species to undergo long periods of summer aestivation and occasionally bypass breeding in years when seasonal pools do not fill make them well suited for this region.

Two lizard species endemic to Baja California, the Orange-throated Whiptail (*Aspidoscelis hyperythra*) and the Granite Night Lizard (*Xantusia henshawi*), have ranges that extend into southern coastal California. However, most desert lizard species are not found here, despite what appears to be favorable thermal conditions for them. This trend does not hold true for some of the desert-oriented snake species, and the Coachwhip (*Masticophis flagellum*), Glossy Snake, Western Patch-nosed Snake (*Salvadora hexalepis*), Western Long-nosed Snake, California Lyre Snake (*Trimorphodon lyrophanes*), and Rosy Boa (*Lichenura trivirgata*) may all be encountered here. This region, along with a segment of the Peninsular Range, are the only two areas in which the Red Diamond Rattlesnake (*Crotalus ruber*) can be found outside of Baja California.

Transverse Range, Tehachapi Range, and Peninsular Ranges

Together with the Sierra Nevada, these smaller southern ranges form a continuous montane barrier between the Great Valley and coastal ranges to the west, and the three desert regions to the east. Indeed, it is the blockage of most Pacific storms by these mountains that creates the xeric conditions of southeastern California. The higher peaks in these ranges are for the most part below 2,500 m (8,000 ft), with a few sites in the San Bernardino and San Gabriel Mountains around 2,925 m (9,500 ft).

Long periods of snow and subfreezing temperatures are therefore prevalent in the higher portions of these ranges during the winter months. Summer daily high temperatures at lower elevations may regularly exceed 38 degrees C (100 degrees F), especially in wind-sheltered canyon areas. The widely varying annual precipitation amounts seen in the Coastal Southern Range are mirrored on the northwest-, west-, and southwest-facing slopes of these three ranges. However, even in wet years when annual precipitation on the western slopes approaches or surpasses 100 cm (40 in.), the perennially dry slopes that border the Mojave and Colorado deserts continue to reflect the xeric climate of the adjacent desert areas.

An oak woodland–conifer community dominates the western slopes of these three regions but gives way to mixed chaparral and eventually desert scrub on the lower east-facing areas. Unique desert plants such as the Joshua tree (*Yucca brevifolia*) and Mojave yucca (*Yucca schidigera*) occur sporadically along the transition line between montane and desert vegetation.

The complement of amphibians and reptiles in these three regions differs markedly between the west and east slopes. *Ensatina klauberi*, a former subspecies of Ensatina recently elevated to species status, is found in the wooded areas of these regions along with seven species of slender salamander, including a robust form, the Tehachapi Slender Salamander (*Batrachoseps stebbinsi*). Xeric-adapted anurans such as the Arroyo Toad and Red-spotted Toad (*Bufo punctatus*) inhabit creeks and pool sites in the Transverse and Peninsular ranges.

The lizard fauna on the western slopes is composed mostly of the same temperate species found in coastal southern California. However, the desert-like habitat in the lower reaches of the east and southeast uplands supports a spectrum of true desert forms such as the Common Chuckwalla (*Sauromalus ater*), Desert Iguana (*Dipsosaurus dorsalis*), Desert Spiny Lizard (*Sceloporus magister*), Desert Night Lizard (*Xantusia vigilis*), and Western Banded Gecko (*Coleonyx variegatus*). Desert snakes, including the Rosy Boa, Western Shovel-nosed Snake (*Chionactis occipitalis*), Sidewinder (*Crotalus cerastes*), and Mojave Rattlesnake (*Crotalus scutulatus*) also occur here. This distribution of desert reptiles along the south and east borders of these three ranges and the Sierra Nevada reflects the role that these mountain systems have apparently played in restricting the spread of these species to seemingly acceptable xeric habitats in Coastal Southern California and the Great Valley.

The Mojave and Colorado Deserts

These are the two classic desert regions of California. Some areas may have no rain at all during drought years, and daily summer high temperatures for most sites hover around 43 degrees C (110 degrees F). Given that both regions are situated in the rain shadow of the western mountain ranges, the single most important physical difference between the two deserts is elevation. The Mojave is a relatively high desert, with elevations ranging between 615 m (2,000 ft) and 1,540 m (5,000 ft). One area in the northern portion of this desert differs markedly from this higher elevation norm. This is Death Valley, the lowest point in the Western Hemisphere at –97 m (–282 ft). The average annual rainfall here is less than 5 cm (2 in.), and in some years the valley basin receives no rain at all. Annual rainfall in the rest of the Mojave ranges from 20 cm (8 in.) to 30.5 cm (12 in.). Midday summer temperatures in Death Valley may rise above 54 degrees C (130 degrees F), and in other parts of this desert summer high temperatures frequently reach 48 degrees C (118 degrees F). Because of its comparatively high elevation, winter lows range between –9 and 3 degrees C (16 and 37 degrees F), with occasional snow at upland sites in areas like Joshua Tree National Park.

In contrast, mountain peaks in the California segment of the Colorado Desert rarely exceed 460 m (1,500 ft), and elevations in flatland areas range between 31 m (100 ft) and 62 m (200 ft) above sea level. The shore area of the Salton Sea is the lowest area in this desert, approaching −40 m (−130 ft) at some points. As a result of this lower elevation regime, winter temperatures are very mild. This, along with a low annual rainfall of 4 to 16 cm (1.5 to 6.3 in.) has led to the expansion of former small deserts towns like Palm Springs into major winter vacation sites. Through the availability of Colorado River water for irrigation, most of the original desert habitat of the Imperial Valley area has been converted to cropland. All of this has resulted in the loss of much prime habitat for numerous amphibian and reptile species.

Most of the Mojave Desert supports a desert scrub plant community dominated by creosote bush (*Larrea divaricata*), brittle bush (*Encelia farinosa*), burro bush (*Franseria dumosa*), and several species of cholla. This community gives way to sparse stands of alkali scrub in the eastern portion of the region. Its most famous plant is the Joshua tree. This largest of all yuccas attains a height of 12 m (40 ft) or more, and is found in sprawling groves throughout the region where cool and moist winters are the norm.

This complex of creosote bush, brittle bush, and burro bush also occupies most lower elevations in the Colorado Desert but in less dense stands than in the Mojave. Broad sandy washes are common here, and where these occur near upland areas, clumps of desert willow (*Chilopsis linearis*), smoke tree (*Dalea spinosa*), blue palo verde (*Cercidium floridum*), and several xeric shrub species often grow along their banks. Small stands of desert riparian vegetation occupy narrow canyon floors with seasonal drainages in both deserts. The most spectacular of these are the widely scattered California fan palm (*Washingtonia filifera*) oases.

As already mentioned, only two of the 38 species of salamanders in California occur in this arid region, one in several parts of the Inyo Mountains in the northern Mojave (Inyo Mountains Slender Salamander) and the other in canyons on the east slope of the Santa Rosa Mountains in the southwestern Colorado Desert (Desert Slender Salamander, *Batrachoseps major aridus*). Here also may be found xeric-adapted anurans like the Red-spotted Toad, Arroyo Toad, Western Spadefoot, and Couch's Spadefoot (*Scaphiopus couchii*). In contrast to this sparse amphibian fauna, the lizard complement of these two deserts is greater than that in any other geographic subdivision in California. It ranges from habitat specialists like the Desert Night Lizard, the Common Chuckwalla, and the three fringe-toed lizard species to a broad spectrum of desert habitat generalists including the Western Whiptail, Zebra-tailed Lizard (*Callisaurus draconoides*), Desert Iguana, Desert Spiny Lizard, Long-nosed Leopard Lizard, Great Basin Collared Lizard, and the ubiquitous Side-blotched Lizard.

These deserts are also home to 26 of the state's 37 snake species (68 percent), a complement that includes five species of rattlesnake. Parts of the Mojave Desert also support the Desert Tortoise (*Gopherus agassizii*), one of only two native species of nonmarine turtles in California. It and the Common Chuckwalla are the only herbivorous species of adult reptiles and amphibians in California, and yet they are found in the least vegetated of all geographic subdivisions, the southern deserts of California.

Ongoing Loss of Habitats and Numbers of California's Amphibians and Reptiles

It is with a true sense of regret that the authors address here a subject that we believe is as disturbing to our readers as it is to us: the continuous loss of critical habitat and populations of the amphibians and reptiles of California. Although that amazing handful of habitat and feeding generalist species that we have highlighted in the natural history accounts continue to hold their own in the face of ever-expanding human habitat manipulation, a far greater number of groups remain impacted by the onslaught of detrimental habitat changes that our species persists in imposing on them.

For both the amphibians and reptiles of this state and the world in general, the predominant impact has been habitat destruction. Whether the site in question was a drained and filled prime sag pond feeding habitat in San Mateo County for the San Francisco Garter Snake (*Thamnophis sirtalis tetrataenia*) and breeding habitat for the California Red-legged Frog (*Rana draytonii*) or a sand dune habitat with a thriving population of the Coachella Valley Fringe-toed Lizard (*Uma inornata*) engulfed by expanding desert development, the result is the same: the extinction of yet another local population. And, if such losses are left unchecked, species extinction will eventually occur. Only by the eleventh-hour intervention of the state and federal endangered species acts in the early 1970s has the total annihilation of these and many other species been temporarily thwarted. Even though in the opinion of many conservationists the administration of these acts has not always been in accordance with the true spirit of these laws, the situation today for all imperiled animals and plants would indeed be dire without this legislation.

Beyond this basic cause for habitat and population loss in both groups, the plight of amphibians differs markedly from that of reptiles. The physiology, behavior, and general natural history of California reptiles make them relatively resistant to several decimating factors other than habitat destruction that impact amphibians. Of some 5,700 species of anurans, salamanders, and caecilians (the legless group of amphibians) reviewed to date (2005), about one third are threatened with extinction, and 168 species have already passed over that border of no return. In recent years the plight of amphibians has garnered much public attention through the highly publicized presumed extinction of spectacular species such as the Golden Toad (*Bufo periglenes*) in Costa Rica. Most important is that the scientific community has vigorously addressed this subject. The Declining Amphibian Population Task Force (DAPTF), spearheaded by Dr. David Wake, director emeritus of the Museum of Vertebrate Zoology, University of California, Berkeley, continues to investigate the nature, extent, and causes of worldwide amphibian decline. Books such as *A Natural History of Amphibians* by the senior author and Dr. Nate Cohen (1995), and the 2005 multiauthored *Amphibian Declines* review amphibian reductions and their probable causes throughout North America and the world. In this natural history guide we address this subject in the "conservation notes" for those California amphibian species exhibiting noticeable or marked decline, and here we will briefly review several factors other than direct habitat destruction that appear to be impacting amphibians in the "Golden State."

Given that most California reptile, bird, and mammal species have as yet not exhibited precipitous, wide-scale population reductions, what then is it about amphibians that makes them so much more vulnerable? In attempting to answer that question, we must first ask one additional question: are all types of amphibians in this state being equally impacted? The answer appears to be no. Beyond the universal impact of habitat destruction, approximately one half of all California species have not experienced noticeable species-wide declines. These are the 31 California plethodontid salamander species, which, of course, differ from this state's other amphibians in that they are not dependent on aquatic habitats for any phase of their life history. Given this premise, what is it, then, about the natural history of California's anurans and water-breeding salamanders that puts them at a far greater risk than all of their terrestrial-reproducing vertebrate neighbors?

A Century of Ever-Increasing Aquatic Predation Pressure

Except for that being exerted through hunting and collecting by humans, predation pressure on most terrestrial vertebrate species, including the plethodontid salamanders, appears to have remained relatively constant over the past few centuries. The introduction of feral house cats and occasional dog packs has been offset by a long period of control of wild predators that is only now being curtailed. Besides these two introduced predators, we have not undertaken the purposeful establishment of undomesticated terrestrial vertebrate predators in this state. However, this has not been the case with aquatic vertebrate predators.

A little over a century ago, there were 68 species of fishes that lived all or part of their lives in California freshwater habitats. As of 2007 the number of freshwater and anadromous fishes stands at 118, as a result of the introduction of 52 nonnative species, of which 30 are relatively large, piscivorous gamefish. Nearly all of the latter introductions have been sanctioned by the California Department of Fish and Game (CDFG) and its predecessor agencies for the purpose of enhancing sport fishing in this state. One or more are now present in nearly every permanent California freshwater habitat, where they feed on a wide variety of small aquatic invertebrates and vertebrates, including amphibian larvae. Two of these, the Largemouth Bass (*Micropterus salmoides*) and the Bluegill (*Lepomis macrochirus*), compose a long-standing recommendation by the CDFG for stocking of ranch and farm ponds, thus rendering these sites unusable by most native aquatic-breeding amphibians.

This present-day situation is in great contrast to that of pristine California when there were only two large warm-water native piscivores, the Sacramento Perch (*Archoplites interruptus*) and the Sacramento Pikeminnow (*Ptychocheilus grandis*). At that time there were also numerous high-Sierran "fishless lakes" in which the only vertebrate residents were amphibian species such as the Sierra Nevada Yellow-legged Frog and the Southern Long-toed Salamander (*Ambystoma macrodactylum sigillatum*) in the northern Sierra Nevada. However, beginning with the first small trout hatchery on the fledgling campus of the University of California at Berkeley in the mid-1870s, the state embarked on what is now the largest trout rearing and stock program in the nation. Currently about 40 percent of the CDFG's inland fisheries budget, around 48 million dollars, supports this "put and take" program in which both native and introduced catchable-size trout species are stocked in most waterways throughout California, including the once-fishless lakes of the Sierra Nevada. This has resulted in the near extinction of the Sierra Nevada Yellow-legged Frog. Trout eradication in a few of these lakes has occurred, but is it action that is too little too late? A more aggressive program may be required to prevent the loss of this native California frog.

Among the 22 nonpiscivorous introduced species are several "grubbers" such as the Common Carp (*Cyprinus carpio*), which consume amphibian eggs. There are also reports that the ubiquitous Mosquitofish (*Gambusia affinia*) will peck off the external gills of newly hatched anuran and salamander larvae. However, tests by the senior author suggest that the young "black-stage" larvae of the California Red-legged Frog are distasteful to this fish. Even so, just the gill-picking behavior of this small, seemingly benign fish may be especially dangerous, because mosquito abatement districts often stock it in fish-free seasonal ponds and marshes that support California Red-legged Frog reproduction but dry after larval metamorphosis in early fall. Because aquatic predation occurs beneath the water surface, it is essentially invisible to the casual observer who sees only the end result: a marked decline or complete disappearance of one or more amphibian species in an area. In California, this may be one of the many impacts causing some of the "mysterious amphibian declines" reported over the past few decades.

The presence of one additional introduced semiaquatic predator, the American Bullfrog, can easily be detected by both sight and sound. Its appearance in permanent aquatic habitats has usually been accompanied by a marked decline or complete disappearance of native ranid species such as the California Red-legged Frog, the Northern Red-legged (*Rana aurora*), and Cascades Frogs in permanent aquatic habitats. It also persists even if predatory fishes are introduced, because its larvae have acquired a noxious taste that apparently discourages such predators. This is because the bullfrog has evolved with many of the introduced eastern game fish species that now dominate most California waterways.

Chemical Contamination of Amphibians and Their Aquatic Habitats

Although predation by introduced fishes has had a major impact on aquatic-breeding amphibians, this alone cannot explain the decline of species in fishless habitats. Wet meadow pool breeders such as the Yosemite Toad and central Sierran and western foothill populations of seasonal wetland breeders like the Western Toad, the California Red-legged Frog, and the Foothill Yellow-legged Frog have undergone alarming declines in the absence of introduced predators. These species do have one thing in common: they all live downwind from the Central Valley, one of the most intensely farmed areas in the world. Carried by the prevailing westerly winds that blow across this region to the Sierra Nevada are molecules of most major pesticides and herbicides used in modern agribusiness. Support for this theory of windborne chemical contamination of Sierran frogs and toads comes from comparisons with relatively stable populations of these anurans in the Mount Shasta area and northern coastal foothills, areas that are not downwind from intensive agriculture but where the extent of habitat destruction and aquatic predator introductions is similar.

Because the surfaces of amphibian adults, larvae, and eggs are highly permeable to chemical contaminants in the air, water, and substrate, they are far more vulnerable to these substances than all other terrestrial vertebrate groups. In this sense they are the modern-day "canaries in the mine shaft," which in many situations may serve as indicators of the future environmental health for our own species. Not only do chemical contaminants enter the body of amphibians more readily than in reptiles, birds, and mammals, but some can exert a dramatic impact at very low dosage levels. A case in point is atrazine, one of the world's most widely used herbicides. In the United States it currently is present in about 40 percent of all herbicides and is annually sprayed over three quarters of the country's corn crop. In 2005 endocrinologists at the University of California, Berkeley, found that males of several species of frogs can be "feminized" by a dose of atrazine that is only one thirtieth of that allowed by the Environmental Protection Agency for safe drinking water. Most disturbing has been the finding that feminizing effects such as the growth of ovaries is actually more acute at very low doses of atrazine than at higher ones, possibly because the endocrine system operates on very low hormone dosages and is therefore more attuned to these low levels than to higher amounts.

With respect to the comparison of aquatic versus terrestrial breeding vertebrates, it would appear that anuran and aquatic-breeding salamander species are far more susceptible to windborne chemical contamination than plethodontids that inhabit the same area. The relatively long larval stage of the former often occurs in shallow aquatic habitats that may collect contaminants and make them available to the larvae throughout their growth period. Such materials are often taken up and stored in lipid reserves that are then mobilized during metamorphosis. It is at this time that gonad development and sex determination take place, and therefore the period when the feminization process can occur. Larval development and metamorphosis also occur during the late spring and summer period when most agricultural spraying in the Central Valley occurs. In contrast, plethodontid development takes place entirely within terrestrially sequestered eggs during the fall and winter months.

Chytrid Fungus Infection

The chytrid fungus, *Batrachochytrium dendrobatidis* (Bd), produces a disease called amphibian chytridiomycosis that has emerged as one of the foremost killers of amphibians worldwide. A recent study at a long-term amphibian research site in central Panama followed the invasion of this fungus and found that within four months 64 species of frogs were infected and some had completely disappeared. This fungus invades the epidermal skin cells, which it causes to thicken, thus interfering with the skin's osmoregulatory ability. It is especially dangerous to species that don't spend much time in water but instead rely on the high permeability of the skin to take up small amounts of moisture from vegetation and soil. When this ability is negated by fungal-induced epidermal thickening, these amphibians may desiccate and die in moist terrestrial foraging and retreat habitats that would otherwise ensure their survival. Currently several declining populations of the Western Toad in Colorado and Wyoming have been linked to the fungus, and recent attempts to reintroduce Sierra Nevada Yellow-legged Frogs into seven Sierra Nevada lakes have failed due to chytridiomycosis.

At this writing this fungus has been detected in 10 native and two introduced amphibian species in California. One of the latter, African Clawed Frog (*Xenopus laevis*), is a worldwide carrier of this fungus. It was imported for human pregnancy testing in the 1940s and 1950s and then retained as a preferred laboratory animal in succeeding decades. The unwarranted release of excess laboratory specimens along with those from pet store sales into local aquatic habitats has most likely initiated this epidemic in North America and elsewhere. The numerous introductions of the Bullfrog in five European countries have been linked with the spread of Bd and the demise of several populations of European anuran populations, and it is probably a major vector for this fungus in western North America as well.

Until recently (2007) the transmission of Bd was thought to be only by a clonal form of the fungus that requires a moist medium to remain viable. Such media include moist algae clumps on collecting nets and wet mud on rubber boots, and therefore humans may play a major role in moving Bd from pond to pond. In the section on capturing amphibians and reptiles we stress the disinfection of all equipment and personal gear through the application of a 10 percent bleach solution before entering a new aquatic site where amphibians are known or suspected to occur. However, current studies suggest that, in addition to clonal reproduction, this fungus may also be capable of sexual reproduction of resistant spores that remain dormant for a decade and can be transmitted by dirt on shoes and tires or even on bird feathers, possibly acquired during "dust baths." If so, the future for many amphibian species now appears bleak.

However, experiments with anuran larvae offer some hope. Unlike the situation in juvenile and adult frogs, in which the keratinized tissue just beneath the skin is affected, in infected larvae major symptoms of amphibian chyridiomycosis do not appear until metamorphosis, after which they soon die. Recent tests with infected larvae treated with an antifungal drug (Itraconazole) eliminated Bd and the metamorphs survived. In addition, two independent studies of frog populations that were recovering from Bd infections found that they contained individuals that could withstand future Bd exposure. This finding suggests that natural immunity is being selected for and may eventually give rise to immune populations.

Ultraviolet Radiation

The heavily pigmented upper surfaces of amphibian eggs and larvae that develop in areas exposed to sunlight protect these organisms against excess ultraviolet (UV) absorption. Montane species like the Yosemite Toad and the Sierra Nevada Yellow-legged Frog may be especially vulnerable to UV radiation, because there is less subfiltering effect of the atmosphere below the primary filter, the stratospheric ozone layer. The recent depletion of the ozone layer, partly or solely due to human-generated vapors, has now been well documented, and this raises the question whether the dorsal pigment protection of select amphibians is sufficient to effectively withstand such an increase. Like the feminizing effect of contaminants on male anurans, egg and embryo mortality due to the effects of excess UV absorption may not be apparent in the wild for many years, especially in long-lived species. It most likely has and will continue to affect some California anuran and aquatic-breeding salamander populations but at present can only be viewed as yet another factor in the still unsolved equation for amphibian declines.

In this brief review of just four of the potential causes for the marked decline of California amphibians, one feature stands out: they are all results of both intentional and unintentional interference by humans with the natural history of these animals. It therefore follows that such destructive practices can only be reversed if we earnestly accept the stewardship of these unique creatures from whose distant ancestors we have also evolved. It is the authors' most sincere hope that you, the reader, will join us in actively supporting and pursuing whenever possible those activities and policies that promote the preservation of amphibians and reptiles, both in California and the world.

Protective Status of Some Amphibians and Reptile Species

A fair number of California amphibians and reptiles have been assigned various levels of protective status under federal and state laws. What follows is brief review of the various protective categories and their designation by abbreviations in the natural history accounts.

In 1966 the U.S. Congress enacted the Endangered Species Preservation Act, the first legislation that afforded some protection to species that were exhibiting serious population declines and appeared to be on a path to extinction. A second bill, the Endangered Species Conservation Act, was passed in 1969, but neither act was strong enough to prevent some states (but not California) from ignoring many of their provisions. As a result, a more forceful and comprehensive bill, the Endangered Species Act (P.L. 93-205), was passed and signed into law on 28 December 1973, and today it remains the federal mandate for endangered and threatened species protection and recovery. Among all states, California led the way in protecting vanishing species with its passage of this state's 1970 Endangered Species Act, which prohibited the importation, taking, possession, and sale of endangered and threatened species.

An endangered species is one that is in danger of extinction throughout all or a significant part of its range. In this book a federal-listed endangered species is designated by **FE** and a state-endangered species by **SE**. A threatened species is one that is likely to become an endangered species throughout all or a significant portion of its range within the foreseeable future. A federally threatened species is designated by **FT** and a state threatened species by **ST**. The State of California has two additional protective categories. One of these, that of fully protected species (**FPS**), is actually the most protective category of all, in that unlike the federal or state endangered listings, the fully protected status precludes any taking (killing or collecting) of a so-listed species under any circumstances. The other designation is that of a California species of special concern (**CSSC**). This category includes species whose numbers are beginning to decline or whose habitats are continuously being reduced and thus require special attention and observation. In contrast to the other protective categories, there is no actual state law that affords legal protection to species in the CSSC category. However, when land use changes near or within their habitat threatens to impact such animals, the CDFG usually requests that protective measures be implemented before any such project can proceed.

One final note concerning endangered and threatened species is that some are really subspecies, and thus the state and federal laws governing their protection should more correctly be called the "Endangered Species and Subspecies Acts." This has led to confusion among some people about why we should be concerned about the loss of one small segment of a population (a subspecies) when the species, represented by other subspecies, is doing fine elsewhere. The answer lies in the basic concept of evolution, whereby natural selection will determine which characteristics within a species' total gene pool will persist through time to eventually emerge as a new species. By having a number of subspecies, each adapted to a different habitat situation, a species has a far better chance of persisting in the face of the rapid environmental change that our own species continues to create in nearly all of California's major habitats today.

The subspecies listed in this book differ from their parent species by relatively subtle morphological and genetic features, the detection of which often requires the capture of an amphibian or reptile for close-up inspection and, in some cases, tissue sampling for DNA analysis. However, as defined in the glossary, a subspecies represents a consistent variation among individuals of a species within a certain segment of that species' geographic range. Indeed, it is the environmental features of such range segments that, through the natural selection process, are responsible for such differences. Given that a primary function of this book is that of a field guide, we have therefore defined all subspecies by giving the geographic area within California where they are found. Thus by simply knowing your location in the field, you will know what subspecies of a given species complex you are viewing.

A number of these subspecies have been assigned to one or more of the previously mentioned protective categories. Because a sound knowledge of their ecology is vital when making decisions about their protection and survival, these have also been accorded a discussion of select natural history features similar to that presented for species.

Common and Latin Name Changes

In recent years there has been a notable increase in proposed name changes for amphibian and reptile species as well as those in other taxa. This is often a source of confusion to the field guide reader who has grown up, so to speak, with one name and is now asked to abruptly switch to another. The following is a summary of how most name changes come about.

A proposed name change that does not also entail a change in taxonomic status (subspecies elevation to species, etc.) is usually initiated by a petition to the International Committee for Zoological Nomenclature (ICZN), which then calls for written responses in support of or against the change. After the various opinions have been gathered and discussed, the Committee then acts as judge and jury to accept or deny the proposed change. As with election results, there are sometimes many who are reluctant to accept a newly sanctioned change and continue favoring their preferred choice. Further complicating this matter is the current trend of adopting a name change before it has completed an ICZN review. Such action appears to be based mainly on new names used in one or more current publications and/or on herpetological websites. The one certainty in all of this is that the scientific and common names of a number of amphibian and reptile species are presently in a constant state of change and will probably remain so for some time to come. Our review of recent books that address some of the amphibian and reptile species found in California and adjacent areas revealed that none are in full agreement with the others with respect to taxonomic nomenclature. Indeed, the senior author now feels that, as a young curator wishing to keep up to date the amphibian and reptile collections at the Museum of Vertebrate Zoology, UC Berkeley, he was often too quick to change specimen names, because many such changes did not persist.

Given these realities, we feel that this problem can best be addressed in our complete coverage of California's herpetological fauna by using those new names that appear most often in such books. When a new species, genus, or family name is introduced in this book, we have attempted to lessen confusion by following it with the former common or Latin name in parentheses. In most cases we have also included a brief explanation of how or why each change has occurred.

New Species, Genera, and Families

Recent advances in DNA sequencing technology have led to a rash of proposed new species, genera, and family groups. At the species level these often do not involve a name change but instead the elevation of one or more subspecies to full species status. One claim by some molecular taxonomists for such changes is that newly detected variations in selected DNA sequences amplify the differences between groups that were originally designated as subspecies based on their external morphology, and taken together the two now merit species status. However, those who question such status changes often wonder just how many DNA sequence differences are required for new species status, and must these occur in both the chromosomes and the mitochondria to be considered significant? In other words, just what are the rules? And because it is rarely known what phenotypic characteristics the various DNA differences may determine, it is possible that such DNA "evidence" may simply reflect the morphological characteristics on which the original subspecies status was based.

Noticeably absent from most new species proposals is whether or not the long-standing prerequisite for species status—that of being a reproductively isolated population—has been met. Can the proposed new species and its close relatives cross-breed and produce viable fertile offspring? If so, they are at best subspecies or races and should remain so. Unfortunately, such evidence is rarely a part of a new species "proposal package." Given this fact, plus the apparent lack of any universally agreed-upon genetic threshold beyond which a new taxon is justified, personal opinion and judgment will probably dominate this area for some time to come. For this book we have adopted the same policy for proposed new taxon acceptance as that for proposed nomenclature changes by attempting to be consistent whenever possible with recently published books that address California amphibians and reptiles and briefly commenting on the opinion(s) that support such changes.

Field Identification Keys

Separate keys to the families and genera of California salamanders, anurans, lizards, snakes, and turtles and tortoises are presented at the beginning of the natural history accounts for each of these groups, where final identification to species can then be made.

The keys are based on external features such as color and scale patterns, numbers and lengths of toes, and so on, so that in most cases identification can be achieved on the spot in the field, often without capturing the animal. Because highly active species may retreat from view before the appropriate key can be consulted, an initial digital camera snapshot will also aid greatly by allowing you to examine and reexamine features long after the animal has disappeared from view.

Each step in a taxonomic key requires a choice between two contrasting sets of features. One will direct you to the pages containing the appropriate natural history accounts for species identification, and the other directs you to a further set of choices. Almost all individual species accounts contain a color illustration of each California amphibian and reptile along with a verbal description of its identifying features and measurements, its preferred habitat, and a range map showing where it occurs in this state. The latter provides one of the most important clues to a specimen's identity, because many species have a relatively small geographic range, and one or more alternative choices may readily be eliminated if a given range does not include your location in the field. In a very real sense, working through a taxonomic key is similar to participating in a treasure hunt in which one discovery leads to another and eventually to the prize, which in this case is the correct identification of the species at hand. This process can be especially entertaining when people in a family or outing group all see the same animal in the field or view it on one person's digital view screen, and then together discuss and debate their way through a key.

AMPHIBIANS

Taxonomy, Anatomy, Physiology, and Behavior of Amphibians

A fair amount of confusion still persists among the general public as to what is an amphibian and what is a reptile. Some of this uncertainty may stem from the name of the discipline that addresses these two groups: herpetology. This term is derived from the Greek word herpeton, which translated means "a crawling animal." Snakes, legless lizards, and caecilians (the legless group of amphibians) do indeed crawl, and some salamanders and lizards may also appear to do so. However, by present-day taxonomic standards this is certainly not a sound feature by which to justify the clumping of these two very different vertebrate groups under one heading. The general body form of a salamander and lizard may also be confusing to a first-time observer, because they are both small, elongate, nonfurred vertebrates with four somewhat equal-size limbs and a long tail. However, upon closer inspection one will note that a lizard has dry, scale-covered skin, claws, and an ear opening, whereas a salamander has unscaled skin and lacks claws and ear openings. Further exploration of the internal anatomy, physiology, and behavior of these two groups reveals numerous major differences that show that many aspects of amphibian biology are similar to those of fish while those of reptiles are more closely allied with birds. With this in mind, discipline titles such as "amphib-ichthyology" and "orno-reptology" would seem more appropriate than herpetology. However, the authors will not attempt to promote such revelations in this book but instead will endeavor to enumerate in this and the following reptile section some of the many differences between these two vertebrate classes, using for the most part our own California species as examples.

Class Amphibia (as of 2010) contains about 6,187 recognized species, but that number continues to rise as a result of the ongoing discovery of new species, especially in the tropics, and the reclassification of existing species groups based on significant difference in DNA sequences. Frogs and toads (order Anura) far outnumber the other living amphibian orders, with around 5,459 species, compared to about 558 species of salamanders (order Caudata) and 170 current species of caecilians (order Gymnophiona). The latter legless tropical amphibians do not occur in California. In this state the lopsided worldwide anuran-to-caudatan ratio is reversed. At present California supports 42 salamander and 27 frog and toad species.

Despite this rather impressive number of species for our state, amphibians are often considered a scarce and relatively unimportant segment of the vertebrate species composition of an area. This is partly because many species are quite small, cryptically colored, and active mostly at night, and in many habitats this may lead to the incorrect assumption that they are scarce or absent. Although the total number of amphibian species is greatly outnumbered by those of birds (about 9,100), there are actually more species of amphibians on earth than those of mammals, and a fair number of field surveys have revealed that the number and biomass of some amphibian species in a given habitat site surpasses that of any resident reptile, bird, or mammal. One such study by the senior author in a Berkeley Hills redwood forest community concluded that the total number and perhaps biomass of either the salamander Ensatina (*Ensatina eschscholtzii*) or the California Slender Salamander (*Batrachoseps attenuatus*) was greater that that of any other resident vertebrate. The ability of amphibians to successfully compete with other groups in many habitats stems from the great range of variability in many major aspects of vertebrate biology that this class possesses. The following brief review of some of the anatomical, physiological, and behavioral characteristics of amphibians highlights several of these features.

Body Form and Locomotion

The body form of most modern-day salamanders approximates that of the first land vertebrates. The forelimbs and hindlimbs are about equal in size and project out from the sides of their bodies. On land they move slowly with a very deliberate and often awkward gait. The latter is most pronounced in aquatic-breeding species where the horizontal sculling action of the body and tail when swimming persists while walking and creates an undulating forward movement. The body of most salamander species is sculptured by vertical folds of skin (costal folds) separated by costal grooves (figure 2). Salamander feet are relatively small compared with those of anurans, and the toes of most species are unwebbed. An exception is the genus *Hydromantes*, in which the short toes are moderately webbed to form a platelike foot that provides maximum flat surface contact with often moist rock faces. This arrangement produces a high degree of surface tension between foot and rock, permitting locomotion on steep and often slippery inclines.

Salamanders have a long, muscular tail that propels aquatic-breeding species in water while the limbs are pressed backward along the body to create a more hydrodynamic form. In terrestrial species such as the plethodontid salamanders it serves as a balance organ, and in the genus *Hydromantes* it functions as a climbing stick when ascending smooth rock faces. Here the tough, blunt tip of the short, muscular tail is pressed against the rock surface behind the next hind foot to step forward. Then while that foot is elevated and moved, the tail provides a rigid support to that side of the body (figure 3).

Plethodontid salamander tails also function as a breakaway organ if grabbed by a predator. Some species like *Ensatina* have a predetermined zone of breakage at the base of the tail that facilitates loss, while in others like the slender salamanders (*Batrachoseps* spp.), which lack tail base

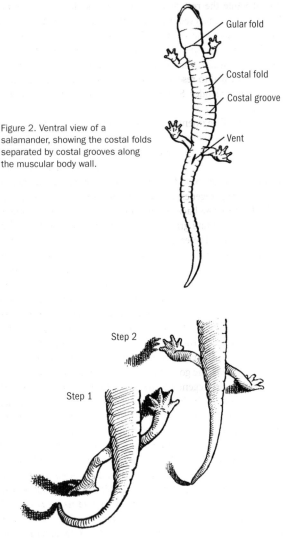

Figure 2. Ventral view of a salamander, showing the costal folds separated by costal grooves along the muscular body wall.

Figure 3. A two-step sequence of a *Hydromantes* spp. salamander that uses its broad, webbed feet and short supporting tail to move on steep, wet slopes.

constrictions, the tail may be severed almost everywhere. However, the severed tail segment does not immediately "die" but instead continues to vigorously wiggle and writhe for several minutes because of the anaerobic nature of its neuromuscular chemistry. This serves to concentrate the attention of the predator, which often proceeds to consume the severed tail segment while the rest of the salamander seeks cover. *Ensatina* and several other species also secrete a very sticky mucus upon being attacked, and while consuming the severed tail a predator's mouth or beak may be temporarily "gummed up," thus allowing even more time for the animal to escape. Once during a food preference study of week-old garter snakes, the junior author had to rescue several of these youngsters that attempted to eat California Slender Salamanders and in the process glued their mouths shut.

The severed tail segment will completely regenerate within a year or two in most species. Unlike lizard tail breakage, where a pseudo tail segment composed of a fibrocartilage rod, nonsegmented muscle, and a modified epidermal covering is produced, the regrown salamander tail contains the vertebrae, blood vessels, and neuromuscular system like the original. Appendage regeneration is not unique to the tail. A severed salamander limb can grow back within several years. Order Caudata is the only terrestrial vertebrate group that possesses this ability, and for many decades it has attracted researchers in the biomedical field, because a complete understanding of this process would be of profound importance to people who have suffered limb loss. At this writing, however, the mechanisms of salamander appendage regeneration remain almost as mysterious as when first observed. However, recent research has revealed that when a salamander loses a limb, cells in the layer where the break occurred differentiate to a stem cell–like state and then reinstitute the development sequence that generated the original limb in the embryo. A genetic database for the Mexican Salamander (*Ambystoma mexicanum*) is currently being developed, with the goal of identifying and fully sequencing the key genes involved in limb regeneration so that comparisons can be made with those in other animals, including humans. Needless to say, the discovery and successful application of this "salamander secret" would produce one of the most astounding medical revelations of all time.

The body and limbs of anurans differ markedly from those of salamanders. Most noticeable is the absence of a true tail in all adult forms. Within California anurans there are also a couple of major differences in body and limb form between the true frogs (*Rana* spp.) and the true toads (*Bufo* spp.). The true frogs have a slender body and narrow waist, whereas the true toad body is stocky and the waist is broad. Frog hindlimbs are also considerably longer than those in front. Their large muscular legs permit a fast "one leap" escape from a pond or creek bank into the safety of the water. Once there, large webs between the five hind foot toes make possible powerful backward thrusting swim strokes while the front legs are addressed backward to streamline the body (figure 4). This method of aquatic hindlimb propulsion is so efficient that it is a part of the required form for one of the four Olympic swimming events, the breast stroke.

One of the most versatile anuran foot designs is found in members of the chorus frog (treefrog) family Hylidae, which have moveable toe tips

Figure 4. The large webbed hind foot of a ranid frog.

Figure 5. The hard, epidermal "spade" on the underside of the hind foot of a spadefoot toad.

called toe pads or disks. The pads contain hundreds of flat-surfaced hexagonal cells on the ventral side. Mucus secreted from glands that open among these cells along with the moist surface of a leaf or rock upon which the chorus frog rests produce a high degree of surface tension between the many flat cell faces and the plant surface. This allows species like the Pacific Chorus Frog (*Pseudacris regilla*) to climb vertical stems and leaves of smooth-surfaced aquatic plants such as cattail and tule, thus adding a third dimension to their world. Anuran toe pads function on a variety of surfaces, including glass.

True toad hindlimbs are considerably shorter than those of most true frogs, and thus are less effective in escape by leaping. Instead these anurans usually execute either rapid short hops or walks, and sometimes even "run." The latter may often be faster than expected, to the frustration of a human pursuer. The undersides of the feet of some California toad species are shod with horny tubercles that protect against skin abrasion and increase gripping power. In the spadefoots (Pelobatidae) the hind foot is equipped with a sharp-edged metatarsal tubercle (figure 5). They use these spadelike structures to loosen and then push soil aside while rotating the body and alternately scraping with the hind feet as they burrow backwards into the ground. Although this method of excavation may seem quite inefficient to us, it certainly works well for these toads, which can burrow to depths of nearly 1 m (about 3 ft) under favorable soil conditions. With the appropriate corrections for the size differences between toad and human, this would be roughly comparable to you digging a hole with two tablespoons to a depth of about 20 times your body width!

Amphibian Skin

Any discussion of the functions of amphibian skin must be preceded by the reminder that it is first and foremost a respiratory organ, supplying up to 85 percent of the oxygen requirement in some lungless salamander species and allowing for the release of 80 percent of carbon dioxide in some mole salamanders. Even in anurans and salamanders with lungs the skin usually shares the oxygen acquisition function with the lungs, and during long periods of submergence the skin is the only active respiratory organ. To function as a respiratory organ, any tissue must not only be thin, moist, and highly vascularized, but it must also have a relatively large surface area that is in contact with air. These requirements negate the presence of any surface protective structures such as scales, feathers, or fur on amphibians, and thus they are the truly "naked" vertebrate group.

Given this overriding prerequisite of filling the role of a respiratory organ, the number and variety of other functions performed by amphibian skin are quite remarkable. One of the most important is the secretions from mucus glands that keep the skin moist so that it can perform its respiratory function. Skin mucus also protects against the entry of bacteria and molds, and it renders the body slippery, thus aiding escape from predators.

Another key function is the secretion of toxic or repellent fluids. Most of these are quite distasteful and can irritate the mouth lining and eyes of a potential predator, including humans. The junior author still vividly remembers his first encounter with the Green Toad (*Bufo debilis*) during a collecting trip to the State of Chihuahua, Mexico. This is a small, beautifully colored toad to which the descriptive term "cute" could justifiably be applied. However, shortly after making a hand capture, he adjusted his glasses with his capture (right) hand, and soon the soft tissue around his right eye swelled so that the lids completely closed and remain in this condition for nearly 12 hours. The potency of some amphibian skin toxins go well beyond the irritation stage. One of the best examples here is found in the three California newt species (*Taricha*), which produce "tarichatoxin," a substance that closely resembles "tetrodotoxin," the most poisonous nonprotein substance known. It acts by blocking sodium transfer across nerve and muscle cell membranes, which can swiftly lead to respiratory failure. This can occur only if a newt is eaten or the poison somehow gets into one's bloodstream, but it is still a good idea to rinse your hands after handling one. Before doing so, smell your hand or even the newt itself. You will note a mild pungent odor that is probably far more readily detectable by predators like the Raccoons, Striped Skunks, and Gray Foxes. If such predators have had a prior disagreeable encounter with a newt, the

Figure 6. An *Ensatina* salamander exhibiting its defensive posture with elevated body and waving tail.

association of bad taste and irritation with that odor alone may deter any further attempts at capture.

The behavioral presentation of toxic material to a predator is also an important component of this defense mechanism. A good example of this is found in *Ensatina* (*E. eschscholtzii*), in which the primary site of toxic secretion is the tail skin. Upon initially being disturbed it elevates and often waves its tail slowly back and forth while secreting a milky-appearing toxin, presumably to ensure that the predator's initial bite will be there and not on the body or head (figure 6). Once the bite or strike does occur, the tail breaks off, leaving the predator with what would literally be a bad taste in the mouth. Because most predators are residents of the same general habitat in which salamanders occur, the musky odor of many toxic secretions may be remembered and deter any further predation attempts by that individual upon all other individuals of that salamander species.

A more passive form of defense is provided by skin coloration. The dorsal and lateral portions of most amphibian bodies are cryptically colored so that they blend well into the predominant background color of their preferred habitat (figure 7). Unlike most other vertebrate groups, amphibians tend to remain motionless for long periods of time, which adds to the effectiveness of their camouflage. The cryptic coloration of amphibians may be enhanced by the pituitary gland secretion melanocyte-stimulating hormone, which controls dispersal or concentration of the dark pigment melanin within specialized skin cells called melanophores. By this method a species like the Pacific Chorus Frog can change its skin color from bright green to a dull greenish brown when moving from a growing to a dead cattail stalk. Melanophore-directed color change

Figure 7. The cryptic coloration of the California Chorus Frog blends well with mottled substrate.

is also present in aquatic amphibian larvae, in which the pineal gland hormone melatonin produces skin blanching in darkness. However, when the larvae are present at inshore sites during daylight, melanin is dispersed throughout the melanophores to produce a more cryptic body coloration. In contrast to the camouflage of most amphibians is the disruptive coloration of species like the Tiger Salamander (*Ambystoma tigrinum*), the California Tiger Salamander (*Ambystoma californense*), and two taxa of *Ensatina*. Here bold pale blotches on a black background disrupt the outline of the salamander body form, especially at night when these species and mammalian salamander predators are most active.

The ventral skin of most amphibians is usually not cryptically colored, because it is rarely exposed. An exception is found in the western newt species, where ventral color ranges from yellow to red and is readily displayed if they are tapped several times on the back. At such times they cease all forward movement, throw their head backwards, and elevate the tail forward while often extending the legs as far out as possible. This now presents the bright ventral coloration to a predator who originally saw only the dark dorsal body surface (figure 8). Birds are a major potential predator on newts, because they are more active on land during the day than most other salamanders, sometimes migrating to a breeding site in bright sunlight. The feeding responses of birds are based heavily on instinctive behavior. The body form and color of a potential prey item

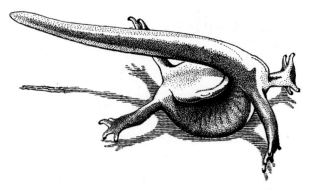

Figure 8. A Coast Range Newt's defensive posture, in which the yellow ventral color is suddenly revealed.

often serves as a sign stimulus that initiates a pecking response. In this case the pecks that initially strike a newt's back suddenly transform the body form and color to one that no longer elicits a peck-and-eat response. Such a color-based defense reaction is especially important, because most birds have excellent color vision but a poorly developed sense of smell and therefore may not react to the odor of toxic skin secretions as mammalian and reptilian predators do.

The surface layer (epidermis) of amphibian skin is shed periodically. People who are familiar with skin shedding in lizards and snakes are often surprised to hear this, because they have never seen shed salamander and frog skins in the field. One reason is that many species eat the shed skin, which serves as a source of moisture and nutrients. In anurans it often splits down the back, the limbs are pulled free, and then the old skin is taken into the mouth and swallowed. In *Ensatina* and presumably other salamanders, the skin first breaks free around the mouth, after which the animal slides forward with forelimbs trailing at its side until the loose skin reaches the chest region. The forelimbs are now pulled free as the body begins to swell anterior to the band of old skin. This swelling then proceeds ventrally, forcing the skin back to the region of the hind legs. Next these legs are pulled free and then used to push the skin backwards on the tail. The clump of old skin near the tip of the tail is now pressed against the ground and removed as the tail is drawn from it. At this point the salamander may perform its own version of recycling by turning and eating the sloughed skin clump (figure 9).

Sensory Input

Olfaction and Chemoreception

The sense of smell or chemoreception is well developed in most amphibians, especially salamanders, which appear to rely on it as much as their

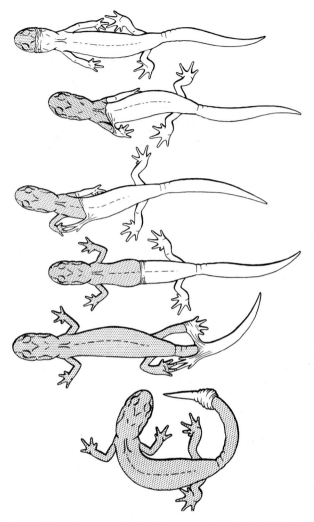

Figure 9. The salamander skin-shedding process, in which the old, light-colored skin is peeled back from head to tail and then usually eaten.

fish ancestors but unlike them have two reception areas instead of one. The first is the olfactory epithelium of the nasal passage, which detects airborne and aquatic odors. The second are the vomeronasal or Jacobson's organs, patches of sensory epithelium located near the internal openings of the nostrils. These detect odors from material in the mouth and inspired air, and also from molecules transported to the nasal passage by the naso-labial grooves (figure 10). The latter structures are unique to plethodontid

Figure 10. The head of an *Ensatina* showing the nasolabial groove passing from the upper lip to the nasal opening.

salamanders and consist of a pair of minute grooves that pass from each side of the upper lip to the nostrils. These salamanders often tap their snouts to the moist substrate as they walk, and water bearing a variety of organic molecules from the surface is transported by capillary action to the nasal openings. From there it is carried through the nasal chamber and on to the Jacobson's organs by cilia. Plethodonts use this system for both food location and in courtship behavior, and they may even employ it to track down a potential mate or detect an intruder.

The sense of smell is best developed in burrowing toads (family Pelobatidae), newts, and many plethodontid salamanders. Larval amphibians appear to rely on olfaction in ways similar to the way fish do. One is by imprinting on the odor of their natal pond or stream and then using that olfactory memory to locate these sites as migrating breeding adults in much the way salmon are able to return to their home spawning creek. Awareness of this behavior is important when relocating amphibians to a new breeding site. Recently the junior author and colleagues constructed a California Red-legged Frog (*Rana draytonii*) breeding pond and then stocked it with several egg clusters from a nearby site that supported a breeding population of this species. They hoped that the future adult residents would be imprinted on the odor of this new pond during larval development and would accept it as "home" and not wander away. To date this new habitat has a thriving resident adult California Red-legged Frog population.

Vision

Most amphibians have large, protruding eyes that provide for very good vision in most species. As a group, frogs have the greatest field of vision of all terrestrial vertebrates and can detect moving objects behind them probably as well as they can in front. This is actually one of the first bits of natural history information acquired by a young person who cautiously attempts to hand-catch a frog by sneaking up on it from behind, only to have it leap to safety in a pond as soon as his arm begins to move forward to grab it. Another often observed feature of the anuran eye is the nictitating membrane, a thin transparent sheet that is drawn up over the eye from behind the lower eyelid when a frog or toad dives and swims. In this instance it appears to function much like a swimmer's goggles. When anurans are on land, it serves as a "windshield" or cornea wiper to moisten and clean the eye. Many amphibians also have good depth-of-field perception made possible by the overlap of the visual fields of both eyes. This may be seen in the skill with which those with extendible tongues can direct the sticky tip directly upon an insect. Such binocular vision is best developed in the plethodontid salamanders, which rival many reptile, bird, and mammals in this visual ability.

Color vision tests with selected species of both anurans and salamanders have demonstrated good recognition of up to eight basic colors. This ability may be used to recognize preferred invertebrate prey species before attempting to strike at them. Contrary to the general mammalian rule that species with good color vision cannot see well in dim light, most amphibians see very well in the dark. Indeed, given that many salamanders and toads are highly nocturnal, such an ability is crucial to their well-being. Despite color vision and good visual acuity, it appears that the visual detection of prey movement is the primary way by which prey is located. Often an inanimate object that is within the normal size range of prey for a given species will be grabbed with the mouth or struck with the extended tongue only if it is moved slowly or wiggled. This requirement for prey movement is similar to fish visual detection of prey, where movement is also the primary sign stimulus for a bite or strike. Anglers who fish vegetated inshore waters occasionally experience this firsthand when the artificial fly or small lure they cast is grabbed by a large hungry ranid frog instead of a fish.

Hearing

Hearing as we know it in terrestrial vertebrates is absent in fish, although in some species sound-based vibrations from the swim bladder are transmitted to a rudimentary inner ear structure by small bony ossicles. This absence is probably due to the presence of the lateral line system, which detects pressure changes produced by currents and underwater sounds. Modifications of this system are still present in many amphibian larvae and permanently in aquatic salamanders and frogs. It does not, however, function in air, and thus the evolving amphibians needed a new system to detect the airborne vibrations we know as sound. Thus, in the anurans we find the basic hearing system that, with only a few variations, is the one that we and our fellow mammals have inherited.

In frogs and toads the basic sound reception unit consists of a large tympanic membrane or eardrum on each side of the head that is continuous with and usually colored the same as the rest of the head skin. There is, of course, no external pinna or "ear," because that would reduce the efficiency of the hydrodynamic body design. In the absence of protruding ear flaps, sound waves are received equally from all directions, and this ability complements an anuran's wide visual field in detecting both prey and predators behind them. Vibrations of the eardrum are transmitted to the inner ear by a single bone, the columella, which is homologous to the stapes, one of our three middle-ear bones. Amphibians have a second middle-ear bone, the operculum, which is not connected to the eardrum but instead forms a loose connection between the columella and the shoulder blade. In this manner vibrations received by the forelimbs from the substrate are also transmitted to the inner ear, which has two sensory membranes that convert fluidborne vibrations from the columella into neural impulses that are then sent on to the brain. One is the basilar papilla, which is stimulated by high-frequency sounds, usually above 1 kHz. The other is the amphibian papilla, thus named because it is unique to this group of vertebrates. It is sensitive to middle and low frequencies below about 1 kHz.

This dual system within the middle and inner ear of anurans enables them to hear a wide variety of sounds, ranging from the shrill, high-pitched vocalizations of their own and other species to low substrate vibrations produced by falling raindrops or approaching predator footsteps. However, salamanders do not have the ability to receive such a wide range of auditory information. They have no eardrum, and in the adults the columella is fused to the skull. Some of the more primitive species have both of the inner-ear membranes found in anurans, but in more advanced forms only the low-frequency sensing amphibian papilla is present. This leaves most salamanders with only the vibratory input from their shoulder blades through the operculum bone to the inner ear as their mode of hearing. More simply put, salamanders hear through their feet. This may not be such a disadvantage as it seems, because much of their invertebrate prey lives in the soil surface layer where only the odor or movement vibrations they produce may reveal their presence. The lack of a mechanism to receive middle- and high-frequency sounds is compatible with the observation that, except for a few species that occasionally utter a grunt or squeak, salamanders are a very silent group.

Respiration

The evolutionary history of amphibians from aquatic to land life is perhaps best seen in the transitions of the respiratory system from larval to adult in aquatic-breeding forms. The respiratory organ of all aquatic larvae is the gill. In salamander larvae and permanently aquatic adult forms, there are three pairs of external gills that protrude through gill slits to the outside (figure 11). The base of each gill contains gill rakers similar to those found in fish. They remain in the pharyngeal cavity and prevent small food particles from escaping as water is drawn into the mouth and then pumped out the slits by pulsations of the throat. Gill size and structure vary according to the dissolved oxygen in a species' larval habitat. Larvae like those of the California Giant Salamander (*Dicamptodon ensatus*)

Figure 11. Ventral view of an aquatic-breeding salamander larva, showing its featherlike external gills.

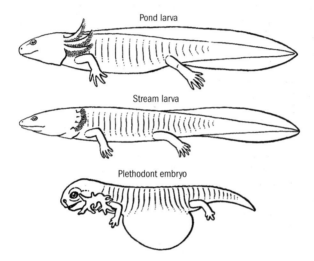

Figure 12. The long, featherlike gills of a pond-dwelling salamander larva, the short, broad gills of a stream-dwelling salamander larva, and the membranous-type gill of a lungless salamander embryo.

that live in cool, well-oxygenated streams have short gills with broad filaments, whereas pond-dwelling larvae of species like the Long-toed Salamander (*Ambystoma macrodactylum*) have long, featherlike gills. A third, membranous-type gill is found in lungless salamander larvae, which pass through their entire larval stage within a gelatinous egg capsule. These broad, flat gills are pressed against the inside of the capsule and receive oxygen by diffusion through it (figure 12).

The gills of the anuran larva or tadpole begin to develop externally, but soon a layer of tissue resembling a soft version of a fish operculum grows over them to form a chamber with usually a single small opening to the

Figure 13. The passage of water into the mouth, over the internal gills, and out the spiracle of an anuran larva, demonstrated through the use of ink-diluted water.

outside, the spiracle. A new set of larger gills now develop within the gill slits inside this chamber as the small, encased original gills regress. Water is drawn in through the mouth, over the gills, and then out the spiracle by the pumping action of the tadpole's throat (figure 13).

During metamorphosis, the gills of most California amphibian larvae are reabsorbed, and adult respiration takes place to varying degrees over the newly created lung surface, body skin, and mouth/pharynx lining. Most anurans and aquatic-breeding salamanders respire somewhat equally through the lung and body skin, with the mouth and pharynx (buccopharyngeal) lining also providing a small additional respiratory surface. However, the plethodontid or lungless salamanders must rely solely on the skin and buccopharyngeal lining for oxygen and carbon dioxide exchange. They are able to accomplish this because of the comparatively large body surface area of the plethodontid body in comparison to its small mass. This, added to the salamander's miniscule oxygen requirements in comparison with those of a small mammal or bird, makes cutaneous respiration for this group a quite logical respiratory mode. A high dependence upon the skin for respiration is also seen in some salamander forms having lungs, especially those like the Southern Torrent Salamander (*Rhyacotriton variegatus*) that spend most of their adult life in cool, well-oxygenated water.

Another reason for the important role of the skin in respiration is that even though lungs may be present, an efficient system for ventilating them is not. Amphibians lack both a rib cage and diaphragm, the essential components of our lung ventilation system. Instead, air is drawn into the amphibian mouth through valvular nostrils as the throat is lowered. Once the expanded pharyngeal area is filled with fresh air, the throat floor is raised as the nasal valves close and the glottis opens, thus allowing the air to pass on to the lungs. Exhalation is accomplished by the elastic recoil of the lung walls that were expanded by the inflow of the air. A mild contraction of the body wall also aids in the exhalation process. Buccopharyngeal respiration is accomplished in the same manner, except that the glottis remains shut, the nasal valves stay open, and the air passes rapidly in and out across the moist, vascularized lining of the mouth and pharyngeal region as the throat floor is rapidly lowered and raised. These continuous throat flutters can be easily observed when viewing an anuran or salamander at eye level in a terrarium. Occasional lung ventilation can be distinguished from buccopharyngeal respiration by noting when a contraction of the body wall occurs, indicating that lung air has just been expelled.

Although these amphibian respiratory mechanisms seem grossly inferior to those of higher vertebrates, they actually support the general activity patterns of salamanders and anurans quite well. As already mentioned, these animals spend much of the time either sequestered or remaining motionless along a pond shore or within the leaf and branch litter of a forest floor. When they do move, it is usually by short bursts of activity followed again by a period of quiescence. Even in water anurans may float motionless at the surface, and salamanders will often hang suspended in the water column. Here the lung takes on an additional role of buoyancy compensator, similar to that of the fish swim bladder, so that no energy expenditure is needed to "tread water." These behaviors are in sharp contrast to the higher degree of activity one sees in most reptiles, mammals, and especially birds. When amphibians require rapid activity for escape or catching prey, it is usually short-lived, as in the case of a frog's leap into a pond. Indeed, the first few such leaps must bring a frog to the safety of the water, because their slow oxygen supply system cannot keep pace with sustained rapid muscular activity. There is also no myoglobin in amphibian muscle to provide for short-term storage of oxygen, which can be made available as rigorous muscle activity begins. Instead, the limited oxygen supplied by the blood is soon consumed, and muscle metabolism switches from an aerobic to an anaerobic mode. The latter is also short-lived, because a product of anaerobic metabolism, lactic acid, soon accumulates to a level at which it causes limb muscle contractions to cease. A practical application of this bit of amphibian physiology is seen in the annual Calaveras Jumping Frog Contest, in which contestants are judged on the distance of their first three jumps. These are usually quite spectacular and greater than most people could jump from a squatting position. However, beyond that initial strong effort, the leap distance dwindles to a point where the frog simply cannot continue jumping because of the accumulation of lactic acid in the muscles.

Osmoregulation

Except for those who occasionally take long desert hikes or run marathons, Californians are rarely concerned about body water loss or gain, especially in this present era of the personal water bottle. However, for amphibians hydroregulation, the prevention of desiccation or excessive hydration through behavioral and physiological adjustments, is perhaps the dominant feature of their existence. Here the vital role of the skin as a respiratory organ dictates to a great extent an amphibian's habitat and behavior. Because oxygen molecules can readily pass through a membrane only if its surface is wet, the skin surface of amphibians must be kept moist. This is most easily accomplished when amphibians restrict their above-ground activities to areas containing wet vegetation or along pond and creek shores where a "quick dip" is always possible. Rain or heavy fog also provides ideal conditions for surface activity, and lengthy upland migrations or wanderings of some species can be successfully accomplished only under such conditions.

When direct contact with water is not possible, skin contact with moist or saturated surfaces may still prevent desiccation. Anyone who has searched for salamanders in upland areas knows that the places to look are under flat ground-cover objects or within rotting logs where moist substrate conditions are maintained even when the surrounding above-ground areas have dried. Many anurans can also replenish body water from moist substrates by absorption through the seat patch, a highly vascularized area on the posterior ventral half of the body. When in a "sitting" position, it is pressed against the ground, and if the substrate is moist, water molecules will pass along an osmotic gradient into the seat patch's rich capillary bed. When on dry substrates that could create a reverse water movement out of the capillaries, the ventral abdominal skin is lifted off the substrate and wrinkled so that adjacent segments of the capillary bed are pressed to each other and not to the ground.

The seat patch is best developed in toads, especially those species that inhabit xeric habitats where the only source of surface moisture for part of the year may be the moist bottoms of drying ponds or creeks. This is just one of several osmotic adaptations that desert-dwelling amphibians have acquired to persist in this osmotically stressful habitat. The primary source of water for desert anurans comes from the abundant invertebrate prey that is abroad each night on the desert floor. For instance, a 10-g (0.3-oz) ground cricket (*Nemobius*) contains about 8 g (0.25 oz) of water, and when eaten by a desert toad, most of this is added to its body water load during the digestive process. Adding to this "preformed water" source is the release of about 0.4 g (0.01 oz) of "metabolic water" when the protein and fat of the cricket body are utilized at the cellular level. Thus, on an average desert foraging night when the higher evening relative humidity decreases the rate of body surface evaporation, several such meals may adequately replace the water lost by a toad through mucus discharge needed to maintain skin respiratory function.

Even when standing water is available in desert pools or creeks, the best foraging for anurans is often achieved by moving through adjacent upland areas. To ensure a body water reserve during such forays, a species such as the Red-spotted Toad (*Bufo punctatus*) leaves a desert pool with its large bladder full of urine. Then as water is lost through skin mucus discharge it is replaced by the reabsorption of this "urine canteen." The large urine load that most land-foraging anurans carry is also discharged if they are grabbed and held in a predator's mouth. Those who have hand-captured toads in the field have experienced the wet-hand version of this protective measure.

In many desert areas there is no surface water available for most of the year, and drought conditions may become so severe that invertebrate prey with their preformed water source also becomes temporarily scarce. Under such conditions, estivation, a retreat from all surface activity due to exceptionally hot, dry conditions, remains the only option for desert anurans. The spadefoots are the most accomplished in this area and can burrow down to moist soil layers that often lie well below the surface. They stop their descent when they arrive at a soil layer where the osmotic pull exerted on water molecules by dissolved materials in their blood is greater than the hydrostatic force that holds water molecules between the soil particles. Here they remain while slowly accumulating just enough water from the surrounding soil to maintain their highly reduced metabolic state. If the water content of the soil layer to which they initially burrowed decreases, these toads resume their digging until an appropriate deeper soil moisture situation is encountered. They may also begin to retain some urea in the blood rather than filtering it all out in the kidneys. Urea is relatively nontoxic at low concentrations in the blood, and its presence there increases the osmotic pressure or pull on water molecules in the surrounding substrate. It is interesting to note that blood urea retention is also the means by which sharks, rays, and skates (Chondrichthyes) osmoregulate in their marine environment.

For those amphibians that spend their larval life and part of their adult existence in water, a completely different osmotic problem exists. The gills and skin readily allow the passage of water molecules, which follow an osmotic gradient from the freshwater of the pond or stream to the capillary blood. If this movement is left unchecked, cellular water balance will surpass acceptable limits, resulting in a cessation of all body functions. This is the same situation that exists in all freshwater fish species, and both groups have a highly functional kidney that extracts excess water from the blood at about the same rate at which it enters at these respiratory surfaces. In freshwater fishes, this high kidney function continues uninterrupted throughout life. In water-oriented amphibians, however, active kidney function can be detrimental when they come out of water to forage or retreat to a burrow system. At such times kidney function is shut down by hormonal action until the next submergence occurs. Indeed, one of the most dynamic shifts in vertebrate kidney function occurs when a frog makes a leap from land into water. In contrast to aquatic and semiaquatic species, the land-based plethodontid salamander species cannot cope with extended periods of emersion in water. Their ability to conduct all life processes on land has apparently relaxed selection for good kidney function to the extent that they will die from cellular hydration if trapped in a flooded burrow or low-lying area.

Body Temperature and Thermoregulation

Amphibians are ectotherms, and except for a small number of species that actively bask, their body temperature closely parallels that of the surrounding air, water, or soil. As classic stenothermic (narrow temperature range) endotherms, it is often hard for us to envision a eurythermic (wide temperature range) ectothermic life style. The thermal profile of the plethodontid salamander Ensatina (*E. eschscholtzii*) as shown in figure 14

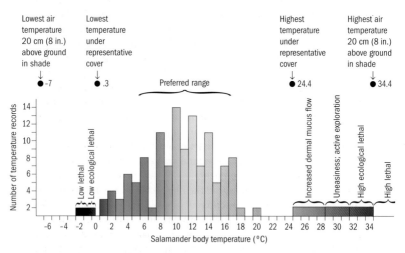

Figure 14. A thermoprofile of the plethodontid salamander Ensatina (*Ensatina eschscholtzii xanthoptica*), recorded from adult individuals over a four-year period in its natural habitat in the Berkeley Hills, California. Bar color denotes potential range of Ensatina body temperature from cold (blues) through preferred range (green–yellow–orange) to the high-lethal point for this species.

provides an introduction to the amphibian thermal world and the various behavioral and physiological responses that occur within it.

At the low end of the thermal range for nearly all animals is the critical minimum (–0.6 degrees C or 31 degrees F), the temperature at which ice crystals begin to form within the cells and eventually rupture the cell walls. Avoidance of this range can only be achieved by retreating to sites such as deep burrows or pond bottoms where freezing does not occur. Immediately above the critical minimum for Ensatina is a narrow, one-degree Celsius ecological lethal range. Here fully coordinated neuromuscular activity is not possible, and therefore the salamander is unable to exercise normal predator avoidance behavior. A test for this condition can be performed by turning a frog or salamander over on its back to see if it can quickly right itself. An amphibian may enter its ecological lethal range when a sudden drop in the ambient temperature occurs as the animal is migrating or foraging away from a safe retreat site. The senior author once observed a vivid example of this situation when he encountered an Ensatina slowly moving across a road during a sleet storm. Its cloacal temperature was 1 degree C (33.8 degrees F), and given the increase of ice water on the substrate surface, an ecological lethal state would soon have set in.

The wide span of body temperatures at which Ensatinas normally function begins at the one-degree Celsius point and extends to about 20 degrees C (68 degrees F). This preferred body temperature range reflects a variety of microhabitat situations that are dictated by day-to-day weather conditions and by the foraging and retreat site choices made by the salamander. Thermal acclimatization, a long-term adjustment to seasonal environmental temperature changes, may also occur. This process, which is also present in temperate freshwater fishes, allows for the efficient function of basic physiological processes throughout a wide range of body temperatures. This is in marked contrast to our high-temperature stenothermic body temperature regime, in which even small variations from the norm such as fever or chills result in malfunctions of our metabolic system.

Because most amphibians do not move far from cool-understory or aquatic habitats during the day, it is unusual for a salamander or anuran to become trapped in a warm, stressful situation. When this does occur, the first defense against hyperthermia is the increase in skin mucus flow. Like sweating in humans, this increases surface evaporative cooling, which may retard body temperature rise for a short time. If it continues to rise, the normally sedate behavior of salamanders such as Ensatina switches to one of rapid exploration of its immediate surroundings as if it were seeking a place to retreat from the mounting heat stress. If either of these measures fails to check body temperature rise, the latter may pass into the high ecological lethal range in which cellular enzymes begin to denature. As in the low ecological lethal range, the righting reflex is again impaired.

However, here the cause is not the partial shutdown of enzyme activity; instead, there begins an irreversible destruction of enzyme molecular structure, and the longer the high thermal state persists, the less likely that recovery will occur. At 35 degrees C (95 degrees F), the high lethal temperature for Ensatina, essentially all enzymes are destroyed and death occurs. Each species and subspecies of adult and larval amphibian has its own thermal profile that dictates to a large extent the habitat(s) in which it can live.

A deviation from classic ectothermy occurs in several anurans, particularly those that inhabit cool habitats. Montane ranid species such as the Sierra Nevada Yellow-legged Frog (*Rana sierrae*) and the Cascades Frog (*Rana cascadae*) regularly elevate their body temperature well above the surrounding cool air and substrate temperatures by basking. Warm-water species like the California Red-legged Frog (*Rana draytonii*) and the American Bullfrog (*Rana catesbeiana*) also engage in extensive basking at or near a pond or stream pool edge. The physiological reward in all cases is a higher degree of neuromuscular coordination, which in turn promotes better feeding- and escape-associated responses. Indeed, throughout the vertebrate world most species that exhibit high neuromuscular coordination are those that maintain a body temperature in the 35- to 40-degree C (95- to 104-degree F) range when active. This is the body temperature range of the endothermic birds and mammals and the heliothermic (basking) reptiles, and here in the basking anuran species we see the first examples of terrestrial ectotherms engaged in behavioral thermoregulation that can elevate their body temperature to the low end of this range.

However, in the complex world of amphibian physiological ecology, an advantage achieved in one area often creates a problem in another. The higher body temperature accrued by basking may stimulate an increase in skin mucus discharge, and when on dry land on a warm, sunny day, an anuran slowly desiccates as it warms. This situation may be prevented if the basking site is a moist shoreline area where the anuran's seat patch can absorb water to offset the dorsal and lateral skin surface loss.

Amphibian Reproduction

Amphibian reproductive biology is more varied than that of any other terrestrial vertebrate class. The name amphibian means "two-lived," and indeed, there is no better example of two contrasting lifestyles than that of the legless, fishlike, herbivorous anuran tadpole and the tetrapod, carnivorous, land-oriented adult frog and toad. In aquatic-breeding amphibians, the larval form, so prevalent in fish and most invertebrates, makes its only appearance in the tetrapod vertebrate classes. Within the major amphibian groups, the larval forms differ greatly, and the majority of California salamander species, the plethodontids, have no free-living larval forms at all. Given this wide spectrum of reproductive biology, the three major modes will be reviewed separately.

Anuran Reproduction

The reproductive season for most anurans commences with a gathering of both sexes at the aquatic-breeding or spawning site. For ranid frog species that inhabit permanent ponds or stream pools, sometimes the adult breeding population is already in place. However, for other species such as the Pacific Chorus Frog or Western Spadefoot (*Spea hammondii*), which usually spawn in seasonal ponds and then forage and/or estivate some distance from such sites, finding a spawning site can be a crucial factor in achieving reproductive success. Locating such sites is often further complicated by the highly variable rainfall and surface runoff patterns in some areas, which may produce an acceptable seasonal pond at one location on one year but then at another site on the next. These temporary wetlands are often used by more than one anuran species, and thus it is important for a given frog or toad to be able to locate a site where its own species is gathering and once there to be able to distinguish individuals of its own kind. These and other reproductive-associated behaviors have brought forth through natural selection one of the most unique features of anuran biology—voice.

Voice

As perhaps the most constantly vocal of all vertebrate species, we may tend to forget just how relatively scarce the use of voice is in many other groups. Except for a few species that emit clicks or grunts during spawning, the fish world is a very silent one. The same holds true for the salamanders and nearly all reptiles. Even within the mammals, only in a few groups such as the marine mammals, bats, elephants, and selected primates is sound production a major factor in their natural history. However, it is the birds and anurans in which vocalization has emerged as a key factor in their respective life histories.

The anuran sound-producing mechanism set the basic anatomical plan for all further vertebrate vocalization. Sounds are made by forcing lung air over vocal cords enclosed in the larynx, an enlarged segment of the trachea. In most frogs and toads, the sound is amplified and often modified by the continued passage of the pulsating air waves from the larynx into a single or paired balloon-like vocal sacs in the throat. The vibrating wall of the sac then functions like the cone of a speaker and transfers the sound energy or waves to the surrounding air, water, or even the ground (figure 15).

Anuran vocalizations are species-specific call notes, consisting of one or two sounds of innately determined frequency and pulse rate that are repeated again and again. This differs from bird song, in which several notes are combined to form a melodic phrase that is partly innate and

Figure 15. Lateral view of a calling male anuran with inflated vocal sac.

partly learned. With the exception of a few baritones such as the California Red-legged Frog and basses like the Bullfrog, the call notes of most California anurans are in the higher frequencies. Some calls, such as those of the Foothill Yellow-legged Frog (*Rana boylii*) are relatively soft and often produced underwater, and thus often go unnoticed by the human listener. Others like those of the spadefoots (Pelobatidae) are very loud and may be heard over a kilometer (0.6 mi) or more away on a calm desert night. The production of such sound energy by a vertebrate that can be held in your closed hand is especially remarkable in that it has no rib cage or diaphragm, the muscular bellows of mammalian sound production. Instead the body wall muscles provide the force for air movement from lungs to vocal sac, and these are well developed in the sex that does all the calling—the male.

The first few males to arrive at a seasonal pond initiate the reproductive season by producing a species-specific "advertisement call," which attracts all other males and the female within hearing distance of the site. During a survey of the amphibians and reptiles of Chihuahua, Mexico, the junior author and his colleagues began to hear the call note of the Red-spotted Toad during the evenings following the first summer rains in their desert camp area. For the next several days they looked for the temporary spawning pool that they were sure had formed nearby but could not find it. Yet each evening as they sat in camp this toad's high musical trill reminded them that once again they had failed to find the "callers." It finally dawned on them that if they were going to locate these toads they would have to behave like toads, and so the next night they set off in the direction of the calls, leaving a lantern on a high rock near camp as a homing signal. After more than an hour of "following their ears," they finally came upon the breeding pond, sequestered within a large boulder area, and when they looked at their camp beacon lantern, it was only a slight glimmer, at least a mile (1.6 km) away.

Male call notes also serve other functions. The well-known two-parted or diphasic "re-bit, re-bit" call of the Pacific Chorus Frog changes to a monophasic "encounter" call when another male approaches and serves to advertise and defend the caller's territory in much the same way a male songbird's vocalization does. The most important respondents to a male's call are the females. If more than one species are present, a female's innate response to her species' specific call note guides her to males of her own kind so that the reproductive effort will not be lost through interspecific spawning. If she finds several males calling nearby, it is often she and not they that makes the choice of a partner, which again is based on voice. This would be a near-impossible task if all males in the area were producing their notes in unison like the tenor section of a men's chorus holding a long full note. Instead, these members of the anuran choral alternate their calls when near one another so that the quality of each may be briefly "appreciated." The result is a chorus that in a species like the Pacific Chorus Frog can be very loud and complex, often consisting of many duets, trios, and quartets, each led by a "chorus master" who initiates each calling period. The eventual choice that a female makes from such groups appears to be based on the forcefulness or other sound qualities of the caller, which often reflects his size and general vigor.

Fertilization

With only one exception, the tailed frogs (*Ascaphus* spp.), in which males deposit spermatoza within the females' cloaca, anurans are an externally fertilizing group, and a behavior called amplexus is the primary breeding act. Here the male grasps the female around her waist (pelvic amplexus) or behind her forelimbs (pectoral amplexus) from the dorsal side so that his cloacal opening is positioned near and above that of his mate's, and when she sheds eggs into the water, he passes spermatozoa (figure 16). The externally fertilized eggs then develop into the familiar anuran tadpole or "polliwog." At hatching it has external gills and an adhesive disk around the mouth with which it attaches to objects while it absorbs the last of its yolk stores and completes its embryonic development. Within a few days the gills are repositioned inside the mouth cavity, and the mouth develops sharp, scraping mandibles. The anuran larva is essentially a small fish. It swims by the horizontal wagging of its tail, the dorsal and ventral fins of which are larger in pond-dwelling species than in stream forms. The head and body contain numerous lateral line organs, the water pressure–sensing system found in most fish. The tadpole mouth area is quite complex, and in addition to the mandibles contains several transverse rows of comblike labial teeth that are used to scrape algae and other plants and decaying matter from objects in the water. Labial tooth counts are often used for larval species identification. Labial tooth formulas are expressed as fractions (2/3, 6/6, and so on), with the numerator indicating the number of rows on the upper lip and the denominator the number on the lower. A fringe of oral papillae rings much of the mouth area and may function as tactile and taste sensors for examining a potential food source (figure 17). The newly hatched larva soon grows a long, coiled intestine, and once

Figure 16. A male California Red-legged Frog amplexing a gravid female. A newly laid and fertilized egg cluster attached to reed stems is in the background.

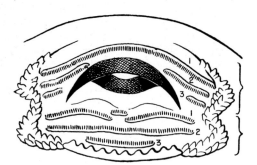

Figure 17. A tadpole mouth showing the dark mandibles surrounded by comblike labial teeth and an incomplete ring of oral papillae. The numbers indicate upper and lower labial teeth rows.

the yolk sac is completely absorbed it assumes a vegetarian-based feeding mode that it retains until metamorphosis.

Metamorphosis

Anuran metamorphosis is one of the most dynamic events of its kind in the animal world. Within the short span of a week or two, first hindlimbs and then forelimbs appear. The forelimbs actually develop first beneath the

Figure 18. The life cycle of an anuran (spadefoot toad), showing the progression from eggs through larval growth stages to newly transformed adult.

anterior body skin but do not break through into view until the hindlimbs begin to lengthen. Next, eyes begin to bulge from the top of the head, and the tail is absorbed through the action of lysosomes, enzymes that reduce protein molecules to amino acids. A tympanic membrane develops on each side of the head, and the small rasping lips are reworked into large, gaping jaws that in several families have rows of fine teeth. Lungs develop while the gills are slowly absorbed, and the long tortuous vegetarian gut is transformed into a short carnivorous digestive system. Figure 18 presents a pictorial review of this entire cycle for the Western Spadefoot.

The anuran larval–adult life cycle entails the most varied utilization of habitat resources in the tetrapod vertebrate world. The primarily herbivorous tadpole occupies the primary-consumer feeding niche, and then at metamorphosis the newly transformed frogs or metamorphs assume the role of secondary consumer and feed on a variety of invertebrate species. As they approach full adult size, large species like the California Red-legged Frog and American Bullfrog often become tertiary consumers and feed on a variety of small vertebrates, including their newly metamorphosed young. Indeed, in some isolated ponds where small vertebrate food such as the Pacific Chorus Frog are scarce or absent, the energy flow from sunlight and aquatic plants to the top of the food chain may take place within one anuran species.

Aquatic-Breeding Salamanders

Like many anuran species, the reproductive year of most aquatic-breeding California salamanders begins with a migration to the spawning pond or creek pool at the onset of heavy winter rains. In this silent group of amphibians voice and hearing cannot be employed to locate a breeding site that may be some distance from the upland summer estivation area. Instead they rely on other navigational guides such as celestial orientation and the recognition of familiar odor of the breeding site upon which they may have been imprinted as larvae or metamorphs. Males usually arrive at the breeding habitat first and recognize the arriving females by their scent.

The most striking difference between breeding in anurans and western North American salamanders is that fertilization is internal and is accomplished when a female picks up with the lips of her vent a sperm packet or spermatophore that the male deposits on the substrate. To facilitate this act the spermatophore sits on a little cone of gelatinous material much like a golf ball sits on top of a tee. Once inside the female's cloaca the spermatophore releases spermatozoa, which then swim to a special chamber on the cloacal roof where they are stored and released to fertilize the eggs as they are laid. Behavior associated with spermatophore deposition and pickup varies between species. The male Tiger Salamander leads the female over his deposited spermatophore as she nudges his tail base, apparently following an attractive scent released in his cloacal region. In other species like the Long-toed Salamander (*Ambystoma macrodactylum*) and the western North America newts, this "tail-nudging walk" is preceded by amplexus, similar to that seen in spawning anurans (figure 19). Male newts are induced to perform amplexus by the female's scent and occasionally are so ardent in this response that an aquatic ball of males may form around one female. However, in these salamanders amplexus has nothing to do directly with the actual fertilization of the eggs.

Like the anuran tadpoles, the newly hatched larvae of water-breeding salamanders look very much like young fish. They have a well-developed tail fin, no legs, and a still-present yolk sac. However, here the similarity

Figure 19. Male newt (top) amplexing a female before the actual fertilization act.

ends. Salamander larvae have external rather than internal gills throughout larval life. Most newly hatched larvae also have a pair of small, laterally projected balancer organs behind the mouth that provide stability on the pond or stream floor until the legs develop. The latter occurs within a few weeks after hatching. Both forelimbs and hindlimbs grow at the same time and enable the larvae to walk along the floor of the aquatic habitat where they do much of their foraging. The feeding niche of salamander larvae differs markedly from larval anurans. They are carnivorous, and their role as a secondary consumer continues in the adult form. There is also an occasional movement to the tertiary-consumer niche in both the larval and adult stages, which often takes the form of cannibalism.

Aquatic salamander metamorphosis is less dynamic than that of anurans, in that the larval body form is quite similar to that of the adult. The external gills are absorbed as lungs develop, and the tail fin also disappears. The legs become more muscular because they now have to support the body on land, and in most species the eyes bulge from the head. The metamorphs usually move from the breeding site to upland feeding or estivation areas, or both, when moist evening weather conditions favor land travel. They return to their "home pond" when sexually mature to breed.

Terrestrial-Breeding Salamanders

Reproduction in the terrestrial members of the lungless salamanders (Plethodontidae) represents one of the most important evolutionary advances of amphibians, in that it eliminates the necessity of a wetland habitat somewhere within a species home range and therefore has allowed for the dispersal of many species into large tracts of upland areas where ponds and streams are absent. In this unique form of reproduction, amplexus is omitted. Instead, "courtship" is conducted with a complex series of olfactory and tactile stimuli as depicted for Ensatina (*E. eschscholtzii*) in figure 20. These culminate in the "tail-straddling" walk during which the male deposits a spermatophore and then directs the female to retrieve it with her vent lips.

Figure 20. "Courtship" sequence of *Ensatina*, in which the female is attracted by the male's pelvic scent gland (top) stimulating her to perform the "tail-straddling walk" (middle), after which spermatophore exchange takes place.

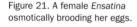

Figure 21. A female *Ensatina* osmotically brooding her eggs.

Unlike most amphibian eggs, which are pigmented on the upper side that faces sunlight, the eggs of our lungless salamanders are unpigmented because they are laid in dark places. They are also quite large in relation to the adult body size than those of water-breeding salamanders because of the large yolk that nourishes the embryo through its complete development to a miniature adult within the gelatinous capsule of the plethodontid egg capsule (plate 17). The nest site is usually a moist rock crevice, under bark of a rotting log or ground cavity, but never in water. Some

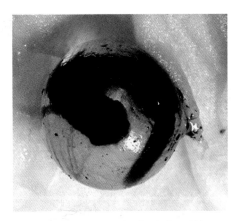

Plate 17. Slender salamander developing within egg.

Plate 18. Recently hatched slender salamander placed on a penny.

lungless salamanders brood their eggs. Here the term "brood" is used in the osmotic sense, because a female salamander's body temperature is that of the surrounding air and substrate. Instead, she secretes copious amounts of water-laden mucus during dry periods to prevent egg desiccation (figure 21). A moist egg capsule surface is also necessary for the passage of oxygen to the large membranous gills of the embryo, which are pressed against the inner surface of the egg capsule. At hatching a precocial miniature adult emerges and begins feeding on small soil invertebrates. Such newly hatched juveniles are among the smallest tetrapod vertebrates in the world. Several hatchlings of smaller species like the slender salamanders (*Batrachoseps* spp.) can fit on a penny (plate 18).

Salamanders (Order Caudata)

Mole Salamanders (Family Ambystomatidae)	69
California Tiger Salamander (*Ambystoma californiense*)	70
Tiger Salamander (*Ambystoma tigrinum*)	75
Barred Tiger Salamander (*Ambystoma mavortium*)	75
Long-toed Salamander (*Ambystoma macrodactylum*)	78
Northwestern Salamander (*Ambystoma gracile*)	82
Giant Salamanders (Family Dicamptodontidae)	84
California Giant Salamander (*Dicamptodon ensatus*)	85
Pacific Giant Salamander (*Dicamptodon tenebrosus*)	88
Torrent or Seep Salamanders (Family Rhyacotritonidae)	89
Southern Torrent Salamander (*Rhyacotriton variegatus*)	90
Newts (Family Salamandridae)	92
Coast Range Newt (*Taricha torosa*)	96
Sierra Newt (*Taricha sierrae*)	96
Rough-skinned Newt (*Taricha granulosa*)	100
Red-bellied Newt (*Taricha rivularis*)	102
Lungless Salamanders (Family Plethodontidae)	104
Woodland Salamanders (Genus *Plethodon*)	106
Dunn's Salamander (*Plethodon dunni*)	107
Del Norte Salamander (*Plethodon elongatus*)	108
Siskiyou Mountains Salamander (*Plethodon stormi*)	109
Scott Bar Salamander (*Plethodon asupak*)	109
Ensatina (Genus *Ensatina*)	110
Ensatina (*Ensatina eschscholtzii*)	110
Large-blotched Ensatina (*Ensatina klauberi*)	110
Climbing Salamanders (Genus *Aneides*)	116
Arboreal Salamander (*Aneides lugubris*)	116
Black Salamander (*Aneides flavipunctatus*)	119
Wandering Salamander (*Aneides vagrans*)	122
Clouded Salamander (*Aneides ferreus*)	122
Slender Salamanders (Genus *Batrachoseps*)	124
Inyo Mountain Salamander (*Batrachoseps campi*)	127
Kern Plateau Salamander (*Batrachoseps robustus*)	128
Tehachapi Slender Salamander (*Batrachoseps stebbinsi*)	129
Kern Canyon Slender Salamander (*Batrachoseps simatus*)	130
Black-bellied Slender Salamander (*Batrachoseps nigriventris*)	131
Gregarious Slender Salamander (*Batrachoseps gregarius*)	132
Channel Islands Slender Salamander (*Batrachoseps pacificus*)	133
Garden Slender Salamander (*Batrachoseps major*)	134
San Gabriel Mountains Slender Salamander (*Batrachoseps gabrieli*)	135
Gabilan Mountains Slender Salamander (*Batrachoseps gavilanensis*)	136
Santa Lucia Mountains Slender Salamander (*Batrachoseps luciae*)	137

Lesser Slender Salamander (*Batrachoseps minor*)	137
San Simeon Slender Salamander (*Batrachoseps incognitus*)	137
Kings River Slender Salamander (*Batrachoseps regius*)	138
Sequoia Slender Salamander (*Batrachoseps kawia*)	138
Relictual Slender Salamander (*Batrachoseps relictus*)	139
Hell Hollow Slender Salamander (*Batrachoseps diabolicus*)	139
California Slender Salamander (*Batrachoseps attenuatus*)	140
Web-toed Salamanders (Genus *Hydromantes*)	142
Mount Lyell Salamander (*Hydromantes platycephalus*)	143
Limestone Salamander (*Hydromantes brunus*)	144
Shasta Salamander (*Hydromantes shastae*)	146

Key to the Families and Genera of Salamanders

California salamander species have four legs and a tail, and may on first appearance resemble some lizard species. However, they lack scales and claws and usually have a moist skin.

1a Nasolabial groove between nostril and lip present; skin smooth, never rough or granular (Lungless Salamanders) 3a
1b Nasolabial groove absent; skin may be smooth or rough 2a

2a Skin rough except in males during breeding season; dorsal color various shades of solid brown; ventral color solid yellow to orange .... Newts
2b Skin smooth; dorsal coloration mottled or spotted ....
.... Mole Salamanders, Giant Salamanders, Torrent Salamanders

3a Base of tail constricted .... Ensatina
3b Tail not constricted at its base 4a

4a Four toes on all four feet; very small legs on a long, wormlike body .... Slender Salamanders
4b Four toes on front feet, five on hind feet; not wormlike in appearance 5a

5a Toes short and webbed: tail relatively short and blunt at tip ....
.... Web-toed Salamanders
5b Toes relatively long with little or no webbing; tail relatively long and usually pointed 6a

6a Adult males usually have projecting upper jaw teeth that can be felt by stroking upward with your finger; toe tips more square than rounded; no distinct stripe on dorsal side ....
.... Climbing Salamanders
6b Upper jaw teeth usually not detectable by stroking upward; toe tips round, not squared off; dorsal stripe usually present ....
.... Woodland Salamanders

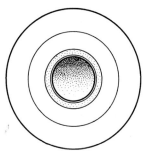

Figure 22. An amphibian egg capsule, showing three concentric jelly capsules surrounding the ovum in the center.

Because of the secretive nature of many salamander species and because some groups such as the mole salamanders are above ground for a very short period each year, the recognition of their eggs and, where appropriate, aquatic larvae is a useful field identification tool. We therefore have included illustrations of eggs for most species and larval illustrations for aquatic-breeding forms with the color plates of the adults at the beginning of each natural history account. Figure 22 shows the capsular chamber and gelatinous envelopes that surround the amphibian ovum and whose position, shape, and number vary between species or similar species groups. The actual shape of the egg and the number and position of ova within the gelatinous envelope also vary.

Mole Salamanders (Family Ambystomatidae)

The mole salamanders or ambystomatids are large, heavy-bodied, smooth-skinned salamanders with small eyes, a laterally flattened tail, and a transverse row of vomerine teeth across the roof of the mouth. Costal grooves are present and usually quite distinct. The name "mole" apparently refers to the habit of most ambystomids spending much of their adult life underground in burrows of ground squirrels, gophers, badgers, and other animals or within decayed logs or stumps. Mole salamanders are usually abroad for only brief periods at night during or following rains. In winter and early spring they often travel in migratory waves to aquatic-breeding sites. Larvae are of the pond type and sometimes fail to transform to the adult form but do develop sexually. Such larvae are referred to as being "neotenic," and if they breed they are said to be "paedogenic" and are called "axolotls."

This family contains only one genus, *Ambystoma*, that occurs only in North America. There are currently 33 recognized species that range from southeastern Alaska, James Bay, and Labrador to the Gulf Coast and the southern edge of the Mexican Plateau. Five species occur in California.

CALIFORNIA TIGER SALAMANDER *Ambystoma californiense*

PROTECTIVE STATUS: FT, ST

IDENTIFICATION: ADULTS snout-to-vent length is 7.6 to 15.5 cm (3 to 6.5 in.). This is a large, stocky salamander with small, protruding eyes, a somewhat flattened head with rounded snout, no parotid glands, and a small tubercle on each side of the rear undersurface of each front and hind foot. The dorsal and lateral body color is black with large pale-yellow or white spots. Dorsal bars of these colors, which are quite prevalent in the Tiger Salamander (*A. tigrinum*), are scarce or absent here. A population in Santa Barbara County that appears to have a greater frequency of bars and reticulations and is molecularly distinctive is currently being considered for species or subspecies status. **EGGS** are attached singly or in small groups to submerged branches, twigs, weeds, and other objects on the pond substrate. They contain three jelly envelopes, the outer of which is 8 to 12 mm (0.3 to 0.5 in.). The diameter of the ovum within is about 2.5 mm (0.1 in.). **LARVAE** are the large-gilled pond type with a broad, flat head, a high dorsal fin that extends to the posterior margin of the head, and a yellowish gray dorsal coloration. The gill rakers number 18 to 24 (often 21) and are located on the anterior face of the third gill arch. These salamanders transform before reaching sexual maturity at about 7.5 to 12.5 cm (3 to 5 in.).

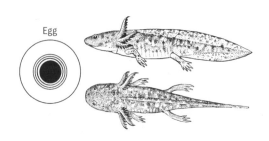

Egg

RANGE: The distribution of the California Tiger Salamander is highly fragmented as a result of the extensive agricultural and urban development throughout most of its original range. At present it occurs from the vicinity of Santa Rosa (Sonoma County) south to the Lompoc area (Santa Barbara County); the central portion of the Great Valley and adjacent foothills; the Sierra Nevada from southern Sacramento County to northwest Tulare County; and in the Inner Coast Range to the Temblor Range.

Plate 19. Migrating California Tiger Salamander.

NATURAL HISTORY: The natural history of the California Tiger Salamander is a prime example of adaptations by an aquatic-breeding amphibian species to the Mediterranean climate of Central California. During heavy night rains in winter and early spring it migrates from burrow retreats to seasonal ponds and slow waters of creek pools to spawn. Breeding follows the pattern described for aquatic-breeding salamanders (see "Taxonomy, Anatomy, Physiology, and Behavior of Amphibians" section), after which the females lay groups of often two to four eggs on submerged objects such as aquatic-plant stalks. The minimum known time from hatching to metamorphosis is around 10 weeks but may be considerably longer if the water temperature remains cool throughout this period. A successful spawning site is therefore one that holds water through late spring. In

Plate 20. California Tiger Salamander breeding pond.

pristine California these were mostly deep vernal pools, but more than 75 percent of such habitats have been filled in during agricultural and urban development. Fortunately, a side product of upland cattle grazing, the stock water pond, has helped buffer the effect of the loss of vernal pools in some areas.

Seasonal ponds and intermittent creek pools are ideal larvae-rearing sites in that they are usually free of fish predators. Here the ambystomid larvae are the secondary consumers instead of small fishes and feed on the spring crop of aquatic insects such as backswimmers (Notonectidae) and water boatmen (Corixidae). The production of such insects and their larvae is often very high in stock ponds as a result of continuous fertilization by wading cattle, which also eat most of the emergent aquatic vegetation. This results in a seemingly barren pond that is often overlooked as a productive amphibian breeding habitat. In reality, the larval biomass in such a pond occasionally exceeds its carrying capacity, and as metamorphosis time approaches cannibalism often sets in. Net sampling at such times usually produces both large, plump larvae along with smaller ones with most of their tail missing. Captive-rearing attempts with groups of larvae in relatively small containers have often resulted in the fastest-growing individuals eating others over half their size. Larval cannibalism, which may have evolved in vernal pool habitats, and the faster growth it promotes may be very significant in a shallow breeding pond, since it could ensure that some individuals will transform and emigrate before pond drying occurs.

The newly transformed larvae (metamorphs) usually remain along the shoreline of a drying pond until a moist spring evening permits safe overland travel to upland retreat sites. At such times they emigrate in mass and may wander randomly or possibly follow historical odor trails until adequate retreat areas are found. At some stock ponds like the one pictured here that usually retain water through summer, larvae will often summer over and emigrate on the evening of the first substantial fall rain. The lungs develop during this time, and throughout the summer the water surface is dotted with small concentric ripples as the larvae gulp air. Observations of this occurrence, especially at sites that are known not to contain fish or newts, constitute an easy way of verifying that a murky stock pond is indeed a California Tiger Salamander breeding habitat. Such an extension of pond residency avoids the risk of desiccation during upland travel if the summer dry season has set in before metamorphosis. Extended pond life in this species does not produce axolotls, as is sometimes the case in *A. tigrinum*.

Rodent burrow systems are most often used for retreats, along with human-created sites such as old culverts and under-house crawl spaces. It was into one of the latter that the junior author once squeezed to explore a plumbing problem soon after he and his wife purchased a small farm with a shallow stock pond. While making his inspection he had the feeling of being watched and began to play his flashlight beam over the wet soil area caused by the plumbing leak. His intuition soon proved correct as the light revealed several pairs of small beady eyes only a few feet from his face. The discovery that he was living in an "upstairs–downstairs" relationship with California Tiger Salamanders was an added real estate bonus that only a herpetologist could fully appreciate.

Under ideal conditions the most important retreat habitat, burrows of the California Ground Squirrel (*Spermophilus beecheyi*), and in some situations the unplugged burrows of Botta's Pocket Gopher (*Thomomys bottae*), are present near a breeding-pond site. However, if appropriate soil and drainage conditions for burrowing mammals do not exist nearby, the larval emigration journey and subsequent annual adult spawning migrations may be amazingly long for such a small, slow-moving vertebrate. The greatest reported to date is 1.6 km (1.0 mi). An unfortunate practice that often creates such a situation is the widespread poisoning of ground squirrels on rangeland where stock pond-breeding sites occur. The pond pictured in this section is located on a ranch where ground squirrels have apparently been completely eradicated, and here California Tiger Salamanders annually migrate across property lines to a neighboring ranch that does not poison squirrels.

These subterranean or moist-surface retreats are the primary habitat of the adult California Tiger Salamander, where they may spend up to 11 months of each year. Burrows not only provide a safe osmotic and thermal environment when weather conditions are unacceptable for surface activity but also serve as a feeding habitat. Many ground-dwelling invertebrates and small vertebrates are also attracted to the microclimate that such sites offer, and here adult tiger salamanders may act as lie-in-wait predators. As in the larvae, cannibalism is most likely a component of the adult diet. On several occasions during a population survey using buried bucket traps, the junior author retrieved large males with the tail of a smaller one still protruding from its mouth. During observations of a captive population maintained in an outdoor artificial-burrow complex, he and a graduate student noted nightly movements by the salamanders to the burrow openings except on very warm, dry evenings. When crickets were released into the enclosure, salamanders at burrow entrances grabbed them with a quick snap of the jaws once contact was made with their snouts. However, without some sort of tactile stimulation of a salamander's head, crickets often passed by apparently undetected. You can observe California Tiger Salamanders at burrow entrances in the field by slowly approaching an occupied retreat burrow on a moist spring or early summer night with a dim flashlight. When your presence is eventually detected, the salamander retreats back into the burrow shaft, slowly walking backwards.

One other important component of the California Tiger Salamander's life history strategy is its longevity, which may average 10 years or more in the wild. Such relatively long life for a small vertebrate mitigates situations during drought years when many seasonal ponds may not retain water through larval metamorphosis. Indeed, in a year when winter rains are exceptionally light, this species may not breed at all. In either situation, good longevity ensures that there may be at least one or two successful reproductive years in the life of each adult.

CONSERVATION NOTE: The California Tiger Salamander is estimated to have disappeared from about 55 percent of its historic range to date. Its geographic range, the former grasslands of the Great Valley and lower foothill regions of the South Coast Range and the Sierra Nevada, has undergone massive agricultural and urban–suburban development during the past century, and the dual habitat needs of this species have often been overlooked. Breeding ponds have sometimes been preserved, but surrounding upland development has negated the burrow retreat habitat, thus eliminating entire populations. Regulatory agencies must be more vigilant in approving land use changes on properties where this species has been reported. New road construction that bisects the burrow-to-pond migratory path can also devastate a tiger salamander population, especially when winter evening traffic is heavy. Under-road culverts that are often proposed as mitigation for such projects are not effective unless a continuous solid "drift wall" is also installed along the outer road shoulder edge to shunt moving salamanders into these safe passage ways. The effect of ranchland ground squirrel poisoning has already been discussed. This practice seems to be based on the assumption that a 0.5 kg (1 lb) California Ground Squirrel seriously competes with a 455 kg (1,000 lb) steer for the annual grass food supply. Ranchers with breeding ponds on their land should be encouraged, possibly through monetary incentives, to maintain poison-free zones around such sites. Breeding-pond owners and mosquito abatement districts should be informed about the effective role of ambystomid larvae in controlling mosquito larvae and that the introduction of Mosquitofish (*Gambusia affinis*) into these breeding habitats must be eliminated. This nonnative fish eats newly hatched larvae, and picks at and destroys the external gills of small ones. For many years the Tiger Salamander (*A. tigrinum*) was imported to California and sold as fish bait. Like several species of imported nonnative bait minnows, these have been released by anglers, and in Monterey and San Benito counties extensive hybridization with the California Tiger Salamander has occurred. The sale of Tiger Salamanders for bait is now illegal, but specimens can still be purchased in pet stores. This trade should also be curtailed, because aquatic pets that are no longer wanted are often released at some nearby pond site.

Family Ambystomidae belongs to a suborder of amphibians that have been around since the Cretaceous, the peak of dinosaur evolution, and during that tenure its members have faced numerous environmental challenges that, unlike their former enormous reptilian neighbors, they have successfully met. Those challenges currently confronted by California's namesake ambystomid species are very likely more formidable than any other yet encountered. Hopefully the source of these impacts can soon begin to prevent further ones.

TIGER SALAMANDER *Ambystoma tigrinum*
INTRODUCED
BARRED TIGER SALAMANDER *Ambystoma mavortium*
INTRODUCED

Tiger Salamander

IDENTIFICATION: ADULT snout-to-vent length ranges from 7.6 to 16.5 cm (3 to 6.5 in.). These species are similar in body form to the California Tiger Salamander but with shorter limbs and more rounded head profile and tail. Color pattern varies from wide yellow "tiger bars" on a black background in *A. mavortium* to black spots and reticulations on a yellowish, yellow-orange, cream, dull-greenish, or grayish background in *A. tigrinum*. As in *A. californiense*, there are two small tubercles on the underside

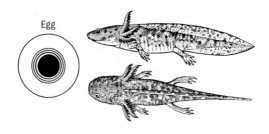

Egg

of each foot. **EGGS** are deposited singly, in rows, or clusters from a few eggs to around 120. The ovum is enclosed in three jelly capsules (see figure 22) with an outer diameter of 4.5 to 12 mm (0.2 to 0.5 in.). The diameter of the ovum is 1.6 to 2.8 mm (0.06 to 0.1 in.). **LARVAE** total length ranges from 7.5 cm (3 in.) to over 20 cm (8 in.), and neotenic forms vary in size from 7.5 to 37.5 cm (7 to 15 in.). Gill rakers number 13 to 24 (often 17 to 21) and are located on the anterior face of the third gill arch. The larval dorsal color ranges from greenish or pale yellow to brown or gray, often with sooty spotting and blotching.

RANGE: These are the most widely distributed salamanders in North America. Subspecies occur in 41 of the lower 48 states (absent in the New England area), five Canadian provinces, and several Mexican states. Until recently, the Barred Tiger Salamander from the Texas–Oklahoma–Kansas area was considered a subspecies of the Tiger Salamander. Several localized populations have been documented in California, and more may be discovered in other areas. These are the result of careless releases by anglers who purchased them for bait. The range map denotes some areas where one or both species have been found, and will unfortunately continue to increase until the importation of these species is curtailed.

NATURAL HISTORY: The natural history of these salamanders closely parallels that of the California Tiger Salamander, which was formerly considered one of the many subspecies of *A. tigrinum*. Its use of a variety of subterranean retreats to "ride out" seasonal climatic extremes, ranging from subfreezing winters to hot, dry desert summers, accounts for much of its ecological success. The ability of its larvae to grow rapidly and transform in ephemeral pools before they dry also accounts for its presence in many contracting habitats. At such sites it feeds on amphipods, copepods, ostracods, midge and mosquito larvae, snails, and the tadpoles of anuran species that may also share the breeding pond. Larval cannibalism is often the key to survival of at least some of a given year class, and in some populations wide-mouthed cannibal morphs develop.

A further feature that buffers the lack of successful reproduction during drought years is its great longevity. Years ago the junior author was given a fully grown Barred Tiger Salamander by a student who had rescued it from a highway while driving through Kansas. It was maintained in a land–water terrarium where it lived for the next 22 years, setting the longevity record for a McGinnis family pet. Its appetite for large earthworms soon added to its girth, resulting in its name of "Fat Albert." As in observations of captive California Tiger Salamanders feeding on crickets, this salamander would only snap at a worm if it contacted its snout. It may be that in the darkness of a burrow chamber, tactile detection of prey items is more reliable than other sensory inputs. In the natural habitat adult food consists of a wide variety of large invertebrates and small vertebrates, particularly those that also use burrow retreats. Cannibalism is also a part of the adult feeding regime and may account for the reported observations of a population in a prairie dog colony in Kansas in which there was only one Barred Tiger Salamander per burrow.

A. tigrinum and *A. mavortium* are the only salamander species that have been introduced into California and established breeding populations. It and five anuran species are the only amphibians that have been introduced into this state. Given that nearly 50 percent of all freshwater fishes in California have been introduced and many of these introductions are now competing vigorously with native species, this number of exotic amphibians is fortunately very low. Although bait shop sales of both species are now illegal, transport of larvae and adults by uninformed anglers and wild pet collectors is still possible. Given the successful hybridization of the Tiger Salamander with the California Tiger Salamander that has already occurred in Monterey and San Benito counties, eradication of all confirmed existing *A. tigrinum* populations should be attempted.

LONG-TOED SALAMANDER   *Ambystoma macrodactylum*
SANTA CRUZ LONG-TOED SALAMANDER   *A. m. croceum*
SOUTHERN LONG-TOED SALAMANDER   *A. m. sigillatum*

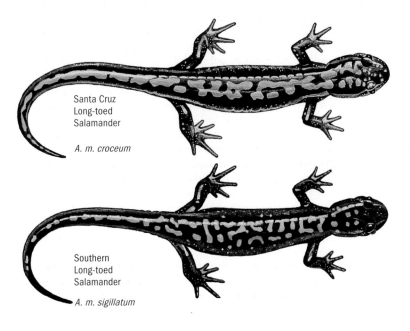

Santa Cruz Long-toed Salamander
A. m. croceum

Southern Long-toed Salamander
A. m. sigillatum

PROTECTIVE STATUS OF *A. M. CROCEUM*: FPS, FE, SE
IDENTIFICATION: ADULT size is 6.3 to 8.3 cm (2.5 to 3.25 in.), total length. They are dusky or black above with bright-yellow broken dorsal stripe (*A. m. sigillatum*) or orange-yellow to orange spots and bars (*A. m. croceum*) from head to tip of tail. The sides are frosted with fine white flecks, and there are tubercles on the underside of the feet as in the tiger salamanders. **EGGS** are contained in two jelly envelopes, outer diameter 12 to 17 mm (0.5 to 0.7 in.), with ovum 5 to 7 mm (0.2 to 0.3 in.) in diameter. **LARVAE** are the pond type, ranging from 6.2 to 7.5 cm (2.5 to 3.0 in.) total length, with 9 to 13 gill rakers on the third gill arch. The dorsal color is light olive gray to brownish gray with dark mottling.

Egg

RANGE: The Southern Long-toed Salamander occurs in the Sierra Nevada and Cascades from Tuolumne County northward and across the northern part of the state from the Siskiyou Mountain Range to the Warner Mountains. The Santa Cruz subspecies inhabits several isolated seasonal wetlands in Santa Cruz County. Three other subspecies occur in six northwestern states and Canadian provinces.

NATURAL HISTORY: The habitat of the Southern Long-toed Salamander varies from sagebrush semidesert east of the Cascade Range through intermediate vegetation types to alpine meadows and the barren rocky shores of high mountain lakes where it spawns in ponds, lakes, and transient pools. It breeds in January and February in lowlands and, after snow melts, in May, June, and probably July in higher mountains. Males migrate to the breeding site and remain during the breeding season. Females usually arrive later, spawn, and then leave. While in amplexus, the male rubs the snout of the female with his chin, where a scent gland is located. He then leads her over his deposited spermatophore with tail extending upward, sometimes almost vertically. This is apparently a signal to her to retrieve the spermatophore with her vent lips.

Eggs are laid singly or in small groups in shallow areas, either on the bottom or attached to grasses, spike rush, or other objects in water. In the Sierra, larvae overwinter at breeding sites above approximately 2,500 m

Plate 21. Southern Long-toed Salamander larva.

MOLE SALAMANDERS (AMBYSTOMATIDAE)

(8,000 ft), transforming during the second summer. Permanent aquatic habitats are therefore necessary for successful reproduction in these montane populations. At lower elevations temporary spawning pools are often used, and the larval development period is only three to four months. During that time the larvae feed on a wide variety of small aquatic invertebrates and tadpoles when present. Adult food includes insects (springtails, flies, mosquitoes, crickets, grasshoppers, caterpillars, beetles), centipedes, pill bugs, earthworms, snails, and slugs. Some also scavenge when small dead prey are encountered.

The Santa Cruz Long-toed Salamander (*A. m. croceum*) is one of the most endangered animal species in California. It breeds at only about a half-dozen ponds in Santa Cruz County. As is the case with most endangered species, it is the species' habitat that is really endangered. In this case it has been dealt a variety of impacts, most of which have negated or seriously altered the breeding pond. The widening of the old two-lane Route 1 south of Santa Cruz to a four-lane freeway in 1969 substantially reduced the size of its largest breeding habitat, Valencia Lagoon. Note that this date was one year before the enactment of the California Endangered Species Act and three years before its federal counterpart came into being. An eleventh-hour lobbying effort by the senior author and his colleagues and the full cooperation of the California Department of Transportation and its engineers fortunately preserved a portion of this site. At a critical hearing concerning the fate of this salamander in the city of Santa Cruz and attended by the senior author, an adolescent lad, fond of salamanders, spoke in its behalf. The impact of his words on the packed listening audience will long be remembered.

Plate 22. Santa Cruz Long-toed Salamander.

Plate 23. Santa Cruz Long-toed Salamander seasonal breeding pond.

Other seasonal breeding ponds were filled for agricultural and urban development. Ellicott Pond, a shallow, heavily vegetated seasonal wetland, was slated for a mobile home park. Fortunately, it was purchased by the CDFG as a reserve for this subspecies. However, seasonal ponds such as this are very shallow and gradually fill with decomposed plant material and silt. As a result, larval development through metamorphosis becomes less likely during dry years. During the winter of 1992–1993 the junior author and John Brode, CDFG herpetologist, studied the Ellicott Pond breeding population and found it greatly reduced from earlier years. The impact here appeared to be the reduction of pond water volume due to problems of organic debris accumulation and siltation caused by illegal off-road vehicle activity on adjacent slopes. Fortunately, an old but still functional well was discovered nearby, and now water is added during the late spring period to prevent early pond dry-out.

CONSERVATION NOTE: As with most highly endangered species whose habitat has been greatly reduced and fragmented, the establishment of well-managed reserves at the few remaining population sites is the best and in many cases only way to prevent extinction in the wild. And as was learned at Ellicott Pond, such reserves cannot just be set aside and then forgotten, but instead must be periodically monitored by qualified biologists so that unforeseen problems can be detected and corrected in time. Extreme caution must also be exercised by all agencies involved in the approval process for land use changes within or near the current known range of the Santa Cruz Long-toed Salamander. The dual critical habitats for ambystomids may sometimes go undetected by unqualified observers, especially during the dry season. The best available biological advice must be sought and followed in such matters, and nothing short of adequate on-site preservation of both the breeding and retreat habitats should be accepted as mitigation.

NORTHWESTERN SALAMANDER *Ambystoma gracile*

IDENTIFICATION: ADULTS range from 7.5 to 11.2 cm (3 to 4.5 in.) in total length. They are brown or nearly black above and have parotid glands (two large swellings on the back of the head) and glandular thickening along the upper border of the tail. There are no foot tubercles as in other ambystomids. **EGGS** are laid in firm, rounded or oval clusters, 5 to 15 by 5 to 10 cm (2 to 6 by 2 to 4 in.) in diameter. The clusters contain from 30 to 270 (often 60 to 140) eggs, with each egg contained in a thick, firm jelly envelope with an outer surface 11 to 15 mm (0.4 to 6.0 in.) in diameter. The ovum diameter is 1.5 to 2.5 mm (0.06 to 1.0 in.). **LARVAE** are 7.5 to 15.0 cm (3 to 6 in.) in total length and of the pond type. There are usually 9 (range 7 to 10) gill rakers on the anterior face of the third gill arch. The dorsal color is gray to brown and often overlaid with yellow, green, or black mottling. Dark spots are often present on the dorsum and upper sides, and the ventral color is cream to pale gray. Young larvae usually have a dark glandular strip along the base of the dorsal fin. Breeding neotenic larvae measure 6.5 to 10.5 cm (2.5 to 4.25 in.) from snout to vent. Their ventral color is slate gray or brown-black in males and light brown in females, and both are flecked with cream.

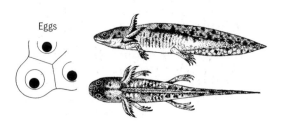

Eggs

RANGE: It occurs from a few miles south of the mouth of the Gualala River, Sonoma County, north through the western portion of the North Coast Range in California to May Islands (also called Mary Islands) in extreme southeastern Alaska.

NATURAL HISTORY: The Northwestern Salamander inhabits grassland, woodland, and forest sites in California's North Coast Range. Like other western ambystomids it spends much of the year in rodent burrows and other underground retreats. While above ground during the breeding migration it can be found under rocks, boards, and logs near the ponds, lakes, and streams where it breeds.

This salamander is well supplied with poison glands, the most prominent of which are the large, toadlike parotid glands behind the eyes. In no other ambystomid is this gland developed to this extent. When disturbed it assumes a butting pose (plate 24) and lashes with its tail as a whitish poison–mucus mixture exudes from its head, back, and upper tail surface. It is also one of the few species that breaks the customary silence of the salamander world. A colleague of the authors, Dr. Paul Licht, described the sound it produces as a "tic," perhaps made by opening and closing the epiglottis. It is heard when an individual is disturbed and often as it assumes its defense posture. During the feeding of captive specimens, the dominant male made this sound, after which subdominant ones exhibited the defense pose. This is one of the few if not the only known examples of intraspecific vocal communication in salamanders.

Plate 24. Northwestern Salamander in defense posture.

In California, breeding takes place once the winter rainy season is well under way in January or February. Spawning sites include permanent or long-lasting seasonal ponds, lakes, and quiet streams void of large predatory fish. Males usually arrive at breeding sites first and often remain during the breeding period, whereas females leave after egg laying. In courting, the male nudges and butts the female, and then engages in amplexus before spermatophore deposition and retrieval. The large, globular, firm-jellied egg clumps are attached to roots, branches, or other firm support in quiet or slowly flowing water. Algae inside egg capsules may impart a greenish cast. The eggs of this salamander are inherently slow in their rate of development, from two to eight weeks depending upon egg size and ambient temperature. Larvae resemble those of tiger salamander species but have a poison gland strip on each side of the base of the dorsal fin that exudes a whitish secretion upon disturbance. Metamorphosis usually occurs in the second year of larval life. Sexual maturity may be reached in about two or three years, but the water temperature during development strongly affects maturation rate. Examples of larval breeding (paedogenesis) may be found throughout this salamander's range but is more prevalent at northern latitudes. In some localities in British Columbia almost the entire "adult" population is paedogenic. Adult food consists of slugs and other invertebrates plus small fish and frogs. Larvae eat small invertebrates, including zooplankton.

Giant Salamanders (Family Dicamptodontidae)

This small family contains four species, two of which occur in California. The term "giant" requires some qualification. The biggest of these, the California Giant Salamander (*Dicamptodon ensatus*), is indeed the largest terrestrial salamander in the world, with a total length of about 30 cm (12 in.). However, it is only about one third the size of the largest aquatic salamander, and in comparison to the spectrum of present-day large land vertebrates, it is a mere midget. There have also been larger land salamanders in the past. The Paleocene fossil record reveals that a related form about twice this size lived in ancestral redwood forests. However, that is still not a very large upper limit for a major vertebrate taxon.

One factor that may put a cap on body size in a group that relies so heavily on cutaneous respiration is the relative decrease in skin surface area as body mass increases. Beyond a certain size range there simply may not be adequate skin respiratory surface to compensate for a poorly developed lung respiration system. It also may be that one key to the success of salamanders is that they have remained small as a group. Most are nocturnal secondary consumers and forage on relatively moist understory floors when above ground. Within this general feeding niche there is relatively light competition from mammals, where only shrews and moles follow

Plate 25. Adult California Giant Salamander and adult slender salamander.

this foraging pattern. It is interesting to note that the smallest species of weasels, the next size up in North American terrestrial vertebrate secondary consumers, are about the same size as the giant salamanders.

CALIFORNIA GIANT SALAMANDER   *Dicamptodon ensatus*

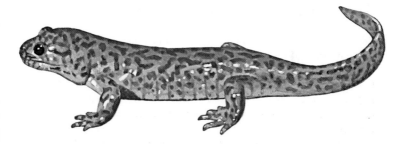

IDENTIFICATION: ADULTS range from 6.3 to 17.3 cm (4.5 to 6 in.) snout-to-vent length and are up to around 30.5 cm (12 in.) in total length. This is a heavy-bodied salamander with sturdy limbs and a massive head. The tail is flattened laterally, especially near the tip, and there are no under-foot tubercles. The dorsal color is tan to light brown with coarse-grained copper or dark-brown marbling. The ventral side is whitish to dull yellow. Costal grooves are present but indistinct. **EGGS** are contained in five jelly envelopes (only two are readily visible). Ovum diameter is 6 to 7 mm (0.2 to 0.3 in.), and each egg is attached singly to the underside of rocks and other underwater objects by short pedicles. **LARVAE** are 15 to 20 cm (6 to 8 in.) in total length at metamorphosis, but neotenic forms range up to

Plate 26. California Giant Salamander larva.

27.5 cm (11 in.). They are of the stream type, with short, bushy, dark-red gills and five to seven gill rakers on the third gill arch. The dorsal body and tail fin are brown with smoky dark and light mottling, and there is a light stripe behind the eye. In streams with a dark substrate the body color is darker with no mottling.

RANGE: It is found in the South and North Coast Ranges from southern Santa Cruz County north to just beyond the Sonoma–Mendocino County border, and from north of San Francisco eastward to western Lake and Glenn counties. It is absent from the East Bay area.

NATURAL HISTORY: The California Giant Salamander inhabits moist forests within and near clear, cold, rocky streams or seepages. When above ground it is found under logs, bark, rocks, and other large objects, usually near water. The cool, humid, shady environment that this species commonly inhabits favors some diurnal activity. Here it feeds on both a variety of large invertebrates plus such unusual salamander prey items as small snakes, lizards, shrews, small rodents (especially young in nests), and other salamanders, including smaller individuals of their own species. On a herpetology course field trip long past, the junior author and his class collected beautiful adult specimens of the California Giant Salamander and an Arboreal Salamander (*Aneides lugubris*). The latter was two thirds the size of the former, and since gallon collecting jars for specimen transport back to the classroom were scarce, these two large salamanders were put into one such container. When unpacking back at the university, it soon became apparent that something was missing: the Arboreal Salamander! Further observation also revealed that the California Giant Salamander had assumed a noticeably greater girth than when captured. Its large mouth and strong jaw muscles can deliver a crushing blow to such species as well as a painful bite to a novice herpetologist's finger.

Above-ground activities occasionally include climbing up as far as 2.4 m (8 ft) in bushes and small trees. The authors find it rather astonishing that such a heavy-bodied salamander can climb well, and we are at a loss for a sound explanation for this behavior. Foraging would seem to be the most likely reason. However, most brush-oriented birds begin nesting activities after early-spring above-ground activity of salamanders has ceased, which would eliminate searching for eggs and nestlings as a motive. Aside from an occasional Arboreal Salamander and our one arboreal anuran, the Pacific Chorus Frog (*Pseudacris regilla*), no other easily catchable small vertebrate prey would seem likely in low, woodland understory branches. The senior author once found a *D. ensatus* in a loggers' notch on a tree trunk where there was a patch of sunlight in an otherwise dark forest. This observation suggests that thermoregulatory behavior may also be involved here. However, for now we must simply add this behavior to the long list of the many intriguing activities of amphibians yet to be fully explored.

This is also one of very few salamanders that can vocalize. Its sounds vary from a low-pitched rattle to a doglike bark or squeal and are sometimes accompanied by a sidewise lunge. These vocalizations may be part of its defensive behavior, and individuals in the laboratory have emitted the barking sound when attacked by captive snakes.

Breeding begins after the first heavy fall rains when adults move to nearby pond or creek pool sites. Breeding may also occur in spring. Pools near springs at the head waters of creeks or beneath waterfalls are usually favored. The most productive waters appear to be those not occupied by fish, especially the young (parrs) of salmonid species. After spermatophore transfer females lay from about 70 to 185 eggs in concealed locations beneath stabilized rock and log jams or other cover. The eggs are attached by their short pedicles to such objects and are attended by the female, possibly as a protection against cannibalism by males and large larvae. After a long development period fall-spawned larvae hatch during winter and live on their yolk for three to four months. After that they feed on the variety of invertebrate prey inhabiting the breeding pond plus smaller larvae

Plate 27. California Giant Salamander breeding pool.

of their own and other amphibian species. Most spawning sites contain shallow, clear water, and larvae can be found by carefully scrutinizing the substrate or turning stones in the shallows. Metamorphosis occurs from June to August of the second summer after hatching, and thus two larval age (or size) classes are usually present at most times. Neotenic individuals may occur in large permanent streams with a strong, cool flow throughout the year.

PACIFIC GIANT SALAMANDER   *Dicamptodon tenebrosus*

IDENTIFICATION: ADULTS are similar in body length to the California Giant Salamander but may tend to have a smaller head. The limbs are shorter, and there are fewer teeth in the upper jaw. The body color, both dorsally and ventrally, is usually darker in this species, and the marbling pattern tends to be finer. **EGGS** and **LARVAE** are also similar, but the tail tip may be black and the upper tail fin is mottled with dark and light more than in *D. ensatus*. The largest neotenic specimen on record is 35.1 cm (14 in.).

RANGE: North Coast Range, from the vicinity of the Gualala River in southern Mendocino County north to just beyond the Washington–British Columbia border; extends eastward into the Cascade Range in Shasta and Siskiyou counties.

NATURAL HISTORY: Until recently the Pacific Giant Salamander and the California Giant Salamander were considered one species. Then a zone of hybridization was found around the Gualala River. Their respective life histories are quite similar, but given its more northerly distribution, the Pacific Giant Salamander usually spawns in May and larvae hatch in late fall or winter.

CONSERVATION NOTE: Clear-cut logging throughout the habitats of both giant salamander species has seriously impacted or even negated many breeding and upland retreat sites. Unfortunately, these unique salamanders have not been afforded any sort of protection to date. Indeed, anyone who buys a California fishing license or is under the age of 16 may legally take four of each species from the wild. Fortunately, these salamanders share some coastal creeks and their tributaries with two anadromous fishes, the Steelhead Rainbow Trout (*Oncorhynchus mykiss*) and the Coho Salmon (*Oncorhynchus kisutch*), which have been granted protection under both the federal and state endangered species acts. As a result, their creek habitats are now also protected, a move that should also help the survival of giant salamanders in this state.

Torrent or Seep Salamanders (Family Rhyacotritonidae)

These are the most aquatic of all California salamanders. They are rarely found on land anywhere but in wet places near water. These include the splash zone of streams and moss-covered talus where water trickles among gravel and large rocks. Boulder clusters in the steeper gradients of streams seem to be favored aquatic sites, because they provide thermal stability, cover, and resting sites amidst strong, turbulent flow. This trend toward a more aquatic life has resulted in greatly reduced lungs, a feature that is important to creek-dwelling salamanders because it reduces buoyancy. Skin respiration in the cold, oxygen-rich waters that they inhabit apparently offsets decreased lung volume in these small salamanders. This family contains four very similar species, which are found only in the Pacific Northwest. The most widely distributed of these, the Southern Torrent Salamander, occurs in the North Coast Range of California.

SOUTHERN TORRENT SALAMANDER   *Rhyacotriton variegatus*

Ventral

PROTECTIVE STATUS: CSSC

IDENTIFICATION: ADULT size is 4.1 to 6.2 cm (1.6 to 2.5 in.). The tail is short and flat-sided. The eyes are large, and the vent lobes in males are prominent and square-cut. The dorsal body is a mottled olive and dusky, and the ventral side is greenish yellow and dark-flecked. **EGGS** are encased in three jelly envelopes and laid singly with no special organs for attachment. The ovum diameter is about 3 to 4 mm (0.1 to 0.2 in.). **LARVAE** are the stream type but with both gills and gill rakers (0 to three per gill arch) reduced to nubbins. Like adults they have large eyes positioned high on the head. The dorsal color is olive with black speckles, and the ventral color is yellowish. These larvae may attain adult size with little apparent change at metamorphosis.

RANGE: In California this species occurs in the humid forests of the North Coast Range from the vicinity of Point Arena, Mendocino County, north to northern coastal Oregon. There is also a population in the upper McCloud drainage, Siskiyou County.

NATURAL HISTORY: The optimal habitat for the Southern Torrent Salamander is humid, old-growth forest with fast, cold streams or seepages. When on land it is found under rocks or other objects usually in the splash zone where the ground is thoroughly wet or in well-shaded environments with moss and ferns. This salamander is highly aquatic and a rapid swimmer, yet it is agile on land and quickly

Plate 28. Southern Torrent Salamander breeding creek.

wiggles into water when discovered under a creek bank rock. Unlike most other western aquatic-breeding salamanders, it does not retreat to underground sites for part of the year. Adults feed on a wide variety of invertebrates, ranging from various types of aquatic insect larvae to terrestrial forms such as beetles, spiders, millipedes, and earthworms. Prey is often captured by a rudimentary form of tongue flipping. Tongue glands produce copious amounts of adhesive mucus that entraps small prey items. Larvae feed on the aquatic prey portion of the adult diet.

Compared with other aquatic-breeding salamanders, little is known about reproduction in this species. Courtship and egg laying may take place throughout the year, but most occurs in spring and early summer. A courting male raises, extends, and tilts his hindquarters upward by extending his hind legs and curls his tail over his back while wagging the tip. This behavior occurs just before spermatophore deposition, at which time it will also chase and bite other males that may come close. Eggs are known only from two clusters of the Columbia Torrent Salamander (*R. kezeri*), each containing 16 eggs, found in mid-December in cracks in sandstone in a spring. No adults were attending these clutches. Because these clutches contained twice as many eggs as the average number (eight) produced by a female, they are presumed to have been the result of communal nesting. Subsequent observations of these sites revealed that it may take a year or more to complete embryonic development. The larval period may be as much as three and a half years, and adults may not breed until they are five to six years old.

CONSERVATION NOTE: Torrent Salamanders are "stay at home" species. One recapture study of individually marked larvae revealed that 75 percent stayed within the same small segment of stream. This sort of behavior has led to isolated populations that are susceptible to local extinction as a result of the large-scale harvesting of old-growth forests and the degrading or destruction of seepages and small springs. The Southern Torrent Salamander is listed as a CSSC by the CDFG, yet at this writing this state agency also allows anyone who purchases a California fishing license to capture and remove from the wild four live specimens. Because its primary habitat is old-growth forests, some populations in California will hopefully be preserved as a spinoff of the federal protection of threatened species like the Spotted Owl (*Strix occidentalis*) and the Marbled Murrelet (*Brachyramphus marmoratum*), which also are residents of this unique habitat.

Newts (Family Salamandridae)

California contains all four Pacific newt species (genus *Taricha*), two of which are endemic to this state. Newts differ from all other salamanders in the West in having rough and rather dry skin when on land, a feature implying that their lung function is more efficient than that of other salamanders. However, as a group many aspects of their natural history are quite similar. We have therefore highlighted several of these in this family introduction and then added species-specific items to the three natural history accounts.

With fall and winter rains newts emerge from underground retreats and rotten logs to forage. In winter and spring they move in large numbers to a pond or creek to breed. Some breeding sites support newt populations that are dispersed throughout a large surrounding upland area. Over 7,000 Coast Range Newts (*Taricha torosa*) were documented in a one-season mark-and-recapture survey at a breeding pond by a former California State University, Hayward (CSUH) graduate student, Beatrice Moore. She recaptured some of the marked individuals over 3.2 km (2 mi) from the pond after the spawning period. In some areas where roads have bisected

Figure 23. Male newt (top) rubbing chin gland on female's nostrils during amplexus.

an ancestral newt migratory pathway, large numbers are often killed during these mass spawning migrations. Indeed, the newts are not the only ones that are at risk in such situations, because the road surface can become so littered with squashed newt bodies that cars may skid and crash. On South Park Drive in Tilden Park in the Berkeley Hills, so many newts were killed and injured by car traffic during their migration to breeding sites that regular road closures have been instituted during the newt migratory season.

Males usually migrate to the breeding site first, at which time they develop a smooth, thick, swollen skin, prominent tail fins, swollen vent region, and rough dark-colored areas on the undersides of their feet called nuptial pads, which aid them in clinging to the slippery body of the female during amplexus. The skin thickening is thought to prevent males from becoming waterlogged during long periods spent in water awaiting females, and fins make the tail a more efficient sculling organ. Females change little and leave the water soon after laying their eggs.

Before spermatophore transfer the male amplexes the female and rubs her snout with his chin, which contains skin glands that secrete a substance to apparently increase her receptivity (figure 23). The recognition of a female by a male and his attraction to her is the result of the scent she produces. This was vividly demonstrated many years ago by Dr. Victor Twitty, a pioneer in newt field research. He placed newt-size pieces of sponge soaked in water containing scent secretions of the female Red-bellied Newt (*T. rivularis*) in a stream containing breeding males, which soon amplexed these objects, confirming that it is the female's "perfume" that is the attraction. It is so potent that one female occasionally attracts several ardent suitors at once, resulting in a gyrating "newt ball", that can exceed 0.6 m (2 ft) in diameter! Her suitors are often so intent on

Plate 29. Red-bellied Newt "ball."

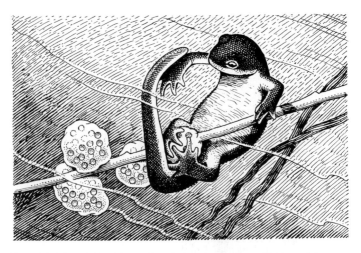

Figure 24. A female Coast Range Newt in the process of attaching her newly produced egg capsule to a submerged stem.

amplexus that you can gently pick up such a ball with cupped hands and return it to the water without having any male let go his grip.

Both authors have occasionally had a finger amplexed when handling a male newt after first holding a female. Amplexus may last an hour or more, after which the pair separates and spermatophore transfer takes place. Females then begin to lay eggs, often at a site where other females have already spawned (figure 24). The eggs, encased in firm-jellied clumps, are attached to sticks, rocks, and other submerged items, either singly or in moderate to large masses.

Newts are more active in daytime than other Californian salamanders and are sometimes seen migrating to breeding sites in bright sunlight. Field studies with displaced Rough-skinned Newts (*T. granulosa*) found that they could orient to their home breeding shore area but only on sunny days. This apparent use of sun compass orientation to find "home" most likely accounts for this daytime activity. However, such diurnal behavior exposes them to a far greater number of potential predators such as birds and selected snake species than does activity at night, and this may account for a newt's exceptionally toxic poison.

Poison-secreting cells of salamandrid skin and several internal tissues produce a type of tetrodotoxin (TTX), the most potent of all neurotoxins. Unlike the neurotoxins of some snakes, it does not block the action of the neurotransmitter acetylcholine at the nerve cell–muscle cell synapse. Instead it negates the basis for the nerve impulse itself: the rapid passage of sodium ions through the nerve cell membrane. However, TTX cannot cross the blood–brain barrier, so that this key organ remains functional as the rest of the victim's body, including the heart and diaphragm

Plate 30. Coast Range Newt embryos within egg capsule.

muscles, are slowly paralyzed. Newt TTX is identical to that found in marine pufferfish, which occasionally kills diners in East Asia who eat incorrectly prepared portions of that type of fish. In California it kills most predators that swallow a newt. A few species, such as the Common Garter Snake (*Thamnophis sirtalis*), have acquired a natural immunity. Humans are not immune, and there have been several poisonings and one death of individuals who for some reason have allowed a newt to pass beyond their lips. However, the poison cannot enter one's system through the skin, provided there are no cuts or abrasions, and thus newts can be safely handled. Even so, washing hands afterwards is advisable, because the secretion can cause severe irritation if you get it into your eyes.

An additional newt protective measure is their unique defensive posture already discussed in the amphibian introductory section (see figure 8). It may be seen in the field by tapping a newt several times on the back before disturbing it in any other way. This harmless procedure mimics the effect of an avian predator pecking on its back and can provide an excellent example of a classic sign stimulus–response action for your hiking companions or field trip class.

Newts feed on a wide variety of terrestrial and aquatic invertebrate prey. Their ability to spend more time above ground than most other aquatic-breeding salamanders most likely exposes them to a wider variety of prey species. They rely on their large eyes to detect surface-active prey and slowly stalk it until close enough for a quick grab with the jaws. When feeding in water they capture prey by suddenly expanding the mouth cavity and sucking the food item in. Many predatory fishes feed in an identical manner. Like several other salamander groups, western newts are long-lived. Marked Red-bellied Newts have been recaptured after 17 years in the field, and the estimated maximum ages for male and female Rough-skinned Newts are 22 and 26 years, respectively.

COAST RANGE NEWT	*Taricha torosa*
SIERRA NEWT	*Taricha sierrae*

Terrestrial

Aquatic

PROTECTIVE STATUS OF *T. TOROSA*: CSSC

TAXONOMIC NOTE: Recent (2007) research using mitochondrial DNA combined with morphological evidence involving color pattern and head shape determined the former subspecies Coast Range Newt and Sierra Newt to be distinct species.

IDENTIFICATION: ADULTS range from 6.9 to 8.7 cm (2.75 to 3.5 in.) in total length. Their dorsal color is tan to dark brown, and the ventral side is orange to pale yellow, with the dorsal color grading into ventral hues on the sides. The lower eyelid and upper lip below the eye are usually without dark pigment. The eyes are large and extend to the outline of the head as viewed from above. Teeth in the roof of the mouth are often in a Y-shaped pattern. Breeding males are as described in the generic account. Occasionally grotesquely warty individuals, caused by a pathogen, occur at some localities in the southern part of the range (Boulder Creek, San Diego County). **EGGS** are laid in spherical masses, 1.2 to 2.5 cm (0.5 to 1.0 in.) in diameter that contain 7 to 39 ova. Each is in its own jelly envelope with an outside diameter 5 to 8 mm (0.2 to 0.3 in.). The ovum diameter is 1.9 to 2.8 mm (0.07 to 0.1 in.). **LARVAE** grow to 5.6 cm (2.25 in.) in total length and are of the pond type. They are the only newt larvae with a dark longitudinal stripe on each side near the base of the dorsal fin. The dorsal and lateral body color is yellowish.

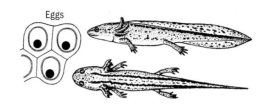

Eggs

RANGE: The Coast Range Newt occurs from the vicinity of Laytonville, Mendocino County, in the North Coast Range, south through the Inner Coast Range to San Francisco Bay and on through the coastal mountains of the South Coast Range and Coastal Southern California region, where the distribution becomes disjunct. It is apparently absent from the outer coastal mountains north of Marin County. There is an isolated population in the southern Sierra tentatively assigned to this species occurring from the Kaweah River, Tulare County, south to Mill Creek, Kern County. It occurs to elevations of 1,280 m (4,200 ft) on Mt. Hamilton, Santa Clara County. The Sierra Newt is found up to about 2,000 m (6,500 ft) in the western drainages of the Sierra from the tributaries of Big Butte Creek and the Feather River south of Lassen Volcanic National Park to the Kaweah River, Tulare County. Isolated populations occur in the Lake Shasta area and in the Squaw Creek drainage, Shasta County.

Plate 31. Male (above) and female (below) Sierra Newt.

NATURAL HISTORY: These two newts are usually associated with woodlands that are often interspersed with grassland and chaparral. Their ranges correspond well with the distribution of the coast and interior live oaks

Plate 32. A typical newt breeding pond.

and the ponderosa pine (*Pinus ponderosa*) zone of the Sierra Nevada and southern mountains. These are the best known of the western newts, because their range includes the areas of highest human population densities in California. Breeding takes place in streams, ponds, lakes, and reservoirs. Stream spawning sites usually have rocky beds and bordering trees with exposed roots immersed in water, which provide cover and attachment sites for eggs. Sierra Newts often frequent pools in clear rocky streams.

Breeding occurs from December to May, but individual newts probably do not breed every year. Some populations finish breeding in February and return to land, but others do not leave the breeding sites until May. Estivation occurs in underground retreats and perhaps in rotting logs from July to early fall. In the San Francisco Bay region, and perhaps elsewhere, populations appear to be composed of early and late breeders. The former primarily use quiet sites with little or no current, while the latter tend to use streams after the subsidence of spring floods in late March and April. The larval period lasts from three to six months, depending on water tempera-

Plate 33. Metamorphs under board at pond edge.

ture and food availability. Those that transform during the late-summer dry season may remain under shoreline cover objects until the first fall rain provides a moist ground surface situation for upland dispersal.

Adult food consists of earthworms, snails, slugs, sowbugs, and a variety of terrestrial and aquatic insects and their larvae. Earthworms are extracted from their burrows by the same method used by robins. The newt grasps the portion of worm protruding from the burrow and then repeatedly applies gentle pulls interspersed with releases of tension, thus avoiding breakage of the body. Larvae eat aquatic insect larvae, including those of mosquitoes, and crustaceans; they also scavenge. Adult newts will also occasionally cannibalize their eggs and larvae. On several occasions the junior author has also observed male Coast Range Newts tearing apart and eating the large egg masses of the California Red-legged Frog (*Rana draytonii*). In a pond with a large breeding newt population but minimal Red-legged Frog numbers, such egg predation pressure may pose a serious threat to this Federal Threatened anuran and should be considered when managing such a site.

CONSERVATION: The Coast Range Newt, *T. torosa*, was once one of the most abundant salamanders throughout the coastal ranges of California. Large-scale development of upland foraging and retreat habitat coupled with the loss of breeding ponds and creek pools through siltation or direct filling have greatly reduced its historic numbers. The junior author has sadly watched the pond in which his former student had documented over 7,000 breeders in one season slowly degrade to the point where it is no longer a viable spawning habitat. This was especially disturbing, because it is located on regional park land, but years of written and verbal requests for meaningful management measures for this site have been ignored. The current policies of the CDFG with respect to this species must also be questioned. As previously mentioned, it is listed as a CSSC, a protective category that requires that any person wishing to collect it must obtain a scientific collection permit containing permission for such an activity from this agency. However, the current (2011) California Freshwater Sport Fishing Regulations permit the take of four specimens from the wild by anyone under 16 years of age, and anyone older who buys a freshwater fishing license can do the same. Given that the Coast Range Newt is perhaps the most easily captured amphibian in this state, many remaining populations could be greatly depleted if only a fraction of the millions of legal anglers exercised their amphibian collection privilege.

ROUGH-SKINNED NEWT　　　　　　　　　　　　*Taricha granulosa*

IDENTIFICATION: ADULT snout-to-vent length of this newt is 7 to 8.9 cm (2.75 to 3.5 in.). It resembles the Coast Range and Sierra Newts but is usually darker above, and the underside is more often orange or reddish than pale yellow. The dark dorsal color stops abruptly on the sides, and the lower eyelid and upper lip below the eye are usually dark-colored. Its eyes are small and do not extend to the outline of the head as viewed from above. The teeth in the roof of the mouth are usually in a V-shaped pattern. When in its defense posture it curls the tip of the tail (California Newt complex usually holds tail straight). **EGGS** are encased in a two- or three-layered gelatinous capsule. The ova are approximately 3.5 mm (0.1 in.) in diameter and are laid singly. **LARVAE** range from 5 to 7.5 cm (2 to 3 in.) in total length and are of the pond type. There is no dark longitudinal dorso-lateral striping as in *T. torosa* and *T. sierrae*. The fins are mottled with dark color, and the trunk has two longitudinal rows of light spots that in older individuals may join to form a single light stripe.

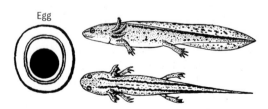

Egg

RANGE: This newt is found from Santa Cruz County in the northern South Coast Range northward through the North Coast Range to southeastern Alaska; inland through the Siskiyou Mountains into the Cascade Range, and south in Sierra Nevada to Butte County. It overlaps the range of the California Newt along the central coast from Mendocino County south. It occurs from sea level to around 2743 m (9,000 ft).

Plate 34. Rough-skinned Newt in breeding pond.

NATURAL HISTORY: In California the Rough-skinned Newt occurs in the humid forests, woodlands, and open grasslands within or near streams, lakes, ponds, and reservoirs of the North Coast and Cascade Ranges. This is our most aquatic newt. It stays in the water long after breeding and sometimes remains there almost permanently with occasional sorties on land during wet weather. It is also the most poisonous newt, with a TTX several times more potent than that of other newt species. It repels most predators, including large piscivorous fish species. This may account for its successful stays in large aquatic sites. At the Cleary Reserve in Napa County the junior author has often observed *T. granulosa* in ponds and creek pools throughout the summer, whereas at the same site *T. torosa* migrated to upland sites soon after breeding. This segregation of the late-spring through early-fall habitat may be one factor in the successful coexistence of the two species between Mendocino and Santa Cruz counties.

The Rough-skinned Newt breeds from December to July, with summer spawning occurring in its northern range and at high elevations. One study of a breeding population in central Oregon indicates that individuals probably breed every other year. Breeding behavior closely follows the generic *Taricha* pattern. The eggs resemble those of the California Newt complex but are usually laid singly and attached to vegetation or other objects in water. Larvae transform in late summer and migrate, often in mass, to upland areas with the first fall rains. At high elevations larvae may overwinter and transform the following summer. Neotenic larvae have also been found in montane lakes in the Cascade Range in southern Oregon, and these may occur in the California segment of the Cascade Range as well. Larvae feed on a wide variety of aquatic invertebrate prey. Adults eat snails, earthworms, crustaceans, spiders, insects, amphibian eggs, and freshwater sponges. Stomach analyses have also revealed some plant material, a most unusual food for an adult amphibian. This species also scavenges, and the intake of plant parts may be a byproduct of that activity.

RED-BELLIED NEWT *Taricha rivularis*

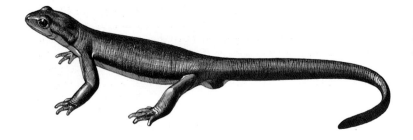

IDENTIFICATION: ADULTS range from 5.9 to 8.1 cm (2.75 to 3.25 in.) in snout-to-vent length and are readily distinguished from other California newts by their dark eyes and tomato-red ventral color. The dark area on undersides of limbs is also more extensive, and the male vent area has a broader dark band than in other newt species. The dark dorsal color stops abruptly on the sides, and the lower eyelid and upper lid are light-colored. **EGGS** are contained in a jelly envelope 8.5 to 10.0 mm (0.3 to 0.4 in.) in diameter, and the ova diameter is 2.4 to 2.9 mm (0.09 to 0.1 in.). They are laid in "streamlined" clusters containing 5 to 16 eggs and usually one to two eggs thick. **LARVAE** measure 4.4 to 5.6 cm (1.75 to 2.25 in.) at transformation. They have some characteristics of the stream type, with balancers reduced or absent and dorsal fin reduced. The body and fin pigmentation is uniform, and there are no dorsolateral stripes or dark body mottling. The eyes are mostly yellow in contrast to dark-eyed adults.

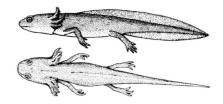
Egg

RANGE: The distribution of this newt is limited to the North Coast Range from the Russian River and vicinity of Agua Caliente, Sonoma County, and Lower Lake, Lake County, to Honeydew, Humboldt County. It coexists with the Rough-skinned Newt and Coast Range Newt in some areas but usually occupies faster water breeding sites.

NATURAL HISTORY: Over much of its range the Red-bellied Newt occurs in mixed-conifer and broad-leafed woodland broken by patches of grassland. The name *rivularis* (river-dwelling) appropriately describes the stream-oriented habits of this species. It breeds from March to May in flowing waters of creeks, streams, and rivers but is aquatic only during that season. Males enter the water as soon as the streams begin to recede from winter flood levels. Both sexes are apparently familiar with and can recognize and orient to a "home" breeding site. Newts that were removed as much as 646 m (2,119 ft) upstream from their home area and released on the stream bank oriented toward and returned to these sites with considerable accuracy. Their initial course direction may have been via sun compass orientation, and then olfaction most likely determined the specific home stream segment. The imprinting of larvae to the odor of a home stream segment may be responsible for this ability.

Males breed every year, but females rarely breed more frequently than at two- or three-year intervals or longer. Mating follows the standard salamandrid pattern, after which eggs resembling those of the California Newt complex but in flattened, streamlined clusters are attached to the undersides of stones or branches in flowing water. At the conclusion of spawning, adults leave the water and are not seen again until the following fall when early rains apparently stimulate above-ground activity. Larvae transform by the end of summer and are rarely seen above ground after that. The scarcity of observations of juveniles when large numbers of adults are active suggests that they live a subterranean life during the five years between metamorphosis and full maturity. This use of flowing creeks for breeding and the relatively long annual subterranean stay of adults and especially juveniles very likely reduces competition between the Red-bellied Newt and the other two California newt species that share its range.

Lungless Salamanders (Family Plethodontidae)

This is the largest family of salamanders in California and the world. It contains about two thirds of all salamander species, and all but a few of these occur in the Western Hemisphere. This group possesses several unique features, some of which will be briefly described in this family introduction. However, the one aspect of their life history that has perhaps contributed more to their success than any other is the ability of most species to reproduce on land. The loss of the aquatic larval form may have especially been important in pristine California with its vast tracts of forest and brush habitats but much less in the way of lakes and ponds. The plethodontids are able to reproduce on land because of their large yolked egg, in which the embryo goes through the gilled larval stage to hatch as a miniature adult. The embryos commonly grow large, three-lobed, leaf-like gills that lie next to the inner surface of the gelatinous egg capsule, through which oxygen and carbon dioxide exchange takes place. Except for a leathery or hard shell and a fluid-filled sack surrounding the embryo, the plethodontid egg is similar to the reptile and bird egg. One habitat requirement is that it must be laid in damp, concealed locations that will remain moist throughout the several-month development period. Except for highly xeric segments of desert areas, most California habitats contain such sites, which can be accessed by very small vertebrates, a category that fits most plethodontid species. Indeed, it was the ability of this group to frequently exploit the "moist-crevice niche," an ecological position unavailable to most other vertebrates, that may be the other major reason for their success.

Most people's introduction to plethodontid salamanders is punctuated by the discovery that this relatively large group of terrestrial vertebrates has no lungs, one of the prime components of our own internal anatomy. Ancestral plethodontids are presumed to have been aquatic, living in cold streams in the Appalachian Mountains of the eastern United States. In the relatively fast-moving water of their ancestral habitat, the buoyancy afforded by lungs may have interfered with locomotion and life in general on the stream substrate. Because these were probably small salamanders and thus had a favorable surface-to-mass ratio for skin respiration in well-oxygenated waters, the lungs could be "evolutionary discarded," and once lost, they were not regained when plethodontids dispersed on land. Because lungs take up considerable space in the vertebrate body, their absence in the plethodontids may have helped promote the small size required by their often crevice-dwelling niche.

In the absence of lungs, plethodontids must depend on skin and buccopharyngeal respiration. Although the capillary lining of the mouth and throat constitute only about 5 to 15 percent of the total vascular surface exposed to air, throat pulsations enhance the skin respiratory role, especially at higher temperatures. The costal folds may also aid in respiration by increasing body surface area. Segmentally arranged short intercostal arteries deliver blood to the costal areas, and lateral cutaneous veins transport it from there to the venter. These artery–vein circuits most likely carry out much of the aeration of the blood. On wet surfaces water flows upward from the substratum by capillary action along the costal grooves between the costal folds, thus keeping the dorsal skin moist and functioning as a respiratory surface.

Another notable feature in many members of this family is the great capacity for tongue protrusion. The tongue pad has no edge attachment to the mouth floor but instead is poised like a mushroom cap on an extendable pediceled base. Glands in the tongue pad secrete a sticky mucus, and the pad can be projected with great accuracy, even at extremely low light levels. The large, forward-focused eyes with their highly sensitive retinas most likely provide the depth-of-field discrimination for such accuracy, even in very dim light. The prey is caught on the sticky pad, and the many teeth in the jaws and roof of the mouth of these plethodontids (Latin for "plethora of teeth") are then employed to hold and crush the food items.

The nasolabial grooves on the snouts of plethodontids are unique structures found in no other postembryonic terrestrial vertebrates. These salamanders regularly tap the substrate with their snouts when they walk, much like snakes do with their tongues as they crawl, and they sometimes lightly probe other individuals. If the surface is moist, fluid passes rapidly up the grooves by capillary action to the external nares, where it is transported through the nasal cavity by ciliary action and then to the chemoreceptors of the Jacobson's organs. The nerves that innervate these organs are considerably more numerous in males, which suggests they may use the grooves in identifying and tracking females. Like fish, salamanders rely heavily on their "nose brain," and odors are probably important in acquiring familiarity with an area, recognizing home range boundaries, intruders into their area, and the territories of other individuals.

In California, surface activity and courtship generally occur during periods of rainfall from fall to spring. In most groups except most slender salamanders, development of the eggs takes place in moist, sequestered sites in late spring and the summer warm, dry period. Hatching occurs in fall and is often timed to coincide with the first rains. Most slender salamanders (*Batrachoseps* spp.) lay their eggs in fall and embryos develop over the winter months. They hatch as miniatures of the adult form.

Plethodontid salamanders seldom form mating aggregations. Adult males are spaced in the habitat, and fighting, which includes biting, has been observed in male encounters. This spacing reduces sexual interference during what is usually a prolonged and energetically costly courtship. This elaborate process begins with the male "injecting" a female with an aphrodisiac-like secretion from his mental gland, located on the underside of his lower jaw near the tip of the chin. In some California plethodontids these "vaccinations" are accomplished by the male pressing the mental gland secretion onto the female's skin with his lower jaw and then puncturing or abrading her skin with his protuberant premaxillary teeth. This procedure, similar to that for smallpox vaccinations of past years, introduces the secretions directly into the rich cutaneous capillary bed.

This process apparently primes females of some species for breeding, while in others such as *Ensatina* it is not necessary. Here the male simply rubs the female's throat with his head (see figure 20, in the "Taxonomy, Anatomy, Physiology, and Behavior of Amphibians" section), after which she follows him with her snout or throat applied to his sacral or tail base area where courtship glands may also be present. The pair now engages in the "tail-straddling walk" in which the female, in her tandem position, walks astride the male's tail. In some species this can last for several hours, after which the male may select a slightly elevated location to deposit a spermatophore. The pair then moves a few steps forward, the female locates the sperm capsule by touch, and picks it up with the lips of her vent as the male lunges backward and strokes her body with his tail.

The internally fertilized eggs are laid in various manners. In the genus *Batrachoseps* they are connected to one another by filaments of jelly like a string of beads. In *Aneides*, eggs are suspended by their stalks like clusters of grapes, while some members of the genus *Plethodon* anchor each egg individually by short basal pedicles. Other *Plethodon* species and *Ensatina* lay nonstalked eggs that simply adhere to each other in an irregular cluster. In many species the eggs are attended by the female, sometimes throughout the entire incubation period. She often engages in what may be called "osmotic brooding" by coiling around her eggs, thus reducing the exposed evaporating surface of the egg mass to air (see figure 21).

Woodland Salamanders (Genus *Plethodon*)

These are primarily woodland or forest salamanders, confined to the United States and Canada. In addition to our western representatives, the genus is widely distributed in humid sections of the eastern United States. Two species occur in the northwest corner of California. These are slim-bodied salamanders with relatively short limbs and a rounded body and tail. Upper jaw teeth extend to the corner of the mouth, the edges of the tongue are free except in front, and many species have a well-defined, broad dorsal stripe of yellow, cream, reddish brown, or orange-brown. Woodland salamanders crawl with sinuous lateral undulations of the body, especially when moving rapidly. When first exposed they may attempt to wriggle head first or backward among rocks, wood fragments, or other cover in which they are found. Western species lay their eggs in spring in damp places on land under stones, in decaying logs, under bark, and in cavities in the ground. Females typically have a biennial, and males an annual, gonadal cycle. Males of most species undergo spermatogenesis in summer and have mature sperm in their sperm ducts in fall and spring.

In some species the eggs are attached to the substratum by short pedicels. It is not known whether western *Plethodon* brood their eggs, but circumstantial evidence indicates that some species do. Hatching occurs in late summer and fall.

DUNN'S SALAMANDER *Plethodon dunni*

IDENTIFICATION: Its snout-to-vent length is 5.0 to 7.5 cm (2 to 3 in.), and it has 14 to 17 (often 15) costal grooves. The dorsal stripe is yellowish tan to dull greenish yellow or olive-yellow, brightening on the tail but obscured with black out toward the tip. The sides are dark brown to black and flecked with white and spotted with yellowish or tan. The upper surfaces of the bases of the limbs are colored like the dorsal stripe. Its ventral surface is slaty with small spots of yellowish or orange. The **EGGS** are laid in grapelike clusters of 4 to 19 and are attached by slender stalks to surfaces in damp hollows within rotting logs.

RANGE: This salamander occurs in ranges west of the Cascade Mountains and south from the extreme southwestern part of Washington to the Smith River area of northwestern California.

NATURAL HISTORY: Dunn's Salamander is most commonly found among moss-covered rock rubble in seepage and spring areas and along permanent, well-shaded small streams. Rocks are almost always present. In the North Coast Range it occurs in outcrops of fine-grained sandstone or shale or other rocks near water. This salamander commonly occurs in wet surroundings and when near streams may escape by swimming, a rare activity for western lungless salamanders.

There are relatively few reports of clutches for this species. A grapelike cluster of nine eggs with an adult female nearby was found in early July in Lincoln County, Oregon. They were attached by a common stalk to a small slab back in a crevice of a shale outcrop. Thirteen eggs laid in the laboratory at UC Berkeley were attached broadly to the side of a jar. Dunn's Salamander feeds on many of the small invertebrates that inhabit the moss-covered substrate of shaded, cool edges of springs and streams.

DEL NORTE SALAMANDER *Plethodon elongatus*

Black phase

Brown phase

PROTECTIVE STATUS: CSSC

IDENTIFICATION: Its snout-to-vent measurement ranges from 6.0 to 7.5 cm (2 to 3 in.), and it usually has 18 costal grooves with 6.5 to 8 costal folds between the toe tips of adpressed limbs. The toes are short and slightly webbed, and the dorsum is dark brown to black, with an even-edged dorsal stripe of brown in specimens from inland areas. **EGGS** are laid in a grapelike cluster suspended by one main strand from the roof of a moist, concealed crevice. The average clutch appears to be about nine.

RANGE: This species occurs from sea level to approximately 900 m (3,000 ft). It inhabits some of the most humid sections of the North Coast Range from the vicinity of Port Orford, Oregon to a few miles south of Orick, Humboldt County, and inland in California along the Klamath and Trinity rivers.

NATURAL HISTORY: The optimal habitat for the Del Norte Salamander appears to be old-growth forest where it occurs in mixed stands of hardwood and conifer. Here it is found under bark, logs, and other ground-cover items on the forest floor. North-facing slopes that retain moisture better throughout the dry summer months seem to be favored. It is most common in stabilized rock slides, including those of road cuts and among rocks and damp soil along stream beds. It may also be found in the interior of decaying logs, and under bark, limbs, and under other surface objects. It appears to prefer moderately to slightly damp, rather than soggy soil.

Males apparently can mate in spring or fall, but peak mating activity probably occurs in early spring. Females are thought to deposit eggs every other year. They remain with the developing embryos in well-hidden nest sites until fall when the eggs hatch. While brooding the eggs they presumably eat little, and it is unlikely they are able to achieve energy levels necessary to "yolk up" eggs for laying the following spring. Females are estimated to mature at around five years of age. Hatchlings probably winter over on their yolk and appear on the surface for the first time in spring.

CONSERVATION NOTE: Here is yet again another species that appears almost totally dependent on old-growth forests for its survival. Unfortunately, small, secretive, and "slimy" salamanders usually take a back seat to feathered fluffy forms such as the Spotted Owl (*Strix occidentalis*), another threatened vertebrate that shares many of the same old-growth forest area with the Del Norte Salamander. Fortunately, this owl's presence has helped preserve some tracts of old-growth forest for less-recognized sympatric species.

SISKIYOU MOUNTAINS SALAMANDER	*Plethodon stormi*
SCOTT BAR SALAMANDER	*Plethodon asupak*

Siskiyou Mountains Salamander

PROTECTIVE STATUS OF *P. STORMI* AND *P. ASUPAK*: ST

IDENTIFICATION: The snout-to-vent length range for both species is 5.0 to 7.5 cm (2 to 3 in.), and they usually have 17 costal folds instead of the 18 normally seen in *P. elongatus*. However, the number of costal folds between the toes of the addressed forelimbs and hindlimbs of *P. stormi* is usually 4 to 5.5, compared to that of *P. asupak* (2.5–3.5). The dorsal color of both is brown and heavily speckled with small light flecks, and the upper surface of the limbs plus the limb bases are usually dusky. Juvenile *P. stormi* are black with an olive-tan dorsal stripe, whereas juvenile *P. asupak* have two orange or red-brown stripes that merge in one in the tail region. Despite these close morphological similarites, *P. asupak* has been determined to be distinct enough genetically from both *P. stormi* and *P. elongatus* to merit full species status. **EGGS** are laid in a grapelike cluster, and the average clutch contains nine eggs.

RANGE: In California *P. stormi* occurs in the Seiad and Horse creek drainages in Siskiyou County. *P. asupak* is found only in a very small area of north Siskiyou County, near the confluence of the Klamath and Scott Rivers.

NATURAL HISTORY: These salamanders live in the driest habitats occupied by any western *Plethodon*. They have been found in old clear-cuts, moss-covered road cuts near seepages, and along streams. The preservation of rock outcrop, talus slope, and seep areas in such areas is especially important. It is interesting to note that unlike their close relative *P. elongatus*, which favors old-growth forest, these salamanders appear to have adapted to sites that are recovering from logging. Their reproductive behavior is presumed to be similar if not identical to that of *P. elongatus*. Both have recently been recognized as distinct species, and with that designation will hopefully come further investigations of their natural history.

Ensatina (Genus *Ensatina*)

At one time genus *Ensatina* was considered to be closely related to genus *Plethodon*. It is now regarded as an ancestral group that diverged from the evolving *Plethodon* line near the beginning of the Tertiary.

The genus, as treated in this book, contains two species. One of these is Ensatina (*Ensatina eschscholtzii*), which consists of a series of subspecies with regions of intergradation between them.

Wake and Schneider (1998) believe that one of these original subspecies, the Large-blotched Salamander, could be elevated to species status based on current molecular studies, but the other original subspecies should remain as such. We concur with this view and present this former subspecies as a new species, *Ensatina klauberi*. In making this designation we also note that hybrids between the Large-blotched Ensatina and the Monterey Ensatina have been found at several locations (Lemm, 2006).

ENSATINA	***Ensatina eschscholtzii***
MONTEREY ENSATINA	*E. e. eschscholtzii*
YELLOW-EYED ENSATINA	*E. e. xanthoptica*
PAINTED ENSATINA	*E. e. picta*
SIERRA NEVADA ENSATINA	*E. e. platensis*
YELLOW-BLOTCHED ENSATINA	*E. e. croceater*
LARGE-BLOTCHED ENSATINA	***Ensatina klauberi***

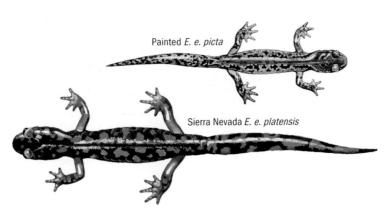

Painted *E. e. picta*

Sierra Nevada *E. e. platensis*

PROTECTIVE STATUS OF *E. E. CROCEATER* AND *E. KLAUBERI*: CSSC

TAXONOMIC NOTE: Because of their great similarity, we discuss the two Ensatina species together. In the account for Ensatina (*Ensatina eschscholtzii*) we deviate from our norm of discussion of California amphibians and reptiles at the species level. This is because its taxonomy is complex, actively under study, and as yet not fully resolved. Furthermore, Ensatina is currently widely recognized as an outstanding example of a Rassenkreis (race circle), examples of which are not common. They are chains of races or subspecies that form more or less a circle, the ends of which overlap and act as distinct species. All but one of the currently recognized taxa occur in California, and all presented here are shown on the accompanying map. The map also indicates hybridization zones and incipient species-level breaks where molecular studies suggest taxa may be approaching the species level of differentiation. It also shows the tentative range of the Large-blotched Ensatina (*Ensatina klauberi*), which, as previously mentioned, we now accept as a full species but urge further study, including molecular, of its relationship to the Monterey Ensatina (*E. e. eschscholtzii*) north of a low-lying geographic break, the San Gorgonio Pass, that separates the San Jacinto and San Bernardino mountains. Hybrids are known from this latter range.

IDENTIFICATION: *Ensatina eschscholtzii* subspecies range in snout-to-vent length from 6.2 cm (1.5 in.) for the Painted Ensatina (*E. e. picta*) to 7.5 cm (3 in.) for the Yellow-blotched Ensatina (*E. e. croceater*). The head and eyes are large, and the latter are noticeably protuberant. The body is rounded with 12 costal grooves, the limbs are relatively long, and the tail is rounded and constricted at the base, a characteristic in which this genus differs from all other western salamanders.

The five California subspecies differ from one another principally in coloration (see color illustrations). Two of the California subspecies, the Sierra Nevada Ensatina (*E. e. platensis*) and Yellow-blotched Ensatina, are blotched dorsally and distributed, in that order, from north to south through the Sierra Nevada and more southerly mountains. Two other subspecies, the Yellow-eyed Ensatina (*E. e. xanthoptica*) and the Monterey Ensatina, are uniformly colored dorsally and distributed in like order through the southern part of the North Coast Range, the South Coast Range, and Coastal Southern California as indicated on the range map. The fifth California subspecies, the Painted Ensatina, has a mottled style of coloration and is limited to a narrow north-coastal area. It is regarded as closest to the ancestral stock of the genus *Ensatina*.

In addition to the five California subspecies, a sixth, the Oregon Ensatina (*E. e. oregonensis*), occurs in western portions of Oregon, Washington, and British Columbia, including the eastern half of Vancouver Island, B.C. It tends to be uniformly colored dorsally and has a fine dark speckling on its ventral surface. An intergrade between it and the Yellow-eyed Ensatina, photographed in northern Napa County, is seen in Plate 35.

Plate 35. *Ensatina* intergrade in northern Napa County.

The Large-blotched Ensatina has blotches, bars, or bands ranging from orange to light-cream color on a blackish background. Its colored markings average larger than those of either of the *E. eschscholtzii* blotched subspecies.

Males of both species have a pronounced forking of the nasolabial grooves at the edge of their swollen, somewhat overhanging upper lip. They also have a longer, more slender tail and a longer, broader snout than females. The female's nasolabial groove is only slightly forked, and the upper lip is not swollen. The **EGGS** are enclosed in two jelly envelopes, about 6 to 8 mm (0.2 to 0.3 in.) in outer surface diameter. The ovum size within the two species ranges from 4 to 7 mm (0.16 to 0.27 in.) in diameter. The embryos have large, flattened, three-lobed gills. Sexual maturity is evidently acquired the fall or spring of the fourth year in the Yellow-eyed Ensatina.

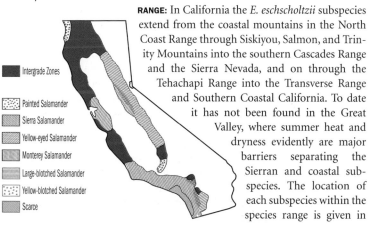

RANGE: In California the *E. eschscholtzii* subspecies extend from the coastal mountains in the North Coast Range through Siskiyou, Salmon, and Trinity Mountains into the southern Cascades Range and the Sierra Nevada, and on through the Tehachapi Range into the Transverse Range and Southern Coastal California. To date it has not been found in the Great Valley, where summer heat and dryness evidently are major barriers separating the Sierran and coastal subspecies. The location of each subspecies within the species range is given in

LUNGLESS SALAMANDERS (PLETHODONTIDAE)

the identification section and is also shown on the range map. This subspecies designation was used by the senior author in his original study of this salamander group along with the term "intergrade" when referring to individuals from zones of color change between subspecies groups. Recent molecular studies have revealed complexities that question the use of this term in some of these zones. However, until the nomenclature is fully resolved, we retain the use of the term intergrade. The large dark blue areas on the range map denote such transitional areas where color patterns merge and intergrades occur.

The Large-blotched Ensatina occurs only in a portion of the Peninsular Range in California. A small isolated population of the Large-blotched Ensatina occurs in the San Pedro Martir Mountains and in a western, coastal volcanic site in Baja California, Mexico.

NATURAL HISTORY: The varied habitats of Ensatina include the redwood zone of the North Coast Range, the chaparral and gray pine–coast live oak woodland of the South Coast Range, and the yellow pine–incense cedar–black oak forests of the Sierra Nevada and higher mountains of southern California. In Coastal Southern California the Monterey Ensatina is widespread in oak woodland and old chaparral, often in remarkably dry places. Despite its wide range and striking color variation, Ensatina appears to be quite uniform in its choice of sites within these plant associations. It tends to avoid steep slopes but instead occurs in more gradual uplands where there is considerable leaf litter and surface objects such as logs, bark, boards, and rocks. It avoids saturated soil and is usually found in only moderately damp surroundings.

Both Ensatina species spend much time underground in the burrows of rodents and other animals and are sometimes found beneath the bark or in the rotten interior of logs and in woodrat middens. Individuals are generally solitary except when breeding or when associated with recently hatched young. With the first rains of late fall, adults, juveniles, and newly hatched young emerge from summer estivation sites. In some areas considerable dehydration can occur during the dry season, thus limiting feeding opportunities. There is therefore a burst of surface activity with the onset of each rainy season, and this is often the best time to conduct field observations and studies. In one such study of a marked population of the Yellow-eyed Ensatina in a redwood habitat in the Oakland Hills, Contra Costa County, the senior author found that many adults evidently spend their entire lives within circumscribed areas ranging from an average diameter of 10 m (33 ft) for females to 20 m (64 ft) for males.

When disturbed by a tap on its back, this salamander often stands stiff-legged and sway-backed with its tail elevated and arched, and its neck erect with the head held horizontally or tipped downward (see figure 6, in the "Taxonomy, Anatomy, Physiology, and Behavior of Amphibians" section). Rarely, it may produce a squeak or snakelike hiss, quite a feat for an animal without lungs! At such times it secretes a milky, sticky fluid, chiefly from

Plate 36. Yellow-eyed Ensatina secreting toxic mucus.

poison glands in its tail, and if the disturbance continues it may swing the tail in the direction of the intruder. If a predator then grabs the tail, it will easily break at its constricted base. In captive feeding experiments, the sticky tail secretion has gummed up the jaws of small snakes and, in some cases, caused a snake to stop feeding.

The courtship of Ensatina has already been presented in the plethodontid salamander introduction as an example of the intricate innate behavior of this family (see figure 20, in the "Taxonomy, Anatomy, Physiology, and Behavior of Amphibians" section). Eggs are usually laid underground in spring and early summer in crevices, burrows of other animals, under bark, and beneath rotting logs and other surface objects. The eggs are laid in grapelike clusters, often loosely adherent to one another and sometimes to the substratum. They are brooded and sometimes moved by the female in a loop of her body (see figure 21, in the "Taxonomy, Anatomy, Physiology, and Behavior of Amphibians" section).

CONSERVATION NOTE: Ensatina is often found in areas of leaf, bark, and branch litter on canopied forest floors. In such areas a total removal of this microhabitat in the name of fire prevention is usually unnecessary. At the least, pockets of such prime understory that have been identified by prior surveys as critical Ensatina habitat areas should be left as refuges for these remarkable animals.

The Yellow-blotched Ensatina and the Large-blotched Ensatina, have been granted CSSC status by the CDFG. They are among the most beautiful North American salamanders and as such are very likely prized by those who collect live amphibians and reptiles. It is most unfortunate that the current (2011) *California Freshwater Sport Fishing Regulations* published by this state agency permit the taking from the wild of four specimens of "Ensatina," regardless of special-status designation.

Climbing Salamanders (Genus *Aneides*)

These salamanders have well-developed lower jaw muscles that protrude in the temporal region and give the head a triangular shape when viewed from above. This feature is especially prominent in adult males. They have large, protuberant eyes and prominent costal grooves. The tail is rounded, tapering from the base, and more or less prehensile. The premaxillary teeth are enlarged in adult males in breeding condition and may be detected by stroking the tip of the snout upward when the mouth is closed. Three of the four species that occur in California (*A. lugubris*, *A. ferreus*, *A. vagrans*) are good climbers and have been found at considerable heights in trees, usually in cavities, under bark, in arboreal rodent nests, and in fern mats within the crowns of old-growth redwoods (*Sequoia sempervirens*). Dr. David Wake, a colleague of the authors, suggests that the jaw specialization of *Aneides* increases the efficiency of seizing and holding prey, and that this increased efficiency is of greater significance to climbing salamanders than to terrestrial ones, which quickly turn their heads to restrain large prey against the ground.

ARBOREAL SALAMANDER *Aneides lugubris*

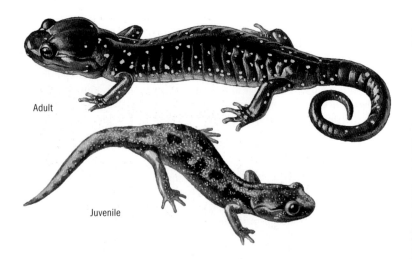

Adult

Juvenile

IDENTIFICATION: This is a relatively large plethodontid salamander with a snout-to-vent length range of 6.2 to 10.0 cm (2.25 to 4 in.). It has a stocky body, and its large jaw muscles are well developed, giving the head a triangular shape when viewed from above. The tail is round, tapered, and somewhat prehensile. The limbs are relatively long, and the tips of toes blunt and broadened. There are usually 15 costal grooves. There is a pronounced dorsal crest on the cranium, especially in males, which serves for the attachment of the massive jaw muscles. The dorsal color is uniformly gray-brown to chocolate-brown, usually spotted with pale yellow. There is great variation in size, number, and positions of spots. Animals from the Sierra Nevada have few small spots or are unspotted, while those from the Gabilan Range of eastern Monterey and western San Benito counties and from South Farallon Island west of San Francisco have numerous large yellow spots. **EGGS** are anchored to the roof of a nesting chamber by individual stalks, ranging from 6 to 20 mm (0.2 to 0.8 in.) in length, one often twisted around one another (figure 25). The ova diameter is approximately 6 mm (0.2 in.). The embryos have three-lobed leaflike gills.

RANGE: The Arboreal Salamander occurs from the vicinity of Eureka in Humboldt County south through the North and South Coast Ranges and Coastal Southern California. Disjunct populations also occur on the lower slopes of the Sierra Nevada from El Dorado to Madera counties and on several islands including South Farallon, Año Nuevo, and Catalina islands off the California coast, and Los Coronados Island off the coast of Baja California, Mexico.

NATURAL HISTORY: In coastal areas it is associated with coast live oaks and gray and yellow pines.

Figure 25. Arboreal Salamanders with their stalked eggs attached to the roof of the nest chamber.

Plate 37. Arboreal Salamander climbing a tree trunk.

In the northern part of its range along the coast it has invaded clearings created by logging and cultivation in the original dense redwood forest. It is also occasionally found in the yards of older homes in the Bay Area where old live oaks and mature shrubs have been preserved. Retreat sites include areas under logs, bark, boards, rocks, and other surface objects, tree cavities, inside rotten logs, wood rat houses, rodent burrows, and mine shafts.

This salamander appears on the surface following the first fall or winter rains and remains active through winter and spring except during cold weather. With the cessation of rainfall and the drying of the surface in late spring, it retreats underground, into moist cavities in trees, rotten logs, or other shelters. It is a good climber and has been found in trees over 18 m (59 ft) above the ground. Aided by its expanded digits and prehensile tail, it often seeks hollows in oaks or other trees. As many as 35 individuals of all ages have been found in a single cavity. Although the common name refers to its climbing propensities, it spends considerable time on the ground.

Upon first encounter when molested, it commonly assumes an Ensatina-like, elevated tail defense pose and threatens with its mouth open. Despite the absence of lungs, it may squeak repeatedly with its mouth closed when tapped, retracting its eyes into their sockets each time a sound is produced. The sound may be created when the retracting eyes force air in the mouth cavity out through the nostrils and/or past the closed lips. The strong jaw and sharp teeth are a protection against some predators. In the laboratory, Ring-necked Snakes (*Diadophis punctatus*) have rejected Arboreal Salamanders while attacking another species of similar color and size. There is also some evidence for use of the teeth in territorial behavior and other forms of intraspecific aggression. The senior author once found a young male with fresh and severe head and neck lacerations, probably inflicted by a large adult male. The injured animal was under one end of a long plank, which harbored at its other end a large adult male and female. This author also experienced interspecific aggression when he sustained a bite on his finger while displaying an Arboreal Salamander during a television interview some years ago. He has also been bitten on the palm of his hand by a large male. The V-shaped laceration in this relatively tough palm-skin area also bled profusely.

After performing a version of the complex Ensatina courtship pattern, spermatophore deposition and retrieval occurs. Eggs are probably laid in late spring or early summer in cavities in ground or hollows in trees or rotten logs. Tree hollow "nests" have been found to heights of at least 4.9 m (16 ft), and nests are often guarded by the adults (see figure 25). From 12 to 23 eggs may compose a clutch. Hatching occurs in fall and early winter, and maturity is probably reached within two years.

The Arboreal Salamander feeds on a wide variety of invertebrates that inhabit both ground surface and tree cavity sites. Like other *Aneides* species, it is able to prey on tree-dwelling insects at night without competition from or predation by birds except perhaps small owls.

BLACK SALAMANDER   *Aneides flavipunctatus*
SANTA CRUZ BLACK SALAMANDER   *A. f. niger*
SPECKLED BLACK SALAMANDER   *A. f. flavipunctatus*

A. f. niger

A. f. flavipunctatus

Juvenile

IDENTIFICATION: The snout-to-vent range is 5.0 to 9.4 cm (2 to 3.75 in.). Its limbs and toes are relatively short, and it has from 14 to 16 costal grooves. As its name denotes, this is indeed a black salamander, although there may be numerous whitish to pale-yellow spots (basis for name flavipunctatus) and considerable numbers of brassy iridophores on the black background color. In some populations the brassy iridophores are so abundant as to impart an olive, greenish gray, or greenish cast over nearly the entire dorsal surface. The ventral surface of all color varieties is deep slate to black, sometimes variously spotted with white iridophores. Males have a larger, more triangular head, an enlarged upper lip, longer legs, and a more glandular ventral surface than females. **EGGS** are suspended from the ceiling of moist retreats by intertwined gelatinous strands. Clutch size ranges from 5 to 25 eggs, based on counts of enlarged ova.

Eggs

RANGE: The Speckled Black Salamander is found from northern Del Norte County and extreme southwestern Oregon in the North Coast Range to Sonoma County. The Santa Cruz Black Salamander occurs as a separate population on the San Francisco peninsula in San Mateo, Santa Cruz, and Santa Clara counties.

NATURAL HISTORY: The Black Salamander may be found in a variety of habitats. They include coastal grassland, open oak and conifer woodlands, redwood and Douglas fir (*Pseudotsuga menziesii*) forests, mixed-evergreen forests, clearings near forest edges, banks of permanent streams, areas around springs, wet to damp rock rubble, and caves and mine shafts. It is generally found outside areas of continuous tree canopy. In many places in the northern part of its range it favors moist rock rubble habitat along river drainages at low elevation. However, in north coastal areas, in the vicinity of Mt. Shasta, and in the Santa Cruz Mountains in the south, it appears to be more strongly attracted to water and commonly occurs under rocks at the edges of streams and in the vicinity of springs and seepages. Areas where this salamander occurs usually receive at least 75 cm (30 in.) of rainfall annually, and some localities may receive as much as 200 cm (80 in.). The gap in the range of this species between the Santa Cruz Mountains south of San Francisco Bay and the vicinity of the Russian River drainage to the north is perplexing, because suitable habitat appears to be present.

For many years the junior author's herpetology course salamander field trip was an overnight affair to the Cleary Reserve in a remote area of the Napa County. The class stayed in an old fieldstone lodge surrounded by an extensive Douglas fir stand. Upon arrival the professor would briefly

Plate 38. Black Salamander on granite outcrop.

disappear into a rock-lined crawl space under the lodge and soon emerge with an adult and juvenile Black Salamander in hand. Although extensive hunting by the class in the adjacent woodland always produced other species of plethodontids, *A. flavipunctatus* was never found outside of this little rock oasis. Although most plethodontid salamanders at lower elevations in California retreat from the surface of the ground during the late spring-to-fall dry period, individuals of *A. flavipunctatus* may be found active in summer in streamside and spring habitats where there is permanent water.

This is apparently a highly sedentary species. When abroad at night it tends to stay near cover such as rock crevices, cracks in dirt banks or in the ground, and rodent burrows. Its coloration, conspicuous by day, is highly concealing at night, and white-spotted individuals, often associated with granite or other speckled rock surfaces, blend with their rock backgrounds with remarkable precision, especially when viewed by moonlight. When roughly handled or injured this salamander may stand with head held high and tail elevated, arched, and waved from side to side, much in the same manner as *Ensatina*.

In captivity courtship and spermatophore deposition has been observed from November to April. Egg laying is presumed to occur in late spring and summer. Only three clusters of eggs have been found to date. In each case a female was in attendance. Hatching occurs in late summer and fall. Hatchlings are found on the surface in fall and early winter. Males appear to reach sexual maturity in their fourth year, females in their fifth.

The Black Salamander feeds on spiders, millipedes, and insects including beetles, ants, termites, and flies. The variability in size of prey taken increase with the size of the salamander.

WANDERING SALAMANDER
CLOUDED SALAMANDER

Aneides vagrans
Aneides ferreus

Wandering Salamander

INTRODUCTORY NOTE: Until recently only one species, the Clouded Salamander, *A. ferreus*, was recognized. However, molecular analysis of specimens throughout its range from central California through northern Oregon and on Vancouver Island, B.C., where it may have been introduced, has shown sufficient differences to merit two species, even though the external characteristics and basic ecology are very similar. Taxonomic revisions such as this pose a challenge to the student of natural history who relies on a field guide rather than a mobile DNA sequencing lab for species identification in the natural habitat. Fortunately, geographic location will often suffice for identification, because it was usually some sort of past geographic isolation that allowed such molecular differences to accrue. We (the natural historians) are fortunate in this particular case, because the geographic division along the North Coast Range falls almost exactly on the California–Oregon border. Here the Smith River flows across the far northwestern corner of the state and may well be the source of the geographical separation that resulted in the current molecular differences.

IDENTIFICATION: Both species have a snout-to-vent range of 4.5 to 7.5 cm (1.8 to 3 in.). Their heads are somewhat flattened, the bodies are rather slim and flattened, the tail rounded, and the limbs slender, and there are 16 or 17 costal grooves. The dorsal color is brown with a clouded pattern of ash gray, light gold, or reddish. The venter is dusky or tan. The dorsal patterning of the Wandering Salamander usually exhibits a more striking pattern of gray and black compared with that of the Clouded Salamander, which is a more diffuse mixture of brown and gold. The dorsal body and limb surfaces of *A. ferreus* may have extensive pale speckling and flecking. Both species may exhibit a dark phase (dorsal body nearly plain brown) and a light phase where the pale gray dominates. **EGGS** are unpigmented and attached by intertwined individual stalks from the roof of a log cavity nesting chamber. Ova diameter ranges from 4 to 6 cm (0.15 to 0.2 in.).

Egg

RANGE: *A. vagrans* occurs from northwest Sonoma County north to the Smith River, Del Norte County. It is also found on Vancouver Island, B.C., where it may have been introduced. *A. ferreus* occurs from the Smith River, Del Norte County north to the vicinity of the Columbia River.

NATURAL HISTORY: These are salamanders of the humid coastal forests and forest clearings, where the dominant large trees are Douglas fir, hemlock (*Tsuga heterophylla*), and redwood. *A. vagrans* is often found under the bark of Douglas fir (sometimes fire-charred), and in stumps and under accumulated leaf litter on the top of sawed stumps of redwoods and other large trees that contain loosened bark and/or vertical fissures. In some areas *A. ferreus* is more common in rock slides and in moist rock crevices. These salamanders are also the most completely arboreal of the western *Aneides*. *A. ferreus* has been found in trees at heights up to 6.5 m (21 ft). Even more impressive is the recent discovery by biologists from Humboldt State University and the Redwood Science Laboratory, both at Arcata, that *A. vagrans* inhabits the large mats of leatherleaf fern (*Polypodium scouleri*) that grow within the canopy of old-growth redwoods. Here it has been observed at heights up to 93 m (302 ft) above ground. Other observations include that of an individual feeding on small winged insects in the canopy and the discovery of a clutch of eggs from high in the crown of a redwood that was felled for lumber. These findings, plus a population estimate of 54 *A. vagrans* for five tree crowns surveyed, strongly suggest that many individuals spend their entire life in these old-growth forest canopies, where water-holding capacity and small insect populations of the large fern mats provide year-round habitat for a plethodontid salamander. At present, only one other group of nonavian vertebrates, the tree voles (*Arborimus* spp.), is known to permanently inhabit this lofty realm.

Eggs are laid in hollows in rotten logs, under bark, and in tree canopy fern mats in late spring and summer. One clutch was found at 30 m (100 ft) in a large coast redwood. Clutch size ranges from six to seven ova (California, *A. vagrans*) to eight to 17 (Oregon, *A. ferreus*) to 14 to 26 (Vancouver, B.C.). The embryos have a single pair of three lobed, leaflike gills with extensive vascularization. Like numerous other plethodontids that live in moist ground surface areas, these species feed on the large variety of small invertebrates that also inhabit such sites. These include ants, beetles and their larvae, spiders, isopods, termites, collembolans, and cockroaches.

CONSERVATION NOTE: The recent finding that populations of the Wandering Salamander permanently inhabit the canopy of old-growth redwood forests provides yet another reason for the preservation of this now scarce botanical resource. Only 4 percent of the original old-growth redwood forests remain, and most of these small stands occur within a few national and state parks. Redwood stands capable of supporting tree-dwelling salamanders are rare outside such parks, and the identification and protection of these is a vital factor in the conservation of these unique salamander species.

Slender Salamanders (Genus *Batrachoseps*)

This genus is confined to the Pacific Coast of North America, with all but one species occurring in California. The number of recognized species has increased dramatically during the past four decades. The senior author's first Peterson *Field Guide to Western Reptiles and Amphibians* in 1966 reported only two species. The number had increased to eight by the time of the second edition (1985), and the current 2003 edition lists 19 species, with 18 occurring within California. One reason for this increase is that several of these are cryptic species that look very similar but that molecular analysis has shown to be sufficiently unique to warrant species status. The main reason for these similarity is that their basic morphology stems from the ecological niche that they occupy. These are the "worm salamanders," so named because they are long and thin with a body girth of mostly less than that of adult earthworms that inhabit the same area. Their costal and caudal grooves look much like the body segmentation seen in earthworms, and the legs of some species are so small that they often go unnoticed when first encountering specimens in the field. One defining feature of this genus in California can be seen in the diminutive hind limbs, which have only four digits, whereas all our other salamander species have five.

The slender salamander body size and form is a classic example of "small" being better, for it permits many members of the *Batrachoseps* genus to enter earthworm and termite burrows plus many narrow crevices where small invertebrate prey can be captured. This allows them to obtain invertebrate prey where no other small vertebrate can, and hence their presence in all 12 of the state's geographical subdivisions. This distribution includes a number of desert fringe and dry-summer habitats, where once again diminutive size permits passage into small crevices and soil retreats to escape seasonal lethal above-ground thermal and osmotic conditions.

Despite this generic conformation to the general slender salamander body plan, there are noticeable interspecific variations in head, body, tail, and limb proportions. The two ends of this spectrum are seen in an Oregon species, the Oregon Slender Salamander (*B. wrightii*) and the California and Gregarious Slender Salamanders (*B. attenuatus* and *B. gregarius*). The former is regarded as the most generalized and ancestral *Batrachoseps*. It has relatively long legs and a robust body form, and it is considered to be close to the ancestral stock that gave rise to the genus. *B. attenuatus* and *B. gregarius*, on the other hand, appear to have the most specialized morphology within the genus. The California Slender Salamander lives in north coastal areas where California lungless salamander species reach their peak in abundance, and thus there is more interspecific competition for plethodontid feeding niches. Its head is little or no wider than its neck, its limbs are minute, and its long body and tail are dwarfed by a plump earthworm, all of which allows it to go into very narrow passageways to search for prey where most other salamanders cannot go.

Between these two ends of the slender salamander spectrum lies a subtle series of interspecific differences that are difficult or in some cases nearly impossible to recognize in the field. This poses a challenge to both the student of natural history and authors of natural history guides. This book therefore relies heavily upon geographic distribution instead of detailed morphological information to help you determine just what species you may encounter in the field. We also present the following generic natural history information for all of California's 18 slender salamander species and stress only significant deviations from some in the individual natural history accounts.

GENERIC DESCRIPTION: The genus snout-to-vent length ranges from 2.5 to 7.0 cm (1 to 2.75 in.). The bodies are slender, rounded, and segmentally grooved. The tail length is about that of the body length in some to over twice the body length in others. The head width is noticeably wider than the neck in some species and about the same width in others. The limb length is also variable, but most have shorter limbs than other salamanders. Many species have a reddish or brown dorsal stripe that extends from the snout to well out on the tail. Juveniles are generally more stocky and have a shorter body and tail length than adults. The best single indicator as to what type of small salamander you have just encountered in the field is its hind feet. If they have four digits, then it must be a species of *Batrachoseps*. Next, browse through the range maps given for each of the species and match your geographic location as closely as possible with one or more. Now read the verbal range description for the one or two species that exist at or near your position, and then consult the color illustration and verbal review of specific features to finally confirm your choice.

GENERIC NATURAL HISTORY: With the first fall or winter rains that end the summer dry period, slender salamanders emerge from their underground retreats and are found on the surface of the ground. The more attenuated forms assume a coil reminiscent of a watch spring when at rest under ground-cover objects, with the head usually resting on a loop of the body. Uncoiled individuals hold the head erect and directed toward the source of light penetrating beneath their protective cover. In this position they may be watching for small prey that are backlit by the crack between the cover and substrate.

Although the limbs of these salamanders are small, they are effective in locomotion, supporting the body well off the ground and providing propulsion without the assistance of marked lateral undulations of the trunk as seen in most other salamander groups. When moving rapidly the limbs are folded back, and forward progress is maintained with lateral undulations of the body and tail that closely resemble snake movement.

At first appearance these small, fragile-looking salamanders seem defenseless, but they resist predation in several rather effective ways. Most observations of defensive behavior have been made on the widely distributed California Slender Salamander. When probed repeatedly, either by a bird's beak or a human finger, this species will at first "play possum," then suddenly engage in rapid fishlike "flip-flops" that may throw it several inches into the air and sometimes over a 1-m- (39-in.)-wide area. This usually takes place on its natural habitat substrate to which its body color is well matched, and upon finally coming to rest it usually lies still and may be readily overlooked. It may also crawl partly under a leaf or other object and then remain perfectly still with its hindquarters still exposed.

If this first line of defense fails, several additional strategies may be employed. If a predator is able to secure a hold, the salamander exudes sticky mucus that is very effective at gumming up the jaws of small amphibian-eating snakes such as the Ring-necked Snake (*Diadophis punctatus*) and juvenile garter snakes (*Thamnophis* spp.). While conducting prey preference tests with young San Francisco Garter Snakes (*Thamnophis sirtalis tetrataenia*), the junior author and former graduate student Sheila Larsen had to perform several emergency mouth cleanings after several of the snakes attempted feeding on *B. attenuatus* adults. This species has also been reported to thwart being swallowed by a young garter snake by forming an overhand loop with its body around the snake's neck. If an initial strike by a snake's jaws or a bird's beak grasps the tail, it will often break off and if not immediately consumed will writhe on the ground for well over three minutes because of the anaerobic nature of amphibian muscle physiology. During this time the rest of the salamander walks slowly away in a smooth, linear movement that is hard to detect in the ground litter of its habitat. The predator, meanwhile, gets a partial meal for its efforts, and the salamander will eventually regenerate a new tail. However, the slender salamander tail is an important fat storage organ, and in *B. attenuatus* this reserve may constitute up to 50 percent of the tail mass of a mature animal. One study has found that tail loss can delay maturity and inhibit reproduction in this species.

Because eggs are probably most often deposited underground, those of only four species have so far been recovered. These have been in clusters under ground-surface objects. The eggs were contained in two jelly envelopes and connected to one another in a beadlike string by a slender strand of jelly. The ova of *Batrachoseps* are unpigmented, and although females have been discovered with their eggs, long-term egg brooding has not been documented in this genus. However, communal "nesting" at favorable sites has been observed for several species and may eventually be found to be a quite common occurrence.

The food of slender salamanders includes a wide variety of small invertebrate species that inhabit surface-litter areas and subterranean openings. Newly hatched earthworms from cocoons deposited within worm burrows are an important food of the very thin, fossorial species. Small slugs, spiders, large mites, juvenile snails, millipedes, sowbugs, and a great variety of ground-dwelling insects and their larvae round out the food preference list.

INYO MOUNTAIN SALAMANDER *Batrachoseps campi*

PROTECTIVE STATUS: CSSC

IDENTIFICATION: This is one of the largest and stockiest slender salamanders with snout-to-vent lengths up to 6.1 cm (2.3 in.).

RANGE: It is restricted to the Inyo Mountains, Inyo County, where it occurs at elevations between 550 and 2,620 m (1,800 to 8,600 ft).

NATURAL HISTORY: Their habitat consists of moist areas adjacent to springs and small water courses mostly below the piñon–juniper belt. They are found along streams in damp crevices and often under surface objects where wild rose and willow are the dominant vegetation.

CONSERVATION NOTE: Water is the most sought-after natural resource in California, and this quest is often most intense in semiarid and desert regions. Springs and seepages are often cleared of all vegetation and rock rubble, and then capped with a pipeline to transport this precious commodity elsewhere for human use. The increase of free-ranging horses and burros also poses an ongoing threat to this salamander by continuously trampling the habitat around these water sources during the long dry season. The relatively few sites where this distinct *Batrachoseps* species occurs require complete protection, and efforts to discover other possible occurrence sites should continue.

KERN PLATEAU SALAMANDER *Batrachoseps robustus*

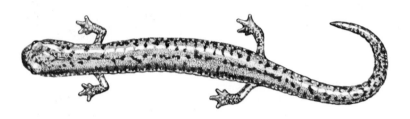

IDENTIFICATION: As its Latin species name indicates, this is a robust, large slender salamander similar in build to *B. campi*, and like it is considered to be close to the ancestral stock that gave rise to modern-day *Batrachoseps*. Its snout-to-vent length varies from 4.4 to 5.7 mm (1.75 to 2.25 in.), its head is broad, and its tail is relatively short. The dorsal coloration differs from other slender salamanders in that it has many dark flecks and spots on a rusty to gray or silvery background, with the latter color more prevalent in drier areas.

RANGE: It occurs from 1,615 to 2,804 m (5,298 to 9,199 ft) in the Sierra Nevada on the Kern Plateau, the eastern Sierran slopes from Olancha Peak to Nine-mile Canyon, and in the Scodie Mountains of Kern County.

NATURAL HISTORY: At higher elevations darker color phase individuals are found in Jeffrey pine and red fir forests, with lighter colored specimens becoming more abundant in drier, lower elevations populated by lodgepole and pinyon pine, black and canyon oak, and sagebrush and rabbitbrush. In all such areas it is active above ground or sequestered under surface-cover objects during their respective wet seasons, after which it apparently seeks deep moist retreats.

TEHACHAPI SLENDER SALAMANDER

Batrachoseps stebbinsi

PROTECTIVE STATUS: ST

IDENTIFICATION: Like *B. campi* and *B. robustus*, this is another robust, short-tailed ancestral-type *Batrachoseps* with a very limited geographic range. Its dorsal body color ranges from light to dark brown or reddish, which sometimes contains light-beige patches. In several areas of the Tehachapi Mountains it coexists with the Black-bellied Salamander (*B. nigriventris*), which represents the other end of the *Batrachoseps* body form spectrum, and thus the two species are rarely confused.

RANGE: Small populations of this salamander are scattered through the Tehachapi Range and the southern end of the Sierra Nevada in Kern County.

NATURAL HISTORY: This salamander inhabits moist ravines and canyons within its range that support an oak or mixed pine–oak woodland. Here it is found under rocks and logs, especially where a good layer of leaf litter has accumulated that holds ground surface moisture long after the annual rains have ceased.

CONSERVATION NOTE: Decades of intense livestock grazing throughout this species' relatively limited range have destroyed much of the ground cover and especially the dense leaf-litter cover that maintains the surface moisture needed by this ST species. Because most of this land is in private ownership, this habitat-degrading practice is hard to control. Ravine and canyon areas should be acquired by the state as mitigation sites whenever possible, and conservation easements should be sought with local ranchers who would then agree to greatly restrict or eliminate stock grazing in these key habitat sites. At this writing this approach to protection is being explored.

KERN CANYON SLENDER SALAMANDER
Batrachoseps simatus

PROTECTIVE STATUS: ST

IDENTIFICATION: This species has the more common *Batrachoseps* body form. It has a narrow head and relatively long body, tail, and legs. Its head and body are somewhat flattened, which may be an adaptation for crevice dwelling. The snout-to-vent range is 4.1 to 5.6 cm (1.6 to 2.2 in.).

RANGE: This species occurs in the lower Kern River Canyon from the northern slopes of the Piute Mountains (Erskine and Bodfish creeks) to Stork Creek.

NATURAL HISTORY: *B. simatus* is found most often in crevices on talus slopes and under ground-cover objects on the north-facing slopes of tributary canyons where interior live oak, canyon oak, and pine species dominate. It also occurs within stands of willow and cottonwoods along canyon streams.

CONSERVATION NOTE: The damming of the Kern River to form Lake Isabella not only flooded former Kern Canyon Slender Salamander habitat but also prompted the construction and expansion of State Route 178 and the ongoing recreation development along it in the southern portion of this species range. All plans for new development along this route should be carefully examined for potential impacts to this salamander by the CDFG, whose charge it is to protect state-listed threatened and endangered species, including those that occur in national forest areas.

BLACK-BELLIED SLENDER SALAMANDER

Batrachoseps nigriventris

IDENTIFICATION: This species has a very similar build as the widely distributed California Slender Salamander: narrow head, long body and tail, and very short legs. Its snout-to-vent range is 3.2 to 4.7 cm (1.25 to 1.9 in.). A dorsal stripe of tan, brown, red-brown, or beige is often present. As its name denotes, the venter is dark with very fine white specks.

RANGE: *B. nigriventris* has the second largest range of all slender salamanders. It extends south from southern Monterey County through the rest of South Coast Range and east into the Tehachapi and Transverse ranges. Disjunct populations also occur in northern Coastal Southern California and on Santa Cruz Island. At several locations throughout this large range it coexists with the following *Batrachoseps* species: Gabilan Mountains Slender Salamander (*B. gavilanensis*), Tehachapi Slender Salamander (*B. stebbinsi*), San Gabriel Mountains Slender Salamander (*B. gabrieli*), Garden Slender Salamander (*B. major*), and the Channel Islands Slender Salamander (*B. pacificus*). Fortunately, all of these are of the robust or partly robust slender salamander body form with head widths greater than that of the body, whereas the Black-bellied Slender Salamander has an extremely narrow head width, long, thin tail, and minute legs. These contrasting features along with its dark ventral side will facilitate field identification in these areas of species overlap.

NATURAL HISTORY: This slender salamander is a habitat generalist whose habitats include mixed oak–pine forests, riparian areas, and open grassland. During the rainy season it can be found under ground-cover objects and in the upper reaches of worm and termite burrows. It is the ability of this species to invade such small passages such as earthworm burrows that seems to allow for the niche separation that permits coexistence with more robust species throughout its range. The use of earthworm burrow retreats has most likely allowed it to persist in some of the older suburban areas of southern coastal California where yard and garden areas received minimal disturbance during the house-building process.

GREGARIOUS SLENDER SALAMANDER

Batrachoseps gregarius

IDENTIFICATION: This is one of the smaller slender salamanders, with a snout-to-vent range of only 3 to 4.6 cm (1.6 to 1.8 in.). It has a short, slim body, narrow head, and short limbs. However, there is a noticeable difference in body form in populations at the two ends of its north-to-south range. In the latter long periods of summer heat and drought has apparently selected for a more slender body with smaller legs and feet that facilitate the use of earthworm and other deep soil burrows as estivation sites during the summer months. A brown stripe and beige to rust-colored streaky patches are often present on the dorsolateral areas of the body and tail.

RANGE: The long and relatively narrow range of this species extends from the south boundary of Yosemite National Park south along the western slopes of the Sierra Nevada between 300 and 1,800 m (984 and 5,905 ft) to near the Kern River. There is a small range overlap with the Sequoia Slender Salamander (*B. kawia*) on the mideastern portion of its range. This latter species has a similar build except for the head, which is broader than that of *B. gregarius* and therefore gives the appearance of having a distinct neck area as opposed to the latter whose narrow head blends with its neck.

NATURAL HISTORY: This is another slender salamander habitat generalist whose habitats span mesic areas of canopy forests of oaks and mixed conifers at higher elevations where high rainfall and mild temperatures prevail to xeric annual grasslands at lower elevations.

CHANNEL ISLANDS SLENDER SALAMANDER

Batrachoseps pacificus

IDENTIFICATION: Although this species' snout-to-vent length range of 4.2 to 7 cm (1.6 to 2.75 in.) is equaled by a few other species, adult specimens have one of the most robust body forms of all the *Batrachoseps*. Given that it is found only on the Santa Barbara Channel Islands, and only the diminutive Black-bellied Slender Salamander (*B. nigriventris*) is found with it on Santa Cruz Island, its large size and your geographic position are the best features for the field identification of this species.

RANGE: The Santa Barbara Channel Islands (San Miguel, Santa Rosa, Santa Cruz, Anacapa).

NATURAL HISTORY: This habitat generalist occurs in oak woodland, coastal scrub, and annual grasslands of the Channel Islands. It may have colonized these islands during the period when they were still connected to the mainland. Because it occasionally uses beach driftwood as ground-cover retreats, individuals may have also rafted to these islands in deep-crevice retreats within driftwood logs.

GARDEN SLENDER SALAMANDER
GARDEN SLENDER SALAMANDER
DESERT SLENDER SALAMANDER

Batrachoseps major
B. m. major
B. m. aridus

Garden Slender Salamander *B. m. major*

Desert Slender Salamander *B. m. aridus*

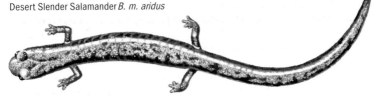

PROTECTIVE STATUS OF *B. M. ARIDUS*: FE, SE

IDENTIFICATION: The subspecies that bears the species name Garden Slender Salamander (*B. m. major*) has an average slender salamander body form, about halfway between the two extreme body plans for this genus. Its snout-to-vent range is 3.2 to 5.9 cm (1.25 to 2.35 in.), and its dorsal color varies from brown or tan to gray or even pinkish. However, some populations in the uplands of the Peninsular Ranges are much darker above, and the venter is gray with small, dark speckles.

The Desert Slender Salamander (*B. m. aridus*) has a noticeably shorter tail than *B. m. major*, and its dorsum is suffused with silvery to brassy flecks that give it a gray, white, or pinkish cast. Its ventral side is dark, but the underside of the tail is a light pink. Some accounts recognize it as a distinct species, based on molecular information. However, until more definitive rules for molecular species recognition are forthcoming, we reserve judgment on species designation.

RANGE: *B. m. major* occurs throughout Coastal Southern California and portions of the Transverse and Peninsular ranges. A population also occurs on Santa Catalina Island. *B. m. aridus* is restricted to Hidden Palm Canyon, a tributary of Deep Canyon west of Palm Desert, and to Guadalupe Canyon, both in Riverside County.

NATURAL HISTORY: The habitat requirements of these two subspecies differ far more than those of many slender salamander species. *B. m. major* is a habitat generalist that can be found in oak woodland, open chaparral, grassland along washes, canyon bottoms, and lower mountain slopes. It still persists in some of the yards and gardens that now occupy its former habitat in Southern Coastal California, and hence its common name. In contrast to this ecological flexibility, the Desert Slender Salamander is found only where it can retreat into moist crevices and soil holes in the two canyon bottoms where it occurs.

CONSERVATION NOTE: The Desert Slender Salamander is a classic endangered species with only a few small populations in a highly specialized desert habitat, a situation very similar to several species of desert pupfish. Its situation has been made more precarious by the massive recreational and suburban development that is ongoing in the greater Palm Desert area. The sprawling 18-hole golf course at the mouth of Deep Canyon (see Plate 16 in the Introduction) possibly destroyed some critical habitat and individuals before the state and federal endangered species acts became law in the early 1970s. Fortunately, the Hidden Palm Canyon site is now a state ecological reserve that may be entered only by permit.

SAN GABRIEL MOUNTAINS SLENDER SALAMANDER
Batrachoseps gabrieli

IDENTIFICATION: Its snout-to-vent range is 3.8 to 5.1 cm (1.5 to 2 in.), and its body proportions are intermediate with respect to members of the genus *Batrachoseps*. However, its head is more flattened than most. One rather unique color feature is the presence in some individuals of relatively large coppery to orange blotches over the pectoral and pelvic regions with additional patches on the tail, while in others this coloration may be in the form of a dorsal stripe. The dorsal background color is black with profuse white flecking, which combines with the bright markings to make this one of the more colorful slender salamanders.

RANGE: This species is restricted to the upper end of the San Gabriel River drainage in the southern portion of the Transverse Range, San Bernardino County.

NATURAL HISTORY: This is a salamander of the talus slopes within stands of oak, big-cone spruce, pine, incense cedar, California laurel, and big-leaf maple. Its flattened head most likely allows it to move deep into narrow crevices where moisture remains as the surrounding areas dry during summer.

GABILAN MOUNTAINS SLENDER SALAMANDER *Batrachoseps gavilanensis*

IDENTIFICATION: This is another species that exhibits many intermediate features within the range of the genus. Even its snout-to-vent range of 3.8 to 6.6 cm (1.5 to 2.5 in.) is near that of several other species. The presence of a tan or ocher dorsal stripe in a portion of the population can be an identifying feature when coupled with a head width that is greater than that of the neck.

RANGE: This species occupies a relatively large portion of the Inner Coast Range in the northern part of the South Coast Range, from southern Santa Cruz and Santa Clara counties south through San Benito County and the eastern part of Monterey County to northern San Luis Obispo County. Throughout most of this area it is the only slender salamander you will encounter. However, there is some overlap with *B. attenuatus* in the northern portion of its range and with *B. nigriventris* in the southeastern area. Fortunately, these two species have the narrowest heads of all slender salamanders, a feature that should permit recognition of the two species in the overlap zones.

NATURAL HISTORY: This species occupies the slender salamander habitat generalist niche. It can be found in oak woodland, gray pine, chaparral, and annual grassland in the Diablo and Gabilan ranges and in redwood and mixed evergreens on the east-facing slopes of the Santa Lucia Mountains. At all sites it is dependent on a good leaf and twig layer on the ground surface along with fallen log and bark cover objects.

SANTA LUCIA MOUNTAINS SLENDER SALAMANDER
Batrachoseps luciae

IDENTIFICATION: This is a moderately robust, short-bodied species, as reflected by its snout-to-vent range of 3.2 to 4.6 cm (1.25 to 1.8 in.). Lighter brown individuals have a broad reddish or brassy dorsal stripe.

RANGE: This species occupies the coastal range segment of the South Coast Range from the southern end of Monterey Bay south to a short distance over the Monterey–San Luis Obispo county border.

NATURAL HISTORY: This is a slender salamander of the moist redwood and mixed-evergreen forests. It is also found in wooded north-facing slopes in areas inland from the coastal redwood belt.

LESSER SLENDER SALAMANDER
Batrachoseps minor

IDENTIFICATION: As its name implies, this species has the smallest average adult size of all slender salamanders, with a snout-to-vent range of 2.5 to 4.6 cm (1 to 2.3 in.). Despite its small size it has a rather robust body and relatively long legs and tail. Its long legs and broad head provide a good contrast with the narrow head and short legs of *B. nigriventris* within whose range it occurs.

RANGE: It occurs in the South Coast Range in a small wooded region of the southern Santa Lucia Mountains in northern San Luis Obispo County.

NATURAL HISTORY: This slender salamander has been found only in wooded habitats containing tanbark oak, coast live oak, blue oak, sycamore, and California laurel. Once common within its small range, its numbers now appeared diminished.

SAN SIMEON SLENDER SALAMANDER
Batrachoseps incognitus

IDENTIFICATION: The Latin species name for this salamander, *incognitus*, is appropriate, because there is really no major feature that sets it apart from the most common body form of the current 19 *Batrachoseps* species. Within its small geographic range it coexists with the Black-bellied Slender Salamander, but the extremely thin body, tiny legs, and narrow head width of this latter species permits recognition of the two in the field.

RANGE: It occurs in a small region of the South Coast Range from the Monterey–San Luis Obispo county line south to Pine Mountain and Rock Butte in northern San Luis Obispo County.

NATURAL HISTORY: This is a closed-canopy woodland species, found primarily in forests dominated by yellow pine. Near the coastal portion of its range the tree canopy is provided by California laurel and sycamore, and more inland by oak woodland.

KINGS RIVER SLENDER SALAMANDER *Batrachoseps regius*

IDENTIFICATION: This is another slender salamander of average proportions except for a relatively short body, reflected by its snout-to-vent range of 3.2 to 3.5 cm (1.25 to 1.35 in.). Its dorsal color also differs from most others in that it is black with little patterning. Once again the isolated location of this new species is perhaps its best identifying feature.

RANGE: It is found only in the Kings River drainage system, from about 335 m (1,099 ft) to 2,470 m (8,104 ft) at Summit Meadows, Kings Canyon National Park, Fresno County.

NATURAL HISTORY: This species prefers north-facing slopes in areas of mixed chaparral, buckeye, California laurel, canyon and blue oak, and ponderosa pine. Crevices beneath rocks in talus areas appear to be a preferred retreat.

SEQUOIA SLENDER SALAMANDER *Batrachoseps kawia*

IDENTIFICATION: This is a rather small and short-bodied species, with a snout-to-vent range of 3.2 to 4.7 cm (1.25 to 1.85 in.). However, it has a relatively broad head and a distinct neck, features that distinguish it from the narrow-headed Gregarious Slender Salamander, which shares a portion of its range.

RANGE: It occurs only in the Kaweah River system in Tulare County, from around 500 to 2,200 m (1,640 to 7,218 ft).

NATURAL HISTORY: As is the case with many of these newly recognized slender salamander species, little is known about their natural history other than the plant communities in which individuals have been collected to date. The Sequoia Slender Salamander appears to prefer habitats with scattered trees and possibly localized mesic sites.

RELICTUAL SLENDER SALAMANDER
Batrachoseps relictus

PROTECTIVE STATUS: CSSC

IDENTIFICATION: With its relatively average snout-to-vent range of 3.5 to 4.7 cm (1.3 to 1.85 in.), short body, and moderately long limbs, this species description could come close to that of several other slender salamanders. With this in mind, the senior author cautioned in his 2003 *Houghton-Mifflin Peterson Field Guide* that molecular information is usually required for certain identification of this species. Indeed, the tongue-in-cheek comment in the introduction to genus *Batrachoseps* concerning the possible need for a mobile DNA sequencing lab for slender salamander field identification could be applied here.

RANGE: This species occurs from the lower Kern River in Kern County to the highlands drained by the Tule and Kern rivers in central Tulare County.

NATURAL HISTORY: At lower elevations where this salamander may be becoming scarce, the habitat consists of rocky terrain with scattered live oak, pines, and buckeye. At higher elevations it occurs in pine, fir, and cedar woodlands, often with deciduous oaks.

CONSERVATION NOTE: With indications that this species is becoming scarce at lower elevations within its range, a logical first conservation step would be to remove it from the CDFG current (2011) list of amphibians that can be legally taken from their natural habitat (possession limit of four) by anyone holding a current fishing license.

HELL HOLLOW SLENDER SALAMANDER
Batrachoseps diabolicus

IDENTIFICATION: This is a small, slim slender salamander with a relatively broad head, moderately long limbs, and relatively large feet for a species with a slim body form. These characteristics serve to separate it from the California Slender Salamander with which it shares part of its range. It is named for "Hell Hollow," a site in Mariposa County on the south side of the Merced River at its junction with Highway 49 where it has been collected.

RANGE: It inhabits the western foothills of the Sierra Nevada from the lower American River drainages in El Dorado County south to Sweetwater Creek, Mariposa County.

NATURAL HISTORY: *B. diabolicus* occurs along the margins of chaparral and open oak and pine woodland, usually in areas that experience long, hot, dry summer weather.

CALIFORNIA SLENDER SALAMANDER
Batrachoseps attenuatus

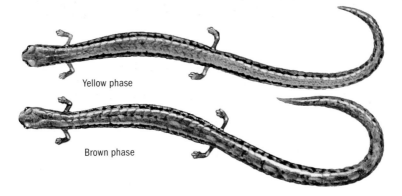

Yellow phase

Brown phase

IDENTIFICATION: This very slender salamander has a narrow head and body and a long tail that may sometimes be nearly twice as long as its snout-to-vent length of 3.2 to 4.7 cm (1.25 to 1.9 in.). The limbs are very thin and short, and most individuals have a dorsal stripe that may vary in color from brown, to brick-red to yellow. These vary with location, with reddish brown occurring most often in the redwood belt of the North Coast Range. The venter is black or dusky, but the underside of the tail is often lighter and tinged with yellow.

RANGE: This species has the largest range of all slender salamanders. It extends from the extreme southwest corner of Oregon south along California's North Coast Range and on into the South Coast Range as far as northern San Benito County. It also occurs in the Sierran foothills from Chico Creek in Butte County south to Calaveras County, and there are scattered populations in the northern part of the Great Valley.

Plate 39. California Slender Salamander with eggs.

NATURAL HISTORY: If you have managed to make it this far through the most challenging genus in this book (both for the reader and the authors), we can at least reward you by having saved the best for last. Not only does this last species have the largest range of all *Batrachoseps*, but much of it encompasses areas with extended wet season and high human populations. Thus, more people encounter this diminutive salamander than perhaps any other amphibian species. One of the most frequent questions that both authors have been asked over the years has concerned the identity of this "tiny legged worm" that was discovered under a garden path rock or border log. It is this conformation that permits the use of earthworm burrow retreats and thus has allowed it to survive in many urban and suburban areas where temporary land surface disturbance has occurred. Such sites include backyards, formal gardens, parks, vacant lots, and occasionally even freeway interchange and off-ramp "islands." The junior author encountered the latter situation recently when he was asked to monitor the clearing of a drainage channel containing the California Red-legged Frog (*Rana draytonii*) that passes through a vegetated triangle between highway on- and off-ramps across from the San Francisco International Airport. The land adjacent to the drainage was covered with ice plant (*Mesembryanthemum*), an introduced species that biologists usually would not equate with wildlife habitats. However, as workmen began removing segments of the dense ice plant cover, California Slender Salamanders were found at densities of approximately four to five per square meter (about one per square yard). The ice plant "understory" was also rich in small invertebrates, and with the good moisture regime it maintained, this "urban salamander" was thriving.

Outside of suburbia *B. attenuatus* occurs in a spectrum of natural habitats including oak savannah, chaparral, woodland, and forests. In such areas it may be found under most types of ground-cover objects and in leaf litter from the first fall rains to the onset of the next dry season. Eggs are laid in late fall or winter, often in a communal nest, and young hatch in late winter or spring. These hatchlings are among the smallest of all terrestrial vertebrates, with several easily fitting on the face of a penny (see plate 18, in the "Taxonomy, Anatomy, Physiology, and Behavior of Amphibians" section).

Egg

Web-toed Salamanders (Genus *Hydromantes*)

In many ways these are the most unique salamanders among the California plethodontids. The use of the webbed feet to adhere to slippery rock surfaces and the muscular tail as a climbing aid has already been reviewed in the section on amphibian anatomy. Other distinct features include a flattened head and body to facilitate retreat into rock crevices and a mushroomlike tongue that can be protruded to over one third the body length. It enables the salamanders to successfully capture insects on slick or slippery rock surfaces where a rapid stalk may not be possible. Given these adaptations for rock crevice, cliff, and talus dwelling, California's three *Hydromantes* species are found at some of the most scenic sites in the state, including Yosemite Valley and the top of Half Dome. They are also the only three *Hydromantes* species in the Western Hemisphere, with the other seven in Mediterranean Europe.

Plate 40. Limestone Salamanders on wet rock.

MOUNT LYELL SALAMANDER — *Hydromantes platycephalus*

PROTECTIVE STATUS: CSSC

IDENTIFICATION: This is a small salamander, snout-to-vent length 4.4 to 9.4 cm (1.75 to 2.75 in.), with a broad, flat head, flattened body, and stout limbs and tail. The toes are short and webbed. The dorsal color is brown to nearly black and obscured in varying amounts by flecks and patches of metallic pale-gold, gray, pinkish or whitish pigment. The dorsal coloration closely resembles the particular texture of the granite rocks among which the salamander lives. Ventral color is brown to sooty, usually with whitish spots on throat and lower sides. Juveniles are black above with a stippling of pale gold, which may give some individuals a greenish cast. No **EGGS** have been found to date.

RANGE: The Mount Lyell Salamander occurs in isolated pockets in the Sierra Nevada from Sierra Buttes, Sierra County to Franklin Pass, Tulare County, Twin Lakes, Silliman Gap, Sequoia National Park, and Mt. Williamson. It has also been found east of the Sierran crest near outflow streams to Owens Valley. This is by far the widest distribution of our three *Hydromantes* species.

NATURAL HISTORY: This unusual species is found in fissures in granite outcrops and cliff faces that are usually associated with patches of soil covered with herbaceous plant growth. Trees are often absent or sparse. The rock surfaces may be wetted by seepages, moisture from melting snow banks, or spray from waterfalls. Openings among talus, crevices associated with exfoliating granite shells, and small-mammal burrows provide places of retreat from winter cold and dry summer weather. It is often found beneath rocks resting on rock surfaces or decomposed granite. Humus may be absent or present in varying amounts, and surfaces may be moist or saturated. Localities in and near Yosemite Valley illustrate the varied conditions under which this salamander occurs. These include (1) under rocks and in crevices in the vicinity of the ice cone that forms at the base of upper Yosemite Falls; (2) spray-drenched talus and cliff areas at the base of Vernal and Bridalveil falls; (3) the talus–cliff interface area at Staircase Falls and other sites on the south side of Yosemite Valley; (4) Firefall Ledge area below Glacier Point; and (5) the top of treeless Half Dome. A deep vertical fissure and crevices between exfoliating granite shells on Half Dome provide moist summer retreats.

The Mount Lyell Salamander has the flattest head and body of all California salamanders and appears to be the most specialized for climbing on rock surfaces and entering crevices. In contrast to most other western plethodontid salamanders, it is active during the late spring and summer, emerging in early May at lower elevations but usually not until June or July at higher elevations. Retreat from the surface occurs from early July (top of Half Dome) to middle or late August or September at lower elevations.

Little is known of the breeding habits of this species. Given the relatively short summer season where it occurs, hatching probably occurs in the fall. Its food consists of small arthropods (insects, spiders) that are supported by the meager herbaceous plant growth and/or wafted onto Half Dome by air currents.

CONSERVATION NOTE: Given that the Mount Lyell Salamander occurs in some large areas that are protected from development such as Yosemite and Sequoia National Parks, one might assume that it is free of serious human-caused impacts. Outside of protected park areas talus deposits are often open to disturbances or removal. Continued vigilance by the CDFG is therefore needed at any sites where such activities are proposed or already taking place, and if surveys reveal the presence of *H. platycephalus*, they should not be allowed to proceed.

LIMESTONE SALAMANDER *Hydromantes brunus*

PROTECTIVE STATUS: ST

IDENTIFICATION: *H. brunus* is similar in snout-to-vent length, 5.0 to 6.9 cm (2 to 2.75 in.), and general body form to the other two California *Hydromantes* species. It differs from both *H. platycephalus* and *H. shastae* in its uniform dark-brown dorsal coloration and pale brown to gray ventral hue. There are other subtle differences such as limb size, but as with other highly localized species, its site of occurrences may often be the best defining feature. A museum specimen contained seven large ova, but no information on eggs in the field is available.

RANGE: This salamander is found in limestone areas of the Sierra Nevada such as those in the Bear Creek drainage to the Merced River, along the North Fork of the Merced River, along tributaries of Bear Creek, at Hell Hollow, and adjacent to Lake McClure.

NATURAL HISTORY: *H. brunus* inhabits rocky north- to east-facing hillsides and shaded gullies, where talus and rock outcrops often are moss-covered. These occur in a woodland–chaparral association characterized by gray pine, live oak, canyon oak, buckeye, California laurel, manzanita, buckbrush, toyon, yerba santa, poison oak, and chamise. It retreats in crevices and under limestone and shale rocks, usually in the vicinity of springs, seepages, and streams. Talus and cracks in cliffs and large rocks, especially when overgrown with moss, are particularly important to this salamander. Shade and cover are needed to survive the warm, dry summer period.

The Limestone Salamander is active during fall, winter, and early spring, especially during or after rains. The watch-spring defense coil assumed by this salamander may sometimes protect it against predation by snakes. Upon coiling, larger individuals have been observed in captivity to resist swallowing by Ring-necked Snakes. Its breeding habits are unknown, but eggs are probably laid in spring and hatch in the fall.

CONSERVATION NOTE: The very small geographic range and specialized habitat requirements make it a classic "endangered species" that should be accorded an official endangered status at the federal level in addition to its current ST rating. An initial blow was most likely dealt to this species when the Merced River Canyon was dammed to form Lake McClure.

Increased gold mining in the area including a planned 720-acre (292 hectare) open-pit and tunnel gold mining project that proposes realignment of Highway 49, which passes through potential habitat area, as well as water developments downstream might also affect known and perhaps as yet undiscovered populations.

A 120-acre (49-hectare) Limestone Salamander reserve was established in 1975 by the State of California. It is located on the south side of Highway 140, east of Briceburg, Mariposa County. The reserve can be entered only by permit. This was a sound protective first step, but additional land should be added and additional localities should be given special protection. The population site adjacent to Lake McClure should be also placed in public ownership and protected against damage to salamander habitat.

SHASTA SALAMANDER *Hydromantes shastae*

PROTECTIVE STATUS: ST

IDENTIFICATION: The Shasta Salamander has a snout-to-vent range of 4.4 to 8.7 cm (1.75 to 2.5 in.). It is similar to *H. platycephalus*, but its head and body are less flattened, its eyes are larger, and its toes are broader and blunter and limbs longer. The dorsum is reddish to beige with white blotching on the venter that often extends across the chest and abdomen. **EGGS** are enclosed in two jelly envelopes with an average outer diameter of about 8 mm (0.3 in.). Most of those observed to date had slender pedicels at each end that were intertwined at the center of a cluster. **EMBRYOS** have relatively small, rounded, unbranched gills with two broad lobes.

RANGE: At this writing 61 sites of occurrence, possibly representing 20 populations. have been located in the Cascade Range in the vicinity of Shasta Lake, Shasta County, and northeast along the drainage of the McCloud River and Squaw Creek to a distance of approximately 80 km (48 mi) from Shasta Dam. Over half of these locations are situated near or immediately adjacent to the shores of Shasta Lake.

NATURAL HISTORY: Vegetation in areas frequented by the Shasta Salamander consists of patchy woodland and forest interspersed with chaparral of manzanita, ceanothus, and other shrubs, plus grasses and other herbaceous growth. Major trees in the area are Douglas fir, gray pine, buckeye, black oak, and canyon oak. Initial field observations of this species were made near limestone caves, outcrops, rocky stream banks, and areas of talus, usually in the vicinity of caves and fissures. However, in the late 1970s a colleague of the authors, Dr. Ted Papenfuss, found this species at a volcanic rock outcrop site away from limestone deposits. More recently others have located populations in forested sites void of any rock outcrops. It now appears that instead of being a habitat specialist as originally assumed, it is much more of a habitat generalist. In limestone outcrop areas it retreats to moist cave and crevice sites during the dry summer period. Outside of such areas it probably uses rodent burrow retreats as do other plethodontid species. When in caves it favors the twilight zone where its arthropod prey is more abundant, rather than the deep recesses of total darkness. Because suitable moist caves are limited, individuals tend to aggregate in such locations.

Two egg clusters have been found in a limestone cave, and each was attended by a female, which curled around the eggs until they hatched. Young hatch in the late fall and come to the surface to feed when the ground is cool and moist from winter rains. They retreat with juveniles and adults when the surface begins to dry in spring.

CONSERVATION NOTE: Without a doubt the greatest impact to the Shasta Salamander to date was the construction of the Shasta Dam in 1949, a time when the term "endangered species" was not part of our vocabulary. This substantially raised the level of the smaller lake at that site, thereby submerging a portion of this species' historic habitat. An increase in Shasta Reservoir's capacity continues to be proposed, and thus future adjacent habitat could be lost. Even if the reservoir's current capacity is not increased, ever-expanding recreational development along parts of its greater shoreline area will continue to threaten this species. The several limestone quarries currently operating in known *H. shastae* population areas could expand their operations. Timber harvesting also continues throughout this species' range and with it additional road construction. Continued vigilance by the CDFG is needed for all proposed land use changes within its range. Those populations on federal lands are also protected by the Northwest Forest Plan, which requires surveys for the Shasta Salamander before any ground-disturbance activities.

As is the case with most special-status species, more field studies are needed to provide sound information on which to base future management and mitigation plans. An important contribution in this area is the recently completed master's degree thesis research of Andrea Herman at Humboldt State University in Arcata. From 2000 to 2002 she conducted a mark-and-recapture survey on a slope populated by California black oak, gray pine, canyon live oak, and California buckeye near a limestone outcrop near the center of this species' range. Within a 0.5-acre

(0.005-hectare) area she captured 399 *H. shastae*, marked 306 with injected fluorescent elastomer pigment, and recaptured 25 percent of these. Her findings support the discoveries of two other recent surveys that found populations of this species outside of limestone or volcanic outcrops, and it strongly suggests that this species may be far more of a habitat generalist than originally assumed.

Frogs and Toads (Order Anura)

Tailed Frogs (Family Ascaphidae)	150
Coastal Tailed Frog (*Ascaphus truei*)	151
Spadefoots (Family Pelobatidae)	154
Western Spadefoot (*Spea hammondii*)	156
Great Basin Spadefoot (*Spea intermontana*)	158
Couch's Spadefoot (*Scaphiopus couchii*)	159
True Toads (Family Bufonidae)	161
Sonoran Desert Toad (Colorado River Toad) (*Bufo alvarius*)	161
Great Plains Toad (*Bufo cognatus*)	164
Red-spotted Toad (*Bufo punctatus*)	166
Arroyo Toad (*Bufo californicus*)	168
Western Toad (*Bufo boreas*)	170
Black Toad (*Bufo exsul*)	174
Yosemite Toad (*Bufo canorus*)	176
Woodhouse's Toad (*Bufo woodhousii*)	178
Chorus Frogs (Family Hylidae)	179
Pacific Chorus Frog (Pacific Treefrog) (*Pseudacris regilla*)	180
California Chorus Frog (California Treefrog) (*Pseudacris cadaverina*)	183
True Frogs (Family Ranidae)	186
California Red-legged Frog (*Rana draytonii*)	186
Northern Red-legged Frog (*Rana aurora*)	186
Cascades Frog (*Rana cascadae*)	193
Oregon Spotted Frog (*Rana pretiosa*)	195
Foothill Yellow-legged Frog (*Rana boylii*)	196
Southern Mountain Yellow-legged Frog (*Rana muscosa*)	199
Sierra Nevada Yellow-legged Frog (*Rana sierrae*)	199
Northern Leopard Frog (*Rana pipiens*)	203
Southern Leopard Frog (*Rana sphenocephala*)	203
Rio Grande Leopard Frog (*Rana berlandieri*)	206
Lowland Leopard Frog (*Rana yavapaiensis*)	207
American Bullfrog (*Rana catesbeiana*)	208
Tongueless Frogs (Family Pipidae)	212
African Clawed Frog (*Xenopus laevis*)	212

Key to the Families and Genera of Frogs and Toads

All frogs and toads have scaleless skin and distinctly larger hind legs than front legs. They also do not have a true tail.

1a	Sharp black claws on the three inner toes of hind foot; no eyelids	African Clawed Frog
1b	No claws on any toes; eyelids present	2a
2a	Fifth outside toe of hind foot broader than others; stubby tail-like structure in males	Coastal Tailed Frog
2b	Fifth toe of hind foot not broader than others	3a
3a	Large oval or round parotid glands behind eyes present	True Toads
3b	Parotid glands absent	4a
4a	Black, sharp-edged "spade" on underside of hind foot; eye pupil vertical in daylight	Spadefoot Toads
4b	No black, sharp-edged "spade" on hind foot; one or more round, light-brown tubercles may be present on hind foot; eye pupil round	5a
5a	Toe pads (expanded toe tips) present; no dorsolateral skin fold on body	Chorus Frogs
5b	No toe pads; dorsolateral body skin fold usually present	True Frogs

Tailed Frogs (Family Ascaphidae)

Tailed frogs are members of an ancient group of anurans. The closest relatives to our Pacific Northwest species are three "tail-less" species in New Zealand. These frogs are named for their tail-like copulatory organ that facilitates internal fertilization by copulation, a method of breeding unique among anurans. It inhabits cold streams, where its larvae live in torrents or quiet water and cling to rocks with their large, suckerlike mouths. Larval development to metamorphosis requires from two to four years, depending on the temperature of the larval habitat. Adults do not breed until their seventh or eighth year.

COASTAL TAILED FROG *Ascaphus truei*

PROTECTIVE STATUS: CSSC

IDENTIFICATION: ADULTS range from 2.5 to 5.1 cm (1 to 2 in.) in snout-to-vent length. The skin is rather rough, more like that of a toad, and the pupils of the eye are vertical. The dorsum color is brown, rust, olive, or gray with a pale triangle on the snout. The male has a short, tail-like copulatory organ with the vent opening at the tip, a unique structure among anurans. The female has no such "tail." The **EGGS** are unpigmented, 4 to 5 mm (0.2 in.) in diameter, surrounded by a jellylike capsule, and connected by jelly strands like a string of beads. **LARVAE** are dark-colored and have a streamlined body with a large round mouth that occupies nearly half of the lower body surface. The labial tooth row formula is 2–3/7–10, and the tail tip is often white or rose-colored, set off by a dark band. Larvae attain a total length up to 5.8 cm (2 1/3 in.) at metamorphosis.

RANGE: In California it inhabits the North Coast Range from the vicinity of Elk, Mendocino County, northward. The inland extent of its range is the McCloud River basin, Shasta County.

NATURAL HISTORY: This is a cold-adapted species, closely tied to water. It inhabits cool coniferous forests or forest edges of redwood, Douglas fir, grand fir, spruce, hemlock, and ponderosa pine, in nearly pure stands or in combinations. Sometimes these stands are interspersed with grassland, chaparral, shrubs, broadleaf deciduous woodland, or several of these.

A number of the characteristics of *Ascaphus* adapt it to life in fast-moving water. These include (1) small lungs that reduce buoyancy, (2) "degenerate" vocal and auditory structures (torrent sounds presumably interfere with hearing), (3) a flattened, streamlined form, (4) internal fertilization, and (5) stream-adapted larvae with a sucking mouth for holding to substrate objects in current flow.

Adults are closely restricted to streams and are usually found in or within a few feet of water, but in cool wet weather, especially in the more humid coastal areas, they may move some distance into damp woods. Where there is prolonged freezing they may hibernate in sand under large rocks in the stream bottom. They are chiefly nocturnal and can be found along stream banks after dark. During the day they remain in spray-drenched crevices or under rocks near waterfalls and along streams.

Mating occurs from May to October, but apparently most takes place in the fall. Fertilization is internal by means of the tail-like copulatory organ. Copulation helps to ensure fertilization in rapidly moving water, which could sweep spermatozoa away from the eggs. Females appear to be able to store viable sperm for at least two years. Typically the male seizes a hind leg of the female with his forelimbs and works forward to assume the pelvic position of amplexus, often with his digits interlocked below her waist. In copulating he flexes his body at the sacral hump, which positions his copulatory organ forward so that it is perpendicular to the pelvic girdle to achieve insertion. The female extends her hindlimbs to form a narrow "V." This may aid the male in guiding his copulatory organ to the cloaca of the female. In observation of a captive pair, separation was initiated by the female by jumping and swimming violently, causing the male to lose his grip. Seventy hours elapsed between the beginning of amplexus and the onset of the female's struggles.

The eggs are the largest of any North American anuran and have the slowest rate of development and narrowest range of thermal tolerance—5 to 18.5 degrees C (41 to 65 degrees F). They are laid in beadlike strings attached to undersurfaces of stones. Embryo development is slow, requiring about six weeks at lower temperatures.

At hatching the larvae are unpigmented and have considerable yolk in the intestine. They tend to remain in "nest sites" under large rocks until most of the yolk is used up and pigmentation has been completed. After that they feed on pollen, diatoms, algae (scraped from rocks), and

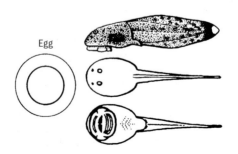

Egg

minute insect larvae. In feeding, larvae commonly face upstream, clinging by their mouths to stones. When attached by their suctorial mouth, they grip with their labial teeth, and hitch along by advancing the upper and lower lips, with the labial teeth and upper jaw aiding locomotion. Organic material on rock surfaces is evidently removed by scouring rather than by biting. Larval development is slow, lasting two years in coastal areas and on the west slope of the Cascades, and up to four years at localities farther inland.

The adults eat a variety of invertebrates associated with stream habitats. These include insects (springtails, stonefly nymphs, beetles, flies, mayflies, and caterpillars), spiders, millipedes, amphipods, and snails.

Plate 41. Female Coastal Tailed Frog.

CONSERVATION NOTE: The Coastal Tailed Frog is closely tied to its cool stream habitat, and any significant variation of its physical properties can greatly affect a population. In some areas one of the most detrimental changes has been the elevation of stream temperature due to the removal of much or all of the stream tree canopy by intense logging. This species has an exceptionally low high lethal temperature range for an anuran: 23 to 24 degrees C (72 to 74 degrees F), and in the upper Sacramento River system the shallow waters of open-canopy streams may reach or surpass that range in summer. Given the slow growth of larvae (two to four years) and the long time needed for sexual maturity for adults (about seven years), repopulation of degraded stream segments from other populations could be extremely difficult, especially if siltation from heavy logging also buries the rock–gravel stream substrate. The CDFG has a long-standing requirement that any land disturbance activity cannot occur closer than 31 m (100 ft) from each side of a permanent stream measured from the outer upper bank edge. Rigorous enforcement of this requirement would do much to preserve Coastal Tailed Frog populations, especially in warm inland areas.

Spadefoots (Family Pelobatidae)

Members of this family are widely distributed and exhibit variable body forms and habits. Species in the New World, including California forms, are squat and short-legged like true toads (Bufonidae), but unlike this family have vertical pupils, lack parotid glands, and have a single black sharp-edged "spade" on the hind foot (see figure 5, in the "Taxonomy, Anatomy, Physiology, and Behavior of Amphibians" section). Many species are adapted to life in arid or semiarid environments, where rainfall is intermittent and often torrential, and therefore breeding sites may vary from year to year. California species breed in temporary pools after heavy rains or in slow streams, reservoirs, and irrigation ditches. Dry periods are spent underground in burrows that these toads make by digging backwards into soil using their foot spades. Rodent burrows are also sometimes used for short-term retreats.

Although spadefoots lose water through their skin more rapidly than some other anurans such as those in genus *Bufo*, they effectively resist desiccation when buried through a marked elevation of the osmotic pressure of their body fluids accomplished by the retention of urea. This creates a situation in which body fluids are at equilibrium with or may actually gain water from the surrounding soil. During such periods of dormancy, which in some areas may last for nine or 10 months, these toads derive their energy needs from the metabolism of fat stored in the large abdominal fat bodies. Oxygen consumption in dormant toads may be one fifth that of those awake and resting on the surface. They have a large stomach capacity, and enough food can be consumed in only a few evenings of foraging during the breeding season to supply the energetic needs during long periods of dormancy. Some individuals with full stomachs and good fat reserves may be able to survive two or more years without eating.

Spadefoots are "explosive breeders" that take advantage of those infrequent periods when substantial rainfall occurs. Buried individuals are quick to sense such events and promptly make their way to the surface. They perhaps hear or "feel' through ground vibrations thunder and the impact of raindrops striking the hard ground surface. Upon emerging, those males that locate a newly formed pond begin a loud evening chorus that attracts females and other males to the site, often from some distance away. There the jelly-coated eggs are laid in cylindrical, grapelike clusters attached to plant stems or other objects (see figure 18, in the "Taxonomy, Anatomy, Physiology, and Behavior of Amphibians" section).

Spadefoot tadpoles are notable in having the fastest rate of development found among anurans (about two weeks). The larvae have special methods for securing food, including moving in schools that stir up nutrients upon or within the bottom muck and cannibalism during the later stages of development. These larvae are much larger at metamorphosis than those of genus *Bufo*, which transform at a very small size and then continue their growth during the following weeks or months in the more mesic habitats that they occupy. However, spadefoots, especially those that inhabit desert areas, must often bury themselves and enter estivation soon after metamorphosis and therefore attain a minimal size for survival until the next emergence period, which could be months or even a year away. Larvae resist desiccation by forming aggregations at the time of pond drying. Active swimming movements occur as the last water evaporates, and the tadpoles become coated with mud. The mud layer slows desiccation, and some may survive long enough after the pond dries to transform. Cannibalism tends to increase with the crowding that develops as ponds dry. Cannibals transform sooner at several times the size of other larvae. They are therefore more resistant to desiccation and also begin their first estivation period at a larger size.

Plate 42. Dying spadefoot tadpoles in a drying pond basin.

WESTERN SPADEFOOT *Spea hammondii*

PROTECTIVE STATUS: CSSC

IDENTIFICATION: ADULTS range from 3.8 to 6.3 cm (1.5 to 2.5 in.) in snout-to-vent length. Their dorsal color is generally olive or gray, often with an hourglass-like marking and orange- or red-tipped skin tubercles. The venter is white to gray and contains no markings. There is no cranial boss or ridge between eyes, the pupils are vertical, and there is a wedge-shaped glossy black "spade" about as wide as it is long on the hind foot. Its **VOICE** is a catlike purr but stronger and hoarser, usually described as a snorelike "kwalk" and lasting about one second. **EGGS** are light olive green or sooty above and whitish below and range from 1.0 to 1.7 mm (0.04 to 0.07 in.) in diameter. They are enclosed in two jelly envelopes with an outer diameter of 3.2 to 5.7 mm (0.1 to 0.2 in). Eggs are deposited in cylindrical clusters containing an average of 24 eggs and are attached to stems and other submerged objects. **LARVAE** may reach nearly 7.5 cm (3.1 in.) total length, but transformation at a smaller size is more common. Their dorsal color ranges from olivaceous to dark brown, gray, or greenish black. The vent is cream to silvery with salmon to copper iridescence, and the tail fins and musculature is often blotched with olive, pale yellow, or copper. The labial tooth row formula is usually 5/5.

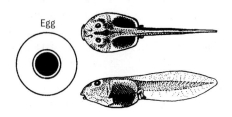

Egg

Plate 43. Western Spadefoot with vertical eye pupil.

RANGE: It occurs at sites within the Great Valley and the Sierra Nevada foothills up to 1,385 m (4,500 ft) and in the South Coast Range south of Monterey Bay and on through Coastal Southern California into Baja California, Mexico. It is absent from the outer cool coast north of Point Conception. It was formerly widespread on the Los Angeles coastal plain but is now apparently extinct throughout much of lowland southern California. It still persists in coastal Orange, western Riverside, and inland San Diego counties. Extensive losses have also occurred in northern California, but it is still present at a handful of sites in Glenn, Butte, and Tehama counties, including near Black Butte Reservoir and at Vina Plains Preserve.

NATURAL HISTORY: This species occurs in open treeless grassland, scrub, or mixed woodland and grassland where temporary pools form or where there are sandy, gravel washes or small streams that are often seasonal (see plate 10, in the "Amphibian and Reptile Distribution throughout California's Habitat Complex" section). When active above ground it is chiefly nocturnal. During the dry season it burrows in loose soil to depths up to about one meter (a little more than three feet), where it avoids temperature extremes and dryness. It may congregate at favorable burrowing sites, which are often well removed from breeding locality. Eggs are laid in late winter and spring. Larval transformation occurs in late spring and summer. Adult food consists of insects, worms, and other invertebrates, which usually appear after soaking rains.

CONSERVATION NOTE: From Ventura County southward, an estimated 80 percent of Western Spadefoot habitat has been lost, and severe habitat depletion has also occurred in the Great Valley. The loss of vernal pools and perhaps rigorous mosquito abatement practices are among the decimating factors. At breeding sites that retain water much of the year and are near permanent ponds or streams, the periodic influx of Bullfrogs (*Rana catesbeiana*) can also affect populations. Seasonal breeding sites may also be overlooked when biological assessments are conducted in response to proposals for suburban or agricultural development, especially if they are made during the dry season. The elevation of several vernal pool crustaceans to Federal Endangered and Threatened status has been important for the Western Spadefoot, because such protection may preserve its breeding habitat as well until it can be elevated to the higher protective status that it deserves.

GREAT BASIN SPADEFOOT *Spea intermontana*

IDENTIFICATION: The **ADULT** snout-to-vent range is 3.8 to 6.3 cm (1.5 to 2 in.). This species closely resembles the Western Spadefoot. There is no satisfactory distinguishing characteristic other than its **VOICE**, which is a low-pitched series of short, rapid "wa-wa-wa" calls. A chorus sounds like sawing with a hand wood saw. Given this species similarity, its locality of occurrence in California (the Great Basin Desert) is the best means of field identification. **EGGS** are laid in small irregular packets of jelly, 1.5 to 2 cm (3.8 to 5 in.) in diameter. Some 10 to 40 are attached to submerged objects or deposited on the bottom of pools. **LARVAE** grow to 7 cm (2.8 in.) in total length and resemble those of the Western Spadefoot. The dorsal body color ranges from dark gray-brown to brown or blackish, often with gold or brassy flecks and patches. Golden highlights may also be present on the abdomen.

RANGE: In California it occurs in the Great Basin Desert east of the Cascade–Sierra Nevada crest from northern Inyo County northward, to around 2,460 m (8,000 ft).

NATURAL HISTORY: *S. intermontana* inhabits sagebrush flats, piñon–juniper woodland, and mountain meadows with scattered conifers. It is less strictly nocturnal than the Western Spadefoot and is active from May to October, especially following rains. It breeds in springs, slowly flowing streams, temporary pools, and canals. Adults feed on a variety of insects including ants, beetles, grasshoppers, and crickets.

COUCH'S SPADEFOOT   *Scaphiopus couchii*

PROTECTIVE STATUS: CSSC

IDENTIFICATION: ADULTS range from 5.7 to 9.1 cm (2.25 to 3.6 in.) from snout to vent. The dorsum is covered with a dark, irregular network of dark blotches on a greenish-yellowish background that is more vivid in females than in males. The eyes are separated by the width of the upper eyelid or more. There is no boss or hump between the eyes, and the hind foot spade is sickle-shaped. Its **VOICE** is a nasal cry or "meow," declining in pitch like

the bleat of a lamb and lasting from one half to one and a half seconds. **EGGS** resemble those of the Western Spadefoot but are smaller. Like other spadefoots they are laid in cylindrical masses. **LARVAE** grow to a total length of 3.1 cm (1.25 in.) and are smaller and darker than those of the Great Basin Spadefoot larva. The dorsum is dark gray, bronze, or nearly black.

RANGE: In California scattered populations occur in the Colorado Desert between Amos and Ogilby on the east side of Algodones Dune; Purgatory, and Buzzard's Peak washes, Imperial County; and in the Mojave Desert 24 km (15 mi) north of Vidal Junction, San Bernardino County.

NATURAL HISTORY: *S. couchii* lives in arid and semiarid regions of short-grass plains, mesquite savannah, and brushy desert. California populations occur along washes in creosote bush desert. When aridity is great breeding may not occur every year. During its short periods of activity above ground it feeds heavily on xeric climate insects such as termites, whose breeding and dispersal also occurs during desert rain periods. Larval cannibalism is often high in this species because of the short life of many desert rain pools.

CONSERVATION NOTE: It has been found that low-frequency sound is the primary emergence stimulus for buried Couch's Spadefoots, whereas an increasing soil temperature and wetting the soil to less than saturation failed to break dormancy in the absence of such sound. In tests with buried specimens, vibrations from an electric motor consistently induced nearly 100 percent emergence from dormancy under very arid conditions, and airborne motorcycle sound had the same effect. Thus, human-caused sounds and vibrations associated with off-road vehicles could induce the emergence of most or all of an adult breeding population at a time when hot and dry surface conditions would be fatal. Because the recreational use of such vehicles continues to increase within the small range of this species in California, known breeding areas should be protected from such intrusion. Based on the finding that breeding now occurs in ponds created as a byproduct of California Department of Transportation (Caltrans) road maintenance along Route 78 in eastern Imperial County, the creation of similar ponds and the continued management of the existing ones should be a major mitigation requirement for the approval of future land use changes in that area. However, a first conservation step for the CDFG, the state agency that asks for and approves such mitigations, is to cease allowing anyone who purchases a fishing license to remove up to four specimens of this State Species of Special Concern from its natural habitat.

True Toads (Family Bufonidae)

These are the anurans that most people think of as toads. They have a chunky body form, short legs compared with most frogs, a "warty" skin, and prominent parotid glands, a feature not present in other anurans. Both the warts and parotid glands secrete a mucus–poison mixture that is potent enough in some species to kill dog-size predators. When molested, most species assume a butting position with the parotids directed toward the intruder. Handling by humans poses no danger if you don't then touch your eyes, and the secretions definitely will not cause warts.

Breeding usually occurs after spring or summer rains. Breeding males develop dark, roughened nuptial pads on their thumbs and inner digits to help them cling to females during pectoral amplexus. Most California species lay eggs in gelatinous strings. Tadpoles are often dark-colored with eyes well in from the outline of the head as viewed from above. The fringe of oral papillae is confined to the sides of the mouth and is indented.

It was recently proposed (2006) that the bufonid toads covered in this guide should be placed in the genus *Anaxyrus*, with the exception of the Sonoran Desert Toad, which was placed in the genus *Incilius*. However, because these proposed changes have been controversial, we have chosen to continue to recognize the genus *Bufo* until there is broader scientific consensus.

SONORAN DESERT TOAD   *Bufo alvarius*
(COLORADO RIVER TOAD)

PROTECTIVE STATUS: CSSC

IDENTIFICATION: ADULTS range from 10 to 19 cm (3 to 6 in.) This is our largest native toad, both in California and North America, with adults ranging from 10 to 19 cm (3 to 6 in.) in snout-to-vent length. It is dark brown or olive above, with smooth skin, long kidney-shaped parotids, and prominent cranial crests. It has several large wartlike structures on its hind legs and smaller ones on its forearm that stand out conspicuously against smooth skin, and a whitish knob at the angle of its mouth. Its **VOICE** resembles a ferryboat whistle given in weak hoots lasting one half to one second. **EGGS** are laid in strings of up to 8,000 ova. **LARVAE** are about 5.6 cm (2.25 in.) in total length. Its body is somewhat flattened and is gray to light golden brown.

RANGE: Historically, in California it occurred in the Colorado Desert in the greater Lower Colorado River area and irrigated lowlands of Imperial County. However, it is now believed to be extirpated from the state.

NATURAL HISTORY: This species is found in brushy desert habitats with creosote bush and mesquite, and also in irrigated farmland. It frequents washes, springs, river bottoms, temporary rain pools, canals, and irrigation ditches. It is nocturnal, more aquatic than most toads, and seeks refuge in burrows of other animals. It is most active from May to July. When molested it assumes the basic Bufo butting pose and may squirt poison over 3 m (10 ft) from its parotid glands. The large "warts" on the limbs also secrete poison that, if swallowed in quantity, is capable of paralyzing and killing a dog. Raccoons disembowel these toads and eat only the internal organs, an area where the poison is apparently not present. Activity is stimulated by rainfall, but it is not dependent on rain for breeding. Development time

Plate 44. Sonoran Desert Toad.

from egg to metamorphosis is relatively short, usually not more than one month. Metamorphs are very small, usually no more than 14 mm (0.6 in.) in snout-to-vent length. Like many other bufonids, they attain much of their initial growth between transformation and the first winter retreat period.

The Sonoran Desert Toad is so large that it is able to eat mice, large scorpions, and other amphibians. In areas where it occurs near human habitation it often comes to outdoor lights at night to catch insects, thus affording an excellent opportunity to observe it.

CONSERVATION NOTE: The ongoing expansion of agricultural development, based on Colorado River water, continues to compromise the small number of California populations of this species. At this writing its status in this state is questionable. More information concerning the existence of any remaining populations and their requirements for coexistence with such activities is needed.

One additional and unusual factor has affected the Sonoran Desert Toad. This centers around its poisonous parotid and leg gland secretion, which has hallucinogenic properties. Within the drug culture this species is known as the "Psychedelic Toad," a title that has led to either direct consumption via "toad licking" or smoking the dried poison squeezed from these glands. This toad's poison has now been classified as a "controlled substance," and well it should be, because it contains 5-MeO-DMT, a potent hallucinogen, psychoactive in humans at doses of three to five milligrams. Up to one gram of poison containing as much as 75 mg of 5-MeO-DMT can be obtained from one toad. This, of course, has kindled much interest in this species outside of the herpetological world, and instructions for finding and capturing *B. alvarius* along with tips on collecting the poison and smoking the dried residue have been in print and available for more than two decades. This has also put a price on this toad's head, and at this writing one out-of-state Internet supplier is selling pairs for $275 and single specimens for $150. One bright spot in all of this is an apparent trend to "raise your own" if your purchased pair cooperates, a pastime that may take some pressure off the remaining wild populations.

Some years ago, before this unusual attribute of this toad was well-known, the junior author and his summer desert biology class stopped at a remote, one-pump gas station in southwestern Arizona to refuel. This place also sold cold drinks, and as they lingered to enjoy them some of the students wandered around to the back of the rundown building, where they found a number of old lounge chairs and, to their delight, a large cage containing Sonoran Desert Toads (then known as the Colorado River Toad). The class was reluctantly granted permission to briefly remove and examine one, all the time being carefully scrutinized by the owner of the establishment. It was not until some years later that the professor realized he had probably taken an official university class into a den of iniquity!

GREAT PLAINS TOAD *Bufo cognatus*

IDENTIFICATION: ADULTS range from 4.6 to 11.4 cm (2 to 4.5 in.) in snout-to-vent length. They have large, well-defined dark blotches in pairs on the back. The cranial crests diverge widely posteriorly and are more or less united on the snout to form a boss. The parotids are oval, and sometimes there is a narrow dorsal stripe down the middle of the back. Its **VOICE** is a harsh, explosive, metallic clatter, resembling a pneumatic hammer, with each call lasting five to 20 seconds. At close range a chorus can be almost deafening. The inflated vocal sac is sausage-shaped and about one half the size of the body. A single female may lay up to 20,000 **EGGS** in long strings, with each string containing two continuous jelly envelopes, which are 2 mm (0.08 in.) at the thickest point when containing a single row of eggs, and 2.7 mm (0.1 in.) when eggs are in a double row. The ova are 1.2 mm (0.05 in.) in diameter. **LARVAE** are approximately 2.5 cm (1 in.) in total length, and the dorsal tail fin is highly arched. Young tadpoles are almost black, but at about 8 mm (0.3 in.) they begin to assume color patterns and become lighter. As they approach metamorphosis the dorsum is usually a mottled brown and gray, dark gray, or blackish, and the vent takes on an iridescent light greenish yellow or reddish hue.

RANGE: This species, which is widespread throughout the Great Plains states, has a small range in the Colorado Desert of southwestern California in Colorado River bottomlands and the irrigated lowlands of Imperial County north to the vicinity of Indio, Riverside County, where it has been

Plate 45. Great Plains Toad.

found in date palm orchards. The long life of these trees and therefore the apparently acceptable orchard floor conditions they create will hopefully permit the continued existence of *B. cognatus* in this area of California.

NATURAL HISTORY: The Great Plains Toad is adapted for life in semiarid lands. It is primarily a grassland species but also frequents creosote bush desert, mesquite woodland, and sagebrush plains. It is chiefly active at night, although it may occasionally be abroad in the daytime, especially during the breeding season. It is an excellent burrower and reduces the hazard of desiccation by going underground. It is capable of storing as much as 30 percent of its gross body weight in the urinary bladder as water, which can be reabsorbed, and while buried it can tolerate the accumulation of a substantial amount of urea from protein metabolism in its body fluids and muscles; this, in turn, creates a favorable osmotic gradient with the surrounding soil. *Bufo cognatus* tends to be opportunistic in breeding and spawns whenever conditions are suitable. Breeding sites include flooded fields, the edges of large temporary pools, permanent springs, and small streams where they often gather in great numbers. Relatively clear, shallow water seems to be preferred. Adult toads feed on moths, caterpillars, flies, beetles, and other insects.

RED-SPOTTED TOAD *Bufo punctatus*

IDENTIFICATION: ADULT snout-to-vent length ranges from 3.8 to 7.6 cm (1.5 to 3 in.). The head and body are flattened, an adaptation for crevice-dwelling. The parotid glands are small and round, the snout is pointed, and it has small cranial crests. Its coloration often has granite-matching pattern, and the warts are reddish-tipped. The young have orange-tipped warts, and the undersides of the feet are yellow. Its **VOICE** is a clear, birdlike trill lasting four to 10 seconds, usually on one pitch but occasionally dropping toward the end. It often calls just after sunset. The vocal sac is round. **EGGS** are laid singly, in short strings or sometimes as loose flat clusters. The single jelly envelope, 3.2 to 3.6 mm (0.12 to 0.14 in.) in diameter, often rapidly accumulates sediment. Ova are 1.0 to 1.3 mm (0.03 to 0.05 in.) in diameter, black above and white below. **LARVAE** grow to about 3.5 cm (1.4 in.) total lateral length. The tooth rows arrangement is typically 2/3. The anterior labium and its tooth rows extend considerably farther ventrolaterally than in our other toad larvae. Tadpoles resemble those of the Western Toad but have coarser spotting on the dorsal fin.

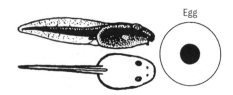

Egg

RANGE: Throughout California's Mojave and Colorado deserts, from northern Inyo County south into Mexico. This species has hybridized with the Western Toad (*B. boreas*) at Darwin Canyon, Inyo County, and Rawson Canyon, Riverside County.

NATURAL HISTORY: This toad inhabits deserts, open grassland, brushland, palm oases, and oak woods. It breeds in springs, seepages, intermittent stream courses, and cattle tanks. The Red-spotted Toad is adapted for life in dry, rocky environments. The flattened head and body enable it to enter narrow crevices, and its low center of gravity and slender body form facilitate climbing. The seat patch is well developed and serves to absorb water from moist surfaces when complete immersion is impossible. In experiments with submerged toads, seat-patch integument accounted for approximately 70 percent of the water uptake through the total integument. It is chiefly nocturnal except when breeding. Breeding takes place when opportunity affords and occurs from April (Death Valley) to as late as September elsewhere. Eggs are encased in one jelly envelope and laid singly in short strings or loose flat clusters on the bottom of small, shallow, often rocky pools. Recapture of toe-marked adults after four years at Deep Canyon, Riverside, County, indicates that these toads may be long-lived, a trait probably important to their survival in arid environments where suitable breeding conditions may not be available for long periods. Adults feed on beetles, ants, bees, and other insects.

Plate 46. Vocalizing male Red-spotted Toad.

TRUE TOADS (BUFONIDAE)

ARROYO TOAD
Bufo californicus

PROTECTIVE STATUS: FE, CSSC

IDENTIFICATION: ADULT snout-to-vent length is from 4.6 to 8.6 cm (1.75 to 3.2 in.). The dorsum is greenish-gray to brown with a light stripe across the head, including the eyelids, and sometimes a light patch on each sacral hump and in the middle of its back. It has no mid-dorsal stripe. The parotid glands are oval, pale anteriorly, and widely separated. Cranial crests are weak or absent. Both male and female have a pale throat, unlike most bufonids where the male's throat is dark. The young have a pale dorsum that often contains dark spots and yellow-tipped back tubercles. Its **VOICE** is a melodious trill lasting eight to 10 seconds, rising in pitch at first and often ending abruptly. **EGGS** are laid in one to three irregular rows and are contained in a continuous gelatinous envelope, about 5.8 mm (0.2 in.) in diameter. The ova average about 1.5 mm (0.06 in.) in diameter. **LARVAE** are about 3.7 cm (1.5 in.) long. They are olive, gray, or tan above, and commonly spotted or mottled with black or brown. Very young tadpoles are black.

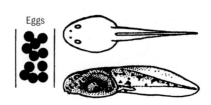

RANGE: *B. californicus* is found in parts of the South Coast Range and Coastal Southern California from the Fort Hunter Liggett area in southern Monterey County south into Baja California. Desert populations occur along the lower Whitewater River, Riverside County, and Mojave River, San Bernardino County. A continuous distribution may have formerly existed when the desert was well watered. Many of the populations are now isolated in the upper tributary streams of larger drainages, apparently as a result of increasing aridity since the last pluvial period.

NATURAL HISTORY: This is a toad of washes, streams (sometimes intermittent), and arroyos of semiarid lowlands, hence the name Arroyo Toad. Trees often present are willows, cottonwoods, sycamores, and coast live oaks. Near the northern part of its range it seems to require large, permanent streams with sandy or gravelly flats that have adjacent stands of oaks and willows. Adults and juveniles require moist shorelines and stable, sandy terraces. Larvae have only been found in broad, very shallow pools with gravel substrate and no emergent vegetation. These are some of the most specialized habitat requirements of all California anuran species.

This toad is chiefly nocturnal but may be active in daytime when breeding. It usually moves by hopping rather than by walking. It breeds from March to July but may be abroad until September. Heavy rainfall marks the highlight of the breeding season. Eggs are laid in tangled strings on the bottom of shallow pools among leaves, sticks, gravel, and mud, or on clean sand bottom. At such sites clutches of several thousand each may be present. Larvae are cryptically colored and blend well with the creek or pond substrate. They seem not to form dense aggregations by day as do those of some other *Bufo* species. At night they are often tightly clumped at the water's edge, often half out of the water. They require between 65 and 85 days to begin metamorphosis. Newly transformed young cannot burrow into moist substrate as do adults and other juveniles, and therefore damp surface retreats such as drying algae mats must be present. Adults feed on snails, Jerusalem crickets, beetles, ants, bees, caterpillars, and moths. Some cannibalism also occurs when newly transformed young appear. Larvae often sift through the aquatic substrate for organic material ranging from algae and fungi to protozoa and bacteria.

CONSERVATION NOTE: As with other amphibian species that depend on a unique and fragile aquatic habitat for breeding, hydrological manipulation and outright destruction of such sites upon which this species depends have had a major impact. Given that water remains California's most sought-after resource, the identification of all sites where this species still exists followed by rigorous protection of same is now vital for its survival.

WESTERN TOAD   *Bufo boreas*
BOREAL TOAD   *B. b. boreas*
CALIFORNIA TOAD   *B. b. halophilus*

IDENTIFICATION: The **ADULT** size range is 5.1 to 12.7 cm (2 to 5 in.) from snout to vent. The dorsum is dusky, gray, or greenish with dark blotches and a narrow, pale, mid-dorsal stripe. There are either weak or no cranial crests, and the parotids are usually oval, well developed, and well separated. There is a sharp-edged skin fold (tarsal fold) on the inner side of the hind foot. The female is larger and stouter than the male. Males usually have smoother skin and a less heavily blotched pattern than females. During the breeding season they have dark-brown nuptial pads on the thumb and inner two digits. The young may lack a dorsal stripe and have bright yellow on the underside of the feet. The male's **VOICE** consists of a continuous series of mellow high-pitched chirps, which are used to trigger the release by another male that may mistakenly attempt amplexus. It may also be emitted when the toad is picked up and gently squeezed, an action that apparently simulates that of an amplexing male. Male calls are not used for long-distance communication as in many other toad species, and thus there is no resonating vocal sac in this species. **EGGS** are contained within strings of continuous jelly consisting of two envelopes. The ova are in single or double (occasionally triple) rows within an inner jelly tube and are 1.2 to 1.8 mm (0.04 to 0.08 in.) in diameter. **LARVAE** are 5.6 cm (2.25 in.) long at metamorphosis. The tip of its tail is rounded, and the eyes are on the dorsal aspect of the head. The labial tooth row formula is 2/3.

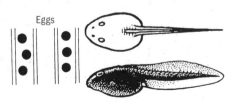

RANGE: It occurs throughout most of California except in most desert areas from Death Valley southward. Its total range extends from southern Alaska to northern Baja California, Mexico. It is also absent from the central high Sierra where the Yosemite Toad (*B. canorus*) occurs. This distribution spans an elevation range of sea level to around 3,600 m (11,800 ft). It coexists and apparently hybridizes with Yosemite Toad at Blue Lakes and perhaps elsewhere in the Sierra Nevada, and with the Red-spotted Toad in Darwin Canyon and near Skinner Reservoir, Riverside County. There are also a number of isolated populations in other arid parts of its range.

NATURAL HISTORY: California habitats for the Western Toad range from grassland, woodland, and meadows in forest areas to gardens and parks in the suburbs. It breeds in ponds, slowly flowing streams, lakes, canals, and reservoirs.

This toad has a very flexible biology, as reflected in its distribution from isolated desert pools to recently deglaciated areas of southern Alaska. The thesis research by two former students of the junior author illustrates this point. Alan Smits recorded deep body temperatures by telemetry from Bay Area Western Toads in a thermal gradient and found a range of 15.0 to 30.1 degrees C (59 to 86 degrees F). When comparing these data with ranges presented graphically in this book for Ensatina (*Ensatina eschscholtzii*) and the Western Fence Lizard (*Sceloporus occidentalis*), the thermal preference of the Western Toad more closely resembles the latter than the former and adapts it well for life in desert fringe areas.

Plate 47. Western Toad in annual grassland habitat.

Mike Taylor found this same species on recently deglaciated islands in Glacier Bay, Alaska, at ambient temperatures only a few degrees above freezing. He also found that in that area this toad was a true "pioneer species," because, except for a stray Moose or two, they were the only terrestrial vertebrates on the newly deglaciated islands that Taylor surveyed. The Moose is a good swimmer, even in ice-cold marine water, but a toad is not. Upper Glacier Bay water temperature is at or below the low ecological lethal for this amphibian, and seawater is a most unfavorable ambient medium. How, then, did this amphibian get there? This puzzle was solved a year or so later when a floating toad was found between newly deglaciated islands and the mainland. Its lungs were fully inflated, and it was in an immobile state but alive. Apparently these toads are occasionally washed into Glacier Bay by the high flows of the mainland streams. These little "bufo bottles" are then randomly dispersed by tides and currents, with a few occasionally landing on ice-free islands to start a new population.

In California, *B. boreas* breeds from January to July, with later spawning occurring at high elevations. Many populations are aggregation breeders that come to suitable pond or marsh spawning sites in large numbers. Here males recognize females through random amplexus. If during this process one male grasps another male, the latter emits a chirping "release call," which signals its sex. The amplexing male responds by releasing his grip and then continues to amplex other individuals until one does not respond by chirping, a sign that his quest for a mate is over. However, his newfound lady must still pass one more test. She must be relatively rotund, indicating that she is gravid and has not as yet spawned. If she has and then resumed her maiden figure, she is also released, and the suitor's search continues.

Eggs number over 16,000 per female and are laid in strings in shallow water. Tadpoles grow up to 5.6 cm (2.25 in.) in total length before metamorphosis but often transform at a smaller size. Western Toad metamorphosis is a very dynamic event. Waves of tiny, newly transformed "toadlets" emerge from the larval habitat over a period of several days and forage while slowly dispersing over the greater shoreline area. Many people who have encountered such an event are often mentally transported to the storybook world of *Gulliver's Travels*, with themselves in the starring role and the metamorphs as the cast of a thousand "Lilliputians." Indeed, one must move with the utmost care lest one of your subjects be crushed. These very small miniature adults forage throughout the post-metamorphosis months, and in most habitats enter winter hibernation at about half the adult size. Larvae feed on the normal complement of algae and detritus found in spawning habitats. Adult food includes insects, crayfish, sowbugs, snails, and spiders plus some cannibalism of metamorphs.

Most Californians are more familiar with the Western Toad than other *Bufo* species, because its range includes many areas of high human population. Extensive gardens and yard landscaping often encourage its presence if a spawning site is nearby, and it is indeed a good neighbor to have. The size of an adult's stomach is stretched to about the size of a man's thumb after a night of feeding on the many insects that attempt to destroy your garden. The large female seen in Plate 48 is a resident of the McGinnis organic garden and is occasionally discovered "hanging out" near one of the garden's many drip irrigation lines. If you find you have these gardener's helpers on your premises, the best reward you can offer for their services is to avoid the use of insecticides and herbicides, which can readily pass through their skin. Instead let them be the pest control specialists. A small garden pond stocked with aquatic plants but *no fish* will ensure their continued presence.

Plate 48. Western Toad in a garden next to drip irrigation line.

BLACK TOAD

Bufo exsul

PROTECTIVE STATUS: ST

IDENTIFICATION: ADULTS are relatively small, ranging from 4.4 to 7.6 cm (1.74 to 3 in.) in snout-to-vent length. The dorsum color is usually solid black with scattered whitish flecks and lines and a narrow white or cream mid-dorsal stripe. The vent is white or cream with spots and blotches. Its **VOICE** resembles that of the Western Toad's but is higher in pitch. Juveniles are olive-colored and have orange-yellow tubercles on the undersides of the feet. The male has nuptial pads on the thumb and inner two digits when breeding and, like *B. boreas*, does not have a vocal sac but does emit a release call when grasped. **EGGS** resemble those of *B. boreas* but are less often in a zigzag or double-row arrangement. They are enclosed in two jelly envelopes, the outer one 4.0 to 6.9 mm (0.16 to 0.27 in.) in diameter. The ova are 1.20 to 1.75 mm (0.05 to 0.07 in.) in diameter. **LARVAE** are similar to *B. boreas*, but the tip of tail is rounded instead of square. The body is nearly solid black above but lighter below, and the tail fins are translucent but with melanophore stippling.

RANGE: This toad is completely isolated in desert spring areas in Deep Spring Valley, approximately 1,524 m (5,000 ft) at the northern tip of the Mojave Desert in Inyo County. Localities are Antelope, Buckhorn, and Deep Springs Corral, the latter with the largest population where springs originate from the base of the southeastern wall of the valley and form a series of sloughs and pools in an area of wet meadows and marshes. A small introduced population persists at a flowing well near Salt Lake in Death Valley National Park.

NATURAL HISTORY: The Black Toad habitat in Deep Springs Valley consists of approximately 10 springs with various associated wetland features (marshes, pools, etc.). The marsh at Buckhorn Springs is typical of such habitats. It is treeless and contains a rich complement of emergent vegetation. The slopes nearby support piñon pine and juniper, and the drier flats contain shadscale, greasewood, and rabbitbrush. The Great Basin Spadefoot Toad (*Spea intermontana*) also occurs in the area.

The Black Toad breeds from March to May, in groups ranging from around 10 to 100 pairs. Breeding usually occurs in the daytime, because night temperatures often fall close to or below freezing during the spawning season. The eggs are deposited in shallow, quiet water to a depth of around 25 cm (10 in.) and are entwined among sedge, watercress, or other aquatic vegetation. Newly transformed metamorphs can be found in June. Sexual maturity is thought to be reached by the end of the second postmetamorphic year. Hibernation occurs in rodent burrows in the banks of the sloughs with as many as 20 to 30 individuals in the same burrow.

CONSERVATION NOTE: On a visit to Buckhorn Springs in June 1949, the senior author estimated the adult population at just this one site to be several thousand. In 1954 one estimate placed the entire species number in excess of 10,000, but by 1971 an estimate of less than 4,000 was proposed.

In 1999 a study using implanted micro-transponders (see the "Recognizing Individual Amphibians and Reptiles in the Field" section) and the mark–recapture format estimated the Deep Springs Corral population to be about 8,400 toads, a number similar to that from a 1980 survey. The most recent (2005) published estimate for the entire Deep Springs Valley area is 24,000 individuals, a number that appears to have remained relatively stable for nearly two decades.

The entire range of the Black Toad is only about 15 hectares (37 acres), which is one of the smallest for North American amphibian species. For such species with a highly restricted range, any major decimating event (severe drought, disease, predator introduction) has the potential of eliminating the entire taxon. Ongoing rigorous protection and management of its remaining habitat can prevent the recurrence of previous impacts such as water diversions from springs, the trampling of shallow streams by cattle, and the introduction of the Common Carp (*Cyprinis carpio*) into some breeding pools. One helpful action has been the delay in diverting water from some springs to pastures until larval metamorphosis has occurred. It is also felt that some limited and carefully regulated grazing by cattle at Buckhorn Springs may benefit this species by removing vegetation that would choke the habitat and by supplying droppings that attract insects that these toads eat.

YOSEMITE TOAD

Bufo canorus

Male

Female

PROTECTIVE STATUS: CSSC

IDENTIFICATION: The **ADULT** snout-to-vent size range of this species, 3.1 to 6.9 cm (1.25 to 2.75 in.), is similar to that of Western Toad, but its dorsal stripe, when present, is only a fine hairline. The skin is smoother than *B. boreas*, and the parotids are large, flattened, and close together. The space between the eyes is narrow, usually less than the width of the upper eyelid. This toad is unusual in that there is a marked color sexual dimorphism. Females are contrastingly marked with numerous black blotches on olive to gray background. The males are yellow-green to dark olive with spots small and scant. The young resemble those of the Western Toad. In addition to color differences, the male is smaller than the female and has fewer and smaller warts. Its **VOICE** is a melodious trill of 10 to 20 or more notes in rapid sequence. **EGGS** are encased in two jelly envelopes, the outer of which forms a scalloped but continuous layer. The inner envelope surrounds each ovum and is usually in contact with the inner envelope of the adjacent ovum. The average diameter of the outer envelope is 4.1 mm (0.16 in.), and the average ovum diameter is 2.1 mm (0.08 in.). The eggs are laid in strings or are variously attached to one another to form a radiating network or cluster four or five eggs deep. **LARVAE** resemble those of *B. boreas* but are usually smaller, with a total length under 3.7 cm (1.5 in.). They also have a shorter, broader snout and shorter tail with a

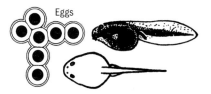

less pointed tip. The labial tooth formula is usually 2/3. Both eggs and larvae are very darkly pigmented, more so than in *B. boreas*. This is perhaps an adaptation to protect against the greater ultraviolet solar radiation at high elevations.

RANGE: It occurs in the Sierra Nevada from an area near Grass Lake, El Dorado County, to the south of Kaiser Pass on the west side of the range and to Evolution Lake in Kings Canyon National Park, Fresno County. This distribution spans an elevation range of 1,460 to 3,630 m (4,800 to 12,000 ft).

NATURAL HISTORY: This is a toad of the high mountain meadows, often where trees are absent or scarce. It seems to be most abundant in open areas with low vegetation, good illumination, good soil moisture, and patches of willows. Scattered lodgepole and white-bark pines may also be present and may form forest borders in some

Plate 49. Male (with head marking) and female Yosemite Toads.

TRUE TOADS (BUFONIDAE)

areas. At lower elevations, surrounding forest cover is denser and consists of red fir, lodgepole, and Jeffrey pines. Breeding habitats consist of water-filled potholes, meandering streams, and shallow meadow snowmelt pools. Favorable habitat is also sometimes present at high mountain lakes. The Yosemite Toad is chiefly diurnal. At high elevations at which it lives, basking in direct sunlight seems to be necessary to maintain the body temperature within a range for efficient neuromuscular coordination. On cool mornings and cold days it remains in rodent burrows or under dense clumps of vegetation. It emerges from hibernation in April and May and ceases surface activity in September and October. At this latter time these toads are usually well laden with fat, and their bodies tend to be enlarged, presumably due to fat storage.

The Yosemite Toad breeds in high mountain lakes, snowmelt pools, and small streams from mid-May through July. The males usually precede the females to the breeding site. Spawning follows the standard bufonid pattern outlined in the genus introduction. Many larvae apparently overwinter and transform the following summer. Juveniles and adults feed on insects, centipedes, and other invertebrates that inhabit moist mountain areas.

CONSERVATION NOTE: The Yosemite Toad is one of many anuran species that has undergone an alarming reduction in numbers in recent years, and as is the case with other such species, the reason(s) for the decrease are poorly understood. At this writing it has disappeared from more than 50 percent of its historic range. Given its occurrence in two national parks, more well-designed and well-executed field surveys and studies of these populations, particularly of egg and larval survival, will hopefully be conducted in the near future. Attention should also be directed to those high mountain lakes that historically supported Yosemite Toad reproduction but no longer do so, possibly because of the presence of introduced trout species.

WOODHOUSE'S TOAD *Bufo woodhousii*

IDENTIFICATION: ADULT snout-to-vent range is 4.4 to 12.7 cm (1.75 to 5 in.). Woodhouse's Toad resembles the Western Toad, but its dorsal color is yellowish brown, grayish, or light olive with a whitish dorsal stripe and a yellow and black network on the posterior face of the thighs. It has prominent cranial crests and elongated divergent parotids but no sharp-edged fold on the inner side of the hind foot. Males have a dark throat. Its **VOICE** is a nasal "wa-a-a-ah" resembling an infant's cry, lasting from one to three seconds and often dropping in pitch at end. **EGGS** are enclosed in a stringlike jelly envelope in one or two rows. These may contain over 28,000 ova and are intertwined around vegetation or other objects in the spawning habitat. The ovum ranges from 1 to 1.5 mm (0.04 to 0.06 in.). **LARVAE** of this species are relatively small, with a maximal length of 23 mm (0.9 in.). The labial tooth rows formula is 2/3.

Eggs

RANGE: If one includes the eastern subspecies (*B. w. fowleri*), this toad has an enormous range that extends from the eastern seaboard to the Colorado Desert in the southeastern corner of California. Here it has been found along the lower Colorado River, in irrigated lands of the Imperial Valley, and in orchards near Coachella Canal east of Indio in Riverside County.

NATURAL HISTORY: Woodhouse's Toad is a classic habitat generalist. It occurs in grassland, sagebrush flats, mesquite plains, river bottoms, flood plains, woodland, forest clearings, and farmland. This toad is chiefly nocturnal. It breeds from March to July in shallow sluggish creeks, freshwater pools, and irrigation ditches. The eggs are laid in long strings in a tangled mass. The tadpoles resemble those of the Western Toad's and transform at under less than 2.5 cm (1 in.) in total length. It feeds on a wide variety of invertebrates, including grasshoppers, crickets, moths, caterpillars, flies, beetles, ants, bees, wasps, sowbugs, scorpions, centipedes, and spiders.

Chorus Frogs (Family Hylidae)

This family is distributed nearly worldwide except for the polar and subpolar regions, Africa south of the Sahara, islands of the Pacific Ocean, and the southern tip of South America. Most are small, slim-waisted, long-legged frogs, usually with adhesive toe pads used in climbing. The toe pads are anchored to the rest of the toe by a small extra joint that gives the toe tip great mobility for climbing. Most species have a well-developed voice, and the calls of chorus frogs are among the most often heard anuran sounds. There are about 600 species in this family, but California contains only two. However, one, the Pacific Chorus Frog (*Pseudacris regilla*), occurs in every geographic subdivision and is perhaps the best-known native amphibian in the state.

PACIFIC CHORUS FROG
(PACIFIC TREEFROG)

Pseudacris regilla

Brown morph

Green morph

IDENTIFICATION: ADULTS range from 1.9 to 5.0 cm (0.75 to 2 in.) in snout-to-vent length. The Pacific Chorus Frog occurs in many color forms, the most common of which are green, gray, brown, reddish, and tan. It can change from dark to light color in few minutes and often has dark spots on its back and legs. The black eye stripe does not change. Males have dark-olive throats with very loose skin. All toes have toe pads, and the hind toes are about two thirds webbed. The margin of the web between successive toes, when spread, is usually curved distinctly inward. There is considerable variation in size and color of *P. regilla* over its extensive geographic range. There may be an optimum size for frogs inhabiting a particular microenvironment that will ensure most effective temperature and body water regulation. Its **VOICE** is typically a loud, two-parted "kreck-ek" with rising inflection, lasting about one second. **EGGS** are encased in two jelly envelopes, the outer of which is sticky, 4.5 to 6.7 mm (0.2 to 0.3 in.) in diameter. Ova are 1.2 to 1.4 mm (0.05 to 0.06 in.) in diameter. **LARVAE** range from 3.1 cm (1.25 in.) to occasionally as much as 4.4 cm (1.75 in.) in total length. The body is globular, not laterally depressed. The eyes are lateral and can be seen bulging out from the outline of the head when viewed from above. The labial tooth row formula is 2/3.

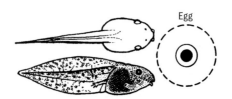

Egg

RANGE: It occurs throughout California but is absent from desert areas except along streams (Mojave River), at oases such as Pushawalla Palms and Zzyzx Springs. It is also found in scattered localities along the lower Colorado River. Its altitudinal range is from sea level to around 3353 m (11,000 ft) in the central Sierra Nevada. *P. regilla* coexists with the California Chorus Frog (*P. cadaverina*) at a number of localities in southern California and Baja California.

NATURAL HISTORY: This frog is a classic example of a habitat generalist. Its habitats vary greatly, from slightly brackish cattail marshes at sea level to mountain meadows above timberline. It frequents ponds, streams, springs, lakes, and reservoirs in open grassland, woodland, or forest, and gardens, golf links, and parks in suburban areas. It is both nocturnal and diurnal. It is also a good climber, but in California it usually stays near ground. As mentioned in the family introduction, more people in California and the Pacific Northwest are probably more familiar with this amphibian than any other. One reason is that it seems to be just as much "at home" in a garden or park with some shallow-water site nearby as in a remote region of the coastal ranges or the Sierra Nevada.

The other reason is that it is more often heard than seen. California has come up short with respect to the number of anuran calls one may hear on a warm spring night. In some southeastern states that number may be

Plate 50. Pacific Chorus Frog climbing a reed stem.

CHORUS FROGS (HYLIDAE)

a dozen or more, but in most areas of the Golden State, that number is usually just one. However, this diminutive species has perhaps the most recognizable anuran voice in America if not the world. This is because years ago when the small community of Hollywood was still surrounded by undeveloped areas (even today "back lots" for many movie studios are located in the Santa Monica Mountains), movie sound technicians apparently went out behind the studio lots on quiet spring evenings and recorded the calls and choruses of the Pacific Chorus Frog for use as background sounds in movies. From that time on, the night scenes from most Hollywood films, whether they were set in California or in many other parts of the world, were accompanied by the voice of our own *P. regilla* or, as many herpetologists call it, "the Hollywood Frog"!

The call that we hear most is its diphasic, two-parted "kreck-ek" advertisement call, which attracts both males and females to a breeding site and also appears to function in the territorial spacing of males throughout a pool or creek site. Its volume is out of proportion to the size of this frog. If you stand near a chorus in full voice, it excludes most other sounds. Another call often heard is the monophasic "encounter call," which sounds somewhat like a growl and is given when one male is approached too closely by another. If an approaching male remains quiet and continues his advance, he may be amplexed by the calling male, which then triggers a "release call" on the part of the intruder. Unlike most salamander species, male anurans in a breeding habitat rely on sound rather than scent to recognize females. The lack of a release call by the frog being amplexed identifies "it" as a "her." We humans are exceptionally good at recognizing subtle variations in musical themes, and with a basic familiarity of these Pacific Chorus Frog voices, one can visualize the interactions taking place in a breeding pond based on the various calls heard on an early spring night.

P. regilla usually calls from November through May, but just what combination of weather conditions initiates calling is still uncertain. One study in the southern part of its range showed that a male's body temperature had to reach about 10 degrees C (50 degrees F) before it began calling. A decline in barometric pressure also seems to stimulate calling, even outside of the breeding season. The junior author recalls one morning in mid-August while listening to the radio in his farm workshop in San Joaquin County, the weather reporter proclaimed it would be "another beautiful day with plenty of sunshine." No sooner had the announcer finished than a Pacific Chorus Frog began his diphasic call from an old stock watering tank nearby. Within an hour the skies had clouded, and later that day there occurred a rare phenomenon for the San Joaquin Valley: a midsummer rain. This author now consults his "resident *regilla* meteorologists" before planning any summer outings.

The Pacific Chorus Frog breeds January to July, with the later spawning dates occurring at high elevations. The eggs are laid in irregular clusters attached to vegetation in shallows of ponds, streams, roadside ditches, rain pools, and marshes. In fact, it is such an opportunistic spawner that any small standing water pools present during the main breeding period may be used. Successful clutches have been produced in such unlikely sites as birdbaths and deep, water-filled tire ruts in clay soil areas.

Females may spawn every other year, but when they do they can often lay up to three clutches. Larval development is quite rapid compared with most other California frog species. Under relatively warm conditions larvae that hatch in mid-March will begin to transform in mid-May. This accelerated development is highly adaptive, because many breeding habitats are seasonal wetlands such as vernal pools.

P. regilla feeds on a wide variety of small invertebrate prey including leafhoppers, springtails, flies, stoneflies, ants, wasps, beetles, caterpillars, spiders, isopods, and small snails. Although it rarely climbs into large trees, it does ascend large forbs and shrubs to feed on small insects that are attracted to blossoms and in some cases fruit and berries. Blackberry thickets along a water course are especially favored foraging sites where these frogs may often be seen sitting on leaves within such stands, apparently not deterred by the sharp thorns on the branches that surround them.

CALIFORNIA CHORUS FROG (CALIFORNIA TREEFROG) *Pseudacris cadaverina*

Two variations

IDENTIFICATION: ADULT snout-to-vent size range is 3.2 to 5.7 mm (1.25 to 2.25 in.). The dorsal surface resembles the color and texture of the rocks upon which this frog normally rests during the day. Frogs on granite substrate tend to be mottled gray and dusky, although sand-colored individuals such as those usually seen in sandstone areas may occur on granite as well. An eye stripe is either vague or absent, and the toe pads are larger and the hind foot webbing is greater than in the Pacific Chorus Frog. Males have a dusky throat. The **VOICE**, usually heard at night, is a ducklike quack, with each call lasting 0.2 to 0.5 second with little or no inflection. **EGGS** are encased in a single jelly envelope, about 4.4 mm (0.17 in.) in diameter. The ova average 2.1 mm (0.08 in.) in diameter. **LARVAL** total length is approximately 3.7 cm (1.5 in.). The body is somewhat depressed, and the snout is rounded. The tail is about two thirds of the total length, and the dorsal fin terminates near the body–tail junction. They differ from *P. regilla* in having a flatter body, more acute snout, and dorsal eyes, set within the outline of the head when viewed from above (eyes are on outline of head in *P. regilla*).

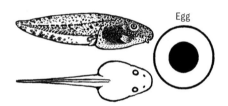

Egg

RANGE: It occurs in the South Coast Range, and Coastal Southern California from central San Luis Obispo County south into Baja California. It is also found in the Transverse Range, Peninsular Ranges, and in desert oases. Its range overlaps that of the Pacific Chorus Frog in many areas, but the two species are seldom found together at the same site.

NATURAL HISTORY: This is a chorus frog of canyon streams and washes where there are quiet pools and some shade. Rocky canyon bottoms with permanent water grown to oaks and sycamores or, at higher elevations, oaks and pines are especially favored. At many localities in southern California the streams occupied by this frog are subject to high water in winter and early spring. At such localities the frogs may disperse in the fall from their exposed streamside habitats to seek shelter in rock crevices in the surrounding hillsides where they spend the fall and winter.

During warm-weather days individuals may be found crouching with limbs tucked close to the body in declivities among the rocks. In this pose

Plate 51. California Chorus Frog on vertical rock wall.

they sometimes rest for long periods, even in direct sunlight. When frightened they often jump into water of nearby pools, swim promptly to the opposite side, and climb out. *P. cadaverina* is more closely bound to the vicinity of water than is *P. regilla*. During periods of low humidity, shelter is sought in crevices among the rocks and in rodent burrows. From two to many individuals are sometimes found sharing a single favorable retreat. Crevices are also used for winter hibernacula, and sites are chosen that are above the high-water mark of the streams inhabited. Following spring emergence, adults return to familiar sites—sometimes even to the particular crevice in the rock used the year before.

Throughout its range the California Chorus Frog breeds from February to early October. Breeding appears to be timed primarily to rainstorms, water flow levels, and temperature conditions in the stream habitats frequented. The season is shorter in high-desert areas subject to temperature extremes and dryness. Eggs are usually deposited singly in quieter water of rocky, clear streams, but they may be clumped when laid in small potholes. They are usually attached to leaves, sticks, and rocks on or near the bottom of rocky pools. Its good climbing ability permits foraging on a wide variety of small invertebrate prey, including ants, caterpillars, beetles, spiders, and centipedes.

True Frogs (Family Ranidae)

Family Ranidae is almost worldwide in distribution and contains more than 700 species. The large, widespread genus *Rana* has about 270 species and occurs in the Western Hemisphere. As the family common name implies, these are the anurans that come to mind when most people think of a frog. They are typically slim-waisted, long-legged, smooth-skinned anurans with well-developed webbing on the hind feet (see figure 4, in the "Taxonomy, Anatomy, Physiology, and Behavior of Amphibians" section). Those species native to western North America have a glandular ridge, the dorsolateral fold, on the upper side of the body extending from the eye region to the rump. They have teeth in the upper jaw, and most are excellent jumpers.

Their jelly-coated eggs are often laid in globular, grapelike clusters. Ranid tadpoles have oral papillae indented at the sides and a wide gap across the top of the upper labium (see figure 17 in the "Taxonomy, Anatomy, Physiology, and Behavior of Amphibians" section). The amplexus position is pectoral. Male ranids can be identified during the breeding season when their thumb bases and forelimbs become enlarged. A dark nuptial pad also appears on each thumb.

It was recently proposed (2006) that a number of ranid frogs covered in this guide should be placed in the genus *Lithobates*, including the American Bullfrog and the leopard frogs. However, because this new taxonomy has been highly debated, we have taken a conservative approach by recognizing all ranids here as members of the genus *Rana* until there is broader agreement among herpetologists.

CALIFORNIA RED-LEGGED FROG   *Rana draytonii*
NORTHERN RED-LEGGED FROG   *Rana aurora*

Ventral

California Red-legged Frog

PROTECTIVE STATUS OF *R. DRAYTONII*: FT, CSSC
PROTECTIVE STATUS OF *R. AURORA*: CSSC
INTRODUCTORY NOTE: Recent studies using molecular, morphological, and behavioral evidence indicate that these former two subspecies should now be accorded species status, and we have accepted this change for our book. However, within California, these former subspecies are nearly identical in appearance and behavior and are therefore discussed together, with the few differences noted where appropriate.
IDENTIFICATION: ADULTS range from 4.4 to 13.3 cm (1.75 to 5.25 in.) in snout-to-vent length. The dorsal color is brown to reddish in *R. draytonii* in the San Francisco Bay region, with small dark flecks and larger blotches. These are more numerous than in *R. aurora* and usually have light centers. Both species usually have a dark mask bordered by a pale stripe on the upper jaw. The groin area is coarsely mottled with black (or dusky) on a reddish or yellowish ground color. The namesake color red (often yellowish in juveniles) occurs on the lower belly and undersides of the legs. Both species also have well-developed dorsolateral folds. Their **VOICE** is a series of stuttering, grating, guttural sounds on one pitch, often ending in a growl, with each call lasting about three seconds. In the South Coast Range *R. draytonii* has small, subgular vocal sacs, and a chorus sounds like continuous low chuckling. The size of the vocal sacs is greatly diminished in populations of both species in the North Coast Range and are absent in *R. aurora* north of the Smith River near the California–Oregon border. This reduction and eventual loss of vocal sacs is accompanied by the increased use of underwater vocalization by both species in the North Coast Range. **EGG** clusters are oval masses with length and width diameters ranging from 7.5 by 10 cm (3 by 4 in.) to 15 by 25 cm (6 by 10 in.). They usually contain from 500 to 1,000 eggs but may occasionally have up to 4,000. The ova are 2.1 to 3.6 mm (0.08 to 0.14 in.) in diameter and are black above and cream below. **LARVAE** grow to over 7.5 cm (3 in.) total length by metamorphosis. The labial tooth row formula is typically 2/3 or 3/4. The dorsum is brown to yellowish with diffuse dark spots on the body and tail, and the vent has a pinkish iridescence. When viewed from above, larvae of both species differ from those of the Pacific Chorus Frog (*Pseudacris regilla*), in that their eyes are inward instead of on the outline of the

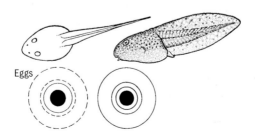

Eggs

Plate 52. California Red-legged Frog egg cluster.

head. They differ from those of the American Bullfrog (*Rana catesbeiana*) in that the latter has a much more rounded snout as viewed from above.

RANGES: The California Red-legged Frog is currently found in segments of the North and South Coast Ranges and the western part of the Transverse Range. It also occurs in the western foothills of the southern Cascade Range and northern Sierra Nevada. It is now apparently absent from most of the floor of the Great Valley and Southern Coastal California. However, several years ago, a young sister and brother from a Calaveras County ranching family discovered adults, tadpoles, and egg clusters on their ranch, discoveries later confirmed by biologists. The range of the California Red-legged Frog overlaps with that of the Northern Red-legged Frog between Point Arena and Elk in Mendocino County. Beyond that point the latter's range extends northward through the coastal half of the North Coast Range in California and on through western Oregon, Washington, and coastal British Columbia and Vancouver Island, B.C., Canada.

NATURAL HISTORY: Red-legged frogs forage in moist woods, forest clearings, stream border vegetation, shrub and grassland communities during the wet season. They frequent permanent ponds, pools along streams, springs, marshes, lakes, and reservoirs. Populations may also be centered around long-term seasonal ponds and marshes that hold water through the late-summer larval metamorphosis season. Such sites must also have rodent burrows or moist debris pile retreats nearby in which both adults and metamorphs can estivate during the period between breeding habitat dry-out and the beginning of the fall–winter rainy season.

Red-legged frogs exhibit a great wanderlust. A segment of both the adult and juvenile population at a given breeding site occasionally roam through acceptable habitats in the adjacent countryside while the rest of the population "stays home," especially if the breeding habitat is a permanent wetland. The junior author has found individual California Red-legged Frogs over a mile from any known breeding pond, and two recent radio-tracking studies in coastal areas north and south of San Francisco Bay recorded overland movements of similar distances. During the rainy season such movements may take place through upland habitat, but during the dry late-spring through early-fall period they are usually confined to moist drainages. The main purpose of these movements would appear to be to forage for large invertebrate and small vertebrate prey. Large ranid frogs are first-class predators, and no predatory species can afford to congregate in large numbers in one small area all year. This reasoning is supported by the findings of Fellers and Kleeman (2007), who report that 66 percent of female and 25 percent of male *R. draytonii* made annual movements from permanent breeding ponds to nonbreeding areas. Such movements are also the means by which new populations may become established, because it only takes a few wide-ranging frogs of opposite sexes to by chance discover an uninhabited wetland and colonize it.

Red-legged frog breeding begins as early as January for *R. draytonii* in the southern part of its range and as late as May for *R. aurora* within its range in California. Eggs are encased in large, irregular grapelike clusters that may be as large as 24.5 cm (10 in.) in diameter. Fairly stiff supports such as grass and weed stalks, stems of sedges, cattails, and submerged branches of willows are used for egg cluster attachment. Larval development may require as long as seven months, perhaps because of the low temperature at some breeding sites and the large size attained by the larvae before transformation. In coastal central California metamorphosis normally occurs in July or early August for the California Red-legged Frog but usually a month or more later for both species in the northern portion of the North Coast Range. The overwintering of larvae may also occur in both species. Two former students of the junior author, Hillary Hodge and Jeff Alvarez, independently observed pre-metamorphosis larvae in February in ponds in San Mateo and Contra Costa counties. These went on to transform in midspring. They may have been the result of a late spawning followed by slow larval growth due to rigorous competition from larger, early-hatched larvae.

Metamorphs remain at permanent wetland breeding sites until the first fall rains moisten upland understory areas, making them osmotically safe for foraging and dispersal. At a small artificial pond at which the junior author documented spawning and development of *R. draytonii* over an eight-year period, metamorphs completely vacated the site within

Plate 53. Adult male California Red-legged Frog.

two nights of the first substantial fall rains. At another habitat site, he and Caltrans biologist Richard Vonarb relocated 22 metamorphs in early November from rodent burrows and beneath ground-cover objects on a half-acre site that was to be graded to form a new pond. The closest permanent breeding habitat was about 100 m (325 ft) away. This large-scale upland dispersal of newly transformed young may be driven by the need both to find foraging sites away from the crowded pond shoreline area and also to escape from being preyed upon by large, cannibalistic adults. Observations such as this illustrate the need for preserving adequate upland area as well as a wetland breeding site when designing protection plans for this special-status species.

Like other large ranid species, adult red-legged frogs feed on a wide variety of large invertebrates and small vertebrates that they are able to capture with either their sticky, extensible tongue or open jaws. Two important prey items in summer and fall are the Pacific Chorus Frog (*Pseudacris regilla*) and their own metamorphs. In some ponds where few chorus frogs and large invertebrates are available in late summer, the entire energy transfer cycle from the sunlight that drives the photosynthesis of green algae to the cannibalistic adult frogs may take place within one animal species.

CONSERVATION NOTE: These comments are directed primarily to the California Red-legged Frog, which to date has experienced far greater human-wrought impacts than the Northern Red-legged Frog within its relatively small northern California range. This has resulted in the designation of the former as a Federal Threatened species, with the result that far more field studies and historical surveys have been directed to it than to the latter.

The California Red-legged Frog was once very abundant, occurring in the Sierra Nevada foothills up to 1219 m (4,000 ft), throughout the Great Valley, and in the coastal ranges from the northern Bay Area south well into Baja California, Mexico. Now it occurs only sporadically in some of these areas. One reason for this decrease was its large size, which made it a target of frog leg hunters for many decades. It is still often mistaken for a small American Bullfrog (*Rana catesbieana*) and illegally taken. The widespread drainage of marshes and ponds during the past century, especially in the Central Valley, has also greatly decreased its numbers.

A third reason for its decrease has a bit of literary history connected with it. California's namesake ranid was made famous by Samuel Clemens who heard about an original "forty-niner" named Leonidas W. Smiley who trained a frog to execute a long jump upon being pinched on its rump. He then carried it around in a box and challenged other "forty-niners" to obtain their own frogs for a high-stakes jumping contest. From this yarn came Mark Twain's "The Celebrated Jumping Frog of Calaveras County," and that frog was the California Red-legged Frog, *R. draytonii*. Today hundreds of spectators flock to Angel's Camp each spring to watch a continuation of this contest. However, the original participant is now very scarce in this area, for the introduced American Bullfrog has replaced the red-legged frog both in the jumping arena and in the latter's breeding habitats.

The introduction of the American Bullfrog into California in 1896 began perhaps the most severe impact to date to both the California and Northern Red-legged Frogs. It is a voracious feeder and grows to nearly twice the size of red-legged frogs, and the highly mobile American Bullfrog has now colonized many former red-legged frog habitats. Within a few years after arriving at a red-legged frog pond, the invading bullfrogs will have eaten all small to medium size red-legs. Eventually the remaining large adults die out, and only a American Bullfrog population remains. The American Bullfrog is also a carrier of the chytrid fungus that is devastating anuran populations throughout the world, and this may account for the rapid decline of *R. draytonii* in many of its former locations.

One of the few effective ways to control an American Bullfrog invasion of an isolated red-legged frog pond habitat is to slowly drain it in late summer when the weather and surrounding terrain are still very dry. These California native ranids have acquired the ability to estivate in rodent burrows and other retreats when seasonal ponds dry, but the introduced eastern species has no such adaptation. Instead, American Bullfrogs usually attempt to bury in the bottom mud, a behavior that they perform in their native range with the onset of cool fall weather. There they remain sequestered in the pond or lake bottom until the spring warming occurs. However, when a California pond is drained in late summer, they desiccate and die within the drying mud substrate. With the beginning of the next rainy season the rightful pond owners emerge from their burrow retreats and regain their home pond. When employing this American Bullfrog control method one must be sure that adequate upland retreats for red-legged frogs are present and that there is no late-developing crop of larvae of these protected species still in the water.

Unfortunately, the American Bullfrog is not the only exotic aquatic species that has compromised both *R. draytonii* and *R. aurora* populations in California. Currently, over 50 species of freshwater fishes have been introduced into California waters, and many of these find their way into aquatic habitats that would otherwise be ideal breeding sites for both red-legged frog species. Impacts from such introductions range from foraging on egg clusters by species like the Common Carp (*Cyprinus carpio*) and Black Bullhead (*Ameiurus melas*) to direct predation on larvae and adults by Largemouth Bass (*Micropterus salmoides*). Perhaps the most destructive of all such introductions has been that of the Western Mosquitofish (*Gambusia affinis*). Millions of these harmless-appearing little fish are stocked annually in a variety of shallow, low-elevation freshwater habitats throughout California by local mosquito abatement districts. However, several studies have shown that they are very adept at picking at and destroying the external gills of newly hatched anuran larvae, including those of red-legged frogs. Many such habitats into which they are introduced are late-drying seasonal wetlands that retain water long enough for red-legged frog larval metamorphosis but also have a fall dry period that precludes permanent fish inhabitancy. However, mosquitofish are usually restocked in the spring and therefore present at or near the time of larval hatching.

In addition to eliminating American Bullfrogs from and preventing fish introduction into red-legged frog breeding sites, and preserving those that have as yet not been invaded, the best recovery effort for this species is to replace some of its many wetland breeding habitats that have been destroyed. The construction and planting of small ponds is a relatively easy and low-cost project. Recently the junior author and Caltrans naturalists Sid Shadle and Richard Vonarb successfully constructed two such breeding habitats, one of which was stocked with egg clusters and the other by wandering adult frogs (see plate 4, in the "Introduction: Amphibian and Reptile Distribution throughout California's Habitat Complex" section). Within three years both had approached carrying capacity for young adults.

CASCADES FROG — *Rana cascadae*

Ventral

PROTECTIVE STATUS: CSSC

IDENTIFICATION: ADULT body length ranges from 4.4 to 7.5 cm (1.75 to 3.0 in.) and is similar to the Northern Red-legged Frog in proportions and external structure, including the presence of lateral folds. The dorsal color is various shades of brown, usually with sharply defined inky-black spots that extend well down on the sides. There is dark mottling or banding on the legs. Vent coloring grades from cream to yellow or orange-yellow on the lower abdomen and underside of the hind legs. It differs from the Oregon Spotted Frog with which it sometimes coexists in being generally smaller and having longer legs with less webbing. Its eyes are also positioned directly laterally, and it has yellow rather than red color on the ventral body and legs. The **VOICE** is a low, grating, chuckling sound on one pitch, resembling that of *R. aurora* with four or five notes per second. **EGGS** are deposited in somewhat flattened, globular clusters at or near the surface in shallows of lakes, roadside ditches, deep pools, or in shallow ponds that may be seasonal. An egg mass found floating free in Emerald Lake, Shasta County, measured approximately 12.5 by 25.0 cm (5 by 10 in.) and contained 425 eggs. Each egg is contained within three jelly envelopes with an outer diameter of about 11 mm (0.04 in.). The ova are 2.2 mm (0.09 in.) in diameter and are very dark above and light cream below. **LARVAE** grow to about 5.0 cm (2 in.). The dorsal body is dark brown, and the upper sides have copper to pinkish flecks on a black ground color. The lower sides are golden, and the tail is finely speckled with pale yellow and copper.

Egg

RANGE: In California, populations occur in the Trinity Mountains, and Shasta and Lassen peak areas of the northern North Coast and Cascade Ranges.

NATURAL HISTORY: This is a montane species that frequents small streams, ponds, and lakes in coniferous forests, especially in areas with open glades and meadows. It is a rather sluggish frog, often allowing close approach. It is mainly diurnal, and when frightened it commonly attempts to escape by swimming, often underwater, to the opposite bank or to the same bank downstream. It may also dive and seek refuge on the bottom. When handled these frogs emit a garlic-like odor much like that produced by the Sierra Nevada Yellow-legged Frog (*Rana sierrae*) and the California Red-legged Frog (*Rana draytonii*) in the southern part of its range. Breeding occurs from early April to mid-July at higher elevations. Males may emerge as soon as ice begins to melt in the breeding pools. Egg clusters contain from 300 to 600 ova and are deposited in shallow pools and lake shoreline sites. They are not attached to plants or other support structures and are often on top of each other. Water temperatures at egg sites during the day may be quite warm, often about 20 to 24 degrees C (68 to 75 degrees F). In the short growing season of montane environments, shallow temporary pools provide warm conditions that hasten development of both eggs and larvae. About two and a half to three months may be required for larval development to metamorphosis, and they sometimes overwinter.

CONSERVATION NOTE: The Cascades Frog is now extinct at about two thirds of the formerly known population sites, and most of them are in the upper Sacramento River drainage. Field surveys by Mark Jennings and Marc Hayes revealed that it is still abundant in lakes and streams of the upper McCloud River system where few or no fish were present. Given the temporary recovery success for the Sierra Nevada Yellow-legged Frog through the eradication of Brook Trout (*Salvelinus fontinalis*) in several high Sierran lakes, fish predation, especially on eggs and larvae, may be the major impact here. A wide variety of introduced fishes has spread throughout the Sacramento River system during the past century. These plus the large number of hatchery-reared "put and take" trout that are annually stocked in these waters could very well be responsible for the mass extinction of the Cascades Frog in this area. In contrast to this situation, the upper McCloud River system has not suffered large-scale fish introductions. It is also one of the few remaining native trout areas and as such receives little or no hatchery trout planting by the CDFG. Here 20 of the 25 historic population sites still support the Cascades Frog.

OREGON SPOTTED FROG — *Rana pretiosa*

Ventral

PROTECTIVE STATUS: CSSC
IDENTIFICATION: ADULT body length ranges from 4.4 to 10.1 cm (1.75 to 4 in.). The dorsal body may be brown, dull olive green, or reddish, with varying numbers of usually black spots. It has a light-colored jaw stripe and bright reddish or salmon-colored sides and under-leg areas. Its head differs from that of the Red-legged Frog in that the eyes are turned slightly upward and it has a light-colored jaw stripe. Dorsolateral folds are usually present. In the young the reddish ventral color is faint or absent. The **VOICE** is a series of faint, rapid, low-pitched clicks that increase in intensity during a call. Calls last up to 10 seconds and are given both in air and occasionally underwater. **EGGS** are laid in globular clusters, 7.5 to 20 cm (3 to 8 in.) in diameter, which may contain anywhere from 150 to 2,000 eggs. The ova are enclosed in either one or two jelly envelopes. **LARVAE** mature at about 10 cm (4 in.) total length. They have long tails and a labial tooth row ratio of 3/4. The dorsal color may be black, dark brown, or greenish, and the ventral side has a silvery to brassy hue.

Egg

RANGE: In California it has been recorded at only five sites in the Great Basin Desert just below the Oregon border. From there it extends north through the Cascade Mountains of Oregon.

NATURAL HISTORY: This frog is usually found in conifer and subalpine forests where it frequents snowmelt pools in mountain meadows, beaver ponds, slow streams and rivers, and lakes. It generally shuns waters lacking vegetation. It may exhibit significant seasonal movements, especially if spawning sites are seasonal wetlands and adults and metamorphs must seek more permanent aquatic sites or rodent burrow retreats. The breeding season begins with males aggregating at a shallow-water habitat with stands of emergent vegetation. Females arrive later and choose laying sites that are sometimes so shallow that the upper part of the egg mass protrudes above the surface as the encapsulating jelly expands. The choice of a very shallow spawning site may be due to the comparatively warmer water found there, which will hasten embryo development. This advantage is often lost, however, by the failure of the eggs above the water surface to hatch because of desiccation or even freezing. Larval development to metamorphosis is about four months. Juveniles take three or more years to reach sexual maturity. Both juveniles and adults feed on a wide variety of insects and the Pacific Chorus Frog (*Pseudacris regilla*), which also breed in shallow, seasonal ponds.

CONSERVATION NOTE: The survival of the isolated populations of this frog in California are threatened by habitat destruction and introduction of nonnative game fishes into former permanent aquatic-breeding sites. The latter situation may have forced some populations to use shallow seasonal spawning sites, which then often dry before larval development is completed.

FOOTHILL YELLOW-LEGGED FROG *Rana boylii*

Ventral

PROTECTIVE STATUS: CSSC

IDENTIFICATION: ADULT body length range is 3.8 to 8.1 cm (1.5 to 3.2 in.). The dorsal body color is gray, brown, reddish, or olive, and is usually flecked and mottled with shades of brown-black. Its general coloration resembles the color of the stream bottom with which the frog blends well. The underside of the hindlimbs and lower belly are yellow and hence its name. There is a triangular light-colored patch on its snout. The dorsolateral folds are indistinct. The skin, including the eardrum surface, is granular. Its **VOICE** is a guttural, grating sound often followed by a "woof." It is rarely heard by humans, because it apparently calls mostly underwater. During calling the throat swells at the sides rather than to the front. **EGGS** are deposited in a grapelike cluster, with individual eggs usually firmly attached to one another and protruding slightly from the surface of the mass. The clusters may measure from 7.5 to 10.0 cm (2 to 4 in.) in greatest dimension and are composed of 100 to more than 1,000 eggs. The eggs are enclosed in three jelly envelopes. The outer one is 3.9 to 6.4 mm (0.15 to 0.25 in.) in diameter. The jelly covering is firm, and all envelopes are distinct. Ova diameter ranges from 1.8 to 2.3 mm (0.07 to 0.09 in.). They are black above and white or light gray-tan below. **LARVAL** total length is 3.7 to 5.6 cm (1.5 to 2.25 in.). The body is somewhat depressed, the eyes are well up on the top of the head, and the dorsal fin peaks near the body midpoint. There are four or more rows of labial teeth in both the anterior and posterior labia. The most common labial tooth row formula is 6/6 or 7/6. Larval dorsal color is olive-gray with rather coarse brown spotting, and the underside is pearly white.

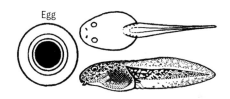

Egg

RANGE: It occurs throughout the North and South Coast Ranges south to the Transverse Range, across northern California to the west slope of the Cascade Range, and south through the foothills of the Sierra Nevada to Tehachapi Creek, Kern County. There are isolated populations in southern California in Elizabeth Lake Canyon and the drainage of the San Gabriel River. It occurs up to 1,830 m (6,000 ft) in the northern Sierra Nevada.

NATURAL HISTORY: *R. boylii* inhabits forest streams and rivers with sunny, sandy, and rocky banks, deep pools, and shallow riffles. It is a diurnal species and may be found basking on the shore or on rocks in streams. It will often sequester beneath loose rocks or in sediment on the stream bottom, and when the current is swift it takes refuge at the downstream base of large boulders. Along intermittent stream courses large aggregations have been observed at locations

Plate 54. Foothill Yellow-legged Frog.

of quiet persistent water during the dry season. Individuals may be abroad throughout the year except in cold weather.

Breeding occurs from mid-March to early June, usually after the high-water mark in streams is past and less sediment is being transported. At any one site spawning lasts only a few weeks. The eggs are attached to stones in stream beds, often partly wedged in between them, or to the stems of sedge or other vegetation. The tadpoles' mottled coloring blends with sand and gravel of stream bottoms, which make it a challenge to detect them. Metamorphosis usually takes place over a few days and temporarily floods the creek or stream shore area with hundreds of young that appear reluctant to enter the main stream current flow or to forage away from the water source. These metamorphs' concentrations are reminiscent of those seen in the Western Toad (*Bufo boreas*). Adult food consists of a wide range of insects, especially those that inhabit water courses. They also eat snails, which may be very abundant in algae-laden streams.

CONSERVATION NOTE: Although *R. boylii* is still moderately abundant in coastal drainages north of Monterey Bay, numerous historic populations appear to have been lost on the western slopes of the Sierra Nevada, especially in the southern part of this range. One contrasting feature between the Sierran and coastal drainages that may be responsible for this difference is water flow management. Fortunately, few coastal streams have as yet been dammed, and thus their spring flows follow a somewhat normal period of gradual subsidence. It is during this period of reduced flow that the Foothill Yellow-legged Frog breeds. It follows this same pattern in Sierran streams, most of which have been dammed upstream from this frog's habitat. However, periodic water releases from upstream reservoirs once spawning has begun temporarily return a stream to its high-flow state which shears eggs from their attachment sites and washes them and newly hatched larvae downstream.

A second ongoing impact during the past century has been the introduction of nonnative predatory game fish species, mostly of the bass–sunfish family (Centrarchidae), into many of the Sierra Nevada foothill streams and rivers. Most of these introductions have been sanctioned by the CDFG, including that of the Redeye Bass (*Micropterus coosae*), which was stocked in numerous drainages between 1954 and 1969. This is an especially voracious predator that is well adapted to foothill stream life and whose appearance in California coincides closely with the reduction of the Foothill Yellow-legged Frog in the lower drainages of the Sierra Nevada.

SOUTHERN MOUNTAIN YELLOW-LEGGED FROG	*Rana muscosa*
SIERRA NEVADA YELLOW-LEGGED FROG	*Rana sierrae*

Sierra Nevada Yellow-legged Frog

Ventral

PROTECTIVE STATUS OF *R. MUSCOSA*: FE
PROTECTIVE STATUS OF *R. SIERRAE*: ST
IDENTIFICATION: Both species are very similar in appearance and until recently were considered one species. **ADULT** body length ranges from 5.0 to 7.6 cm (2 to 3 in.). The skin, including the eardrum, is smooth to tuberculate. The dorsolateral folds are obscure. In *R. sierrae* the heel of the hind foot usually does not reach the nostril when the leg is extended forward, whereas that of the *R. muscosa* does. Inner and outer hind foot tubercles

are present, and the toes are fully webbed and have slightly expanded tips. The dorsal color is usually yellowish brown, occasionally reddish brown, with brown or black spots or lichenlike markings. The eardrums are usually colored like the rest of the head. When handled, its skin secretions smell like garlic. The **VOICE** of both species is a faint clicking or rasping sound, sometimes ending in a loud "woof." However, the mating call of *R. sierrae* differs from that of *R. muscosa* in having transitions between the pulsed and noted sounds. Both species usually call underwater, and there are no vocal sacs. **EGGS** are deposited in globular clusters measuring approximately 2.5 by 3.7 cm (1 by 1.5 in.). The ovum is 1.8 to 2.6 mm (0.07 to 0.1 in.) in diameter and is encased in three gelatinous envelopes with an outer diameter of 6.4 to 7.9 mm (0.25 to 0.31 in.). **LARVAE** grow to over 5 cm (2 in.) and have a labial tooth formula of 2/4 to 4/4. Larval dorsal color is brown with a golden tint, and dark spots are often present on the tail musculature and fins. The ventral surface of the body is faintly yellow. In the Sierra Nevada and the San Bernardino Mountains, tadpoles of both species require more than one summer to transform.

RANGE: *R. sierrae* once ranged throughout the Sierra Nevada above 1,231 m (4,000 ft.) to southern Tulare County and east to Dexter Creek at Crooked Meadow about eight miles northeast of Crestview, Mono County. Disjunct populations of *R. muscosa* were also numerous in southern California on Palomar Mountain and in the San Gabriel and San Jacinto mountains, and still are present in the San Bernardino Mountains. Over the past three decades the Sierra Nevada Yellow-legged Frog has disappeared from about 95 percent of its historic range, and seven out of 10 recent re-introductions at sites where it once occurred in Sequoia and Kings Canyon national parks have failed. *R. muscosa* is also on the brink of extinction in Southern California.

NATURAL HISTORY: The Sierra Nevada Mountain Yellow-legged Frog is the only ranid found in the highlands of the Sierra Nevada in California. It occurs from the chaparral belt to coniferous forests and high mountain meadows, almost to timberline. It inhabits sunny river banks, creeks, meadow streams, isolated pools, and lake borders in the high Sierra Nevada and stream courses. The Southern Mountain Yellow-legged Frog frequents streams that range from rocky, steep drainages to those with a gentle gradient, marshy margins, and sod banks. Large clear pools up to three feet deep are especially favored.

These are diurnal, highly aquatic species that are usually found within one to three jumps of water. When irritated they produce a musky garlic-like odor and sometimes flatten the body, retract the eyes, and place the forefeet at the sides of the head. *R. muscosa* is active throughout the year

Plate 55. Sierra Nevada Yellow-legged Frog in a snowmelt meadow.

along streams in southern California, but at high elevations in the Sierra Nevada the annual activity period of *R. sierrae* may be only three months. Here both larvae and adults may spend six to nine months beneath ice-covered waters. Adult deaths may occur in waters that are low in oxygen content. The tadpoles, however, are more tolerant to low oxygen levels. Body heat is at a premium for high-altitude ectotherms, and these frogs attempt to maximize their body temperature at nearly all times during the day by basking and seeking the warmer locations in water and on land. By such behavioral thermoregulation they are able to elevate their body temperature to as much as 13 degrees C (23 degrees F) above the surrounding air and 18.6 degrees C (33.5 degrees F) above the average water temperature.

In the high Sierra *R. sierrae* breeding occurs in late May, June, and July, and may begin before meadows are free of snow and when ice is still present in parts of streams. Rapid embryo development and larval growth are essential in snowmelt pool spawning sites. Here it is driven by elevated daytime water temperatures accrued to these shallow ponds by high summer solar radiation. In Sierran lakes larvae overwinter and at very high elevations may not transform until their third or fourth larval year. Larval tolerance to the low oxygen levels in the water beneath the ice in these lakes is a key factor in their survival. *R. muscosa* breeding commences when high stream flows subside. Fast larval development is also crucial for populations in seasonal creeks to avoid desiccation when pools dry in late summer. Metamorphs and adults feed on the wide variety of aquatic terrestrial arthropods that inhabit wetland areas.

CONSERVATION NOTE: The Southern Mountain Yellow-legged Frog, along with the Arroyo Toad (*Bufo californicus*), are the only California anurans that have been granted the dubious honor of being listed as endangered under the Federal Endangered Species Act. Unfortunately, this designation is well deserved, because their numbers have diminished greatly during the twentieth century. What has turned out to be perhaps one major problem began in 1871 at, of all places, the campus of the University of California, Berkeley. It was there that the former "California Acclimatization Society" established a small fish hatchery and imported 5,000 Brook Trout (*Salvelinus fontinalis*) eggs from New York State. In 1872 the California Fish Commission began planting this eastern species in California lakes and rivers. Its successor, the CDFG, has continued the process to date, with as many as 1.7 million Brook Trout fingerlings being stocked annually in the mid-twentieth century. One reason that CDFG has been so enamored with this fish is that unlike our native trout species, which require permanent lake feeder creeks in which to spawn, the Brook Trout does not and instead can utilize shoreline gravel beds of lakes for egg deposition. Many of these high-elevation lakes were originally void of native fishes but apparently had thriving Sierra Nevada Yellow-legged Frog populations. However, they soon became overpopulated with stunted Brook Trout, which devoured most available small food items, including frog eggs, and the rest is history. In addition, the nonnative Brown Trout (*Salmo trutta*), the native Rainbow Trout (*Oncorhynchus mykiss*), and even California's "official state freshwater fish," the Golden Trout (*Oncorhynchus mykiss aguabonita*), have also been stocked in great numbers in streams once inhabited by *R. muscosa*, but their affect on this species is unclear. A recent effort to eliminate introduced fishes from a few lakes where these endangered frogs still persist has been encouraging.

Among other potential impacts such as airborne pollution, toxins, and silt from recreational places and mining, the most potentially devastating one now appears to be the introduction of the chytrid fungus *Batrachochytrium dendrobatidis* (Bd) into Sierra Nevada Yellow-legged Frog lakes. This fungus destroys the ability of amphibian skin to absorb water, resulting in lethal dehydration in species that spend much of the year in moist terrestrial habitats void of standing water (see "Introduction: The Ongoing Loss of the Habitats and Numbers of California's Amphibians and Reptiles" section). In the seven re-introduction sites that failed in Sequoia and Kings canyons, healthy introduced frogs were infected with Bd and died within two years. These discouraging findings suggest that the fungus either persisted in these lakes after the initial frog die-off or were transported there since that time by humans and/or birds. However, recent findings by Roland Knapp at the Sierra Nevada Aquatic Research Laboratory of the University of California have revealed that some frogs in fungally infected areas are living with low-level infections, suggesting that frogs in recovery populations can survive future exposure, unlike frogs with no prior exposure to Bd. Also encouraging are the observations, by a colleague of the junior author, of numerous *R. sierrae* along a stream in the Palisades Glacier above Big Pine, California in 2009 and 2010.

NORTHERN LEOPARD FROG
NATIVE AND INTRODUCED
SOUTHERN LEOPARD FROG
INTRODUCED

Rana pipiens

Rana sphenocephala

Northern Leopard Frog
(two color phases)

PROTECTIVE STATUS OF *R. PIPIENS*: CSSC

IDENTIFICATION: The body lengths of mature **ADULTS** of both species varies from 5.1 to 11.1 cm (2 to 4.75 in.). The dorsal color may be green or shades of brown. *R. pipiens* has numerous well-defined round or oval dark spots outlined in pale color or white on its sides, whereas *R. sphenocephala* has only a few spots with no light outline. *R. sphenocephala* also has a distinct light spot in the center of the tympanum. Both have undersides that are white or cream-colored, well-defined tympanum folds that do not angle inward toward the rear, and a light line along the upper jaw. Juveniles have little or no spotting. Breeding males have darkened, swollen thumb bases and loose skin between the jaw and shoulders where the paired lateral vocal sacs protrude. Its **VOICE** is a low rattling "snore." Choruses are a series of moaning, chuckling, and grunting sounds. It often emits a squawk when jumping into water, and may give a loud scream when grabbed. **EGGS** are laid in firm round clusters ranging from 5 to 15 cm (2 to 6 in.) in diameter that may contain up to nearly 6,500 ova. The latter are surrounded by either two or three jelly envelopes. **LARVAE** mature at about 8.7 cm (3.5 in.) total length, and the larval tooth rows are usually 2/3. They are dark brown, olive gray, or gray on the top and sides of the body. The ventral area is an off-white, and the coiled intestine can be seen through the skin.

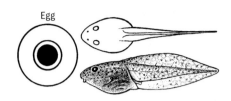
Egg

RANGE: Northern Leopard Frogs were introduced east of Red Bluff at Battle Creek Meadows (near Mineral), Tehama County, in 1918, and at Lake Tahoe, where they were brought in to supply the demands of restaurants. They have been reported at nine sites in California including Santa Catalina Island and are expected to disperse throughout the Great Valley canal system. Historically, presumed native populations occurred in Owens Valley and the northeastern corner of the state. These have been severely reduced, but it is not known if total extinction has occurred. The Southern Leopard Frog was most likely introduced in Riverside County in the early 1990s in a shipment of American Bullfrogs, fish, and crayfish from Louisiana. It spread throughout the Santa Ana river and is now common in the Prado Flood Control Basin.

NATURAL HISTORY: These are two of the most widely distributed anurans in North America and two of only six amphibians that have been introduced into California and have become established. They are a highly adaptable species occurring in a great variety of habitats, including grassland, brushland, woodland, forests, and farmland. In upland and some northern areas *R. pipiens* is often a "meadow frog." It is very cold-adapted and outside of California is widely distributed in areas with winter snow. When not breeding it may move quite far from water into moist fields and pastures.

Despite the Northern Leopard Frog's being an introduced species, Californians may be more familiar with it than any other amphibian. Indeed, they may know more about its internal anatomy than their own! This is because for many decades it has been the "laboratory frog" for high-school and college biology classes. And for children who grow up east of the Sierras these are often their first and perhaps only wild-animal pet. This large group includes the junior author, who reared several Northern Leopard Frog tadpoles or "pollywogs" to adulthood in an old dishpan in the basement window well of the family home in Wisconsin. The project ended sadly for this budding herpetologist but happily for the frogs when on one warm, rainy late-summer night their strong new hind legs sent them soaring over the well wall and on to freedom. However, they had more than adequately paid for their larval room and board by instructing their keeper in one of the most dynamic events in vertebrate natural history—anuran metamorphosis.

For breeding sites these frogs prefer cattail-populated ponds and swamps or other marshy areas, grassy overflows, and shallow, slow streams. They generally shun ponds lacking vegetation. The spring reproductive season begins with males moving to spawning sites and calling. Amplexing pairs will often move to areas where egg clusters have already been deposited and add their own clutch to the communal laying effort. Embryo development takes from one to three weeks, and larvae metamorphosis occurs in two to three months after hatching. Despite their transcontinental distribution, populations of both species have declined in many areas of the United States and Canada. Suspected or confirmed impacts include overcollecting for the biological supply and fishery bait trades, increase in agriculture pesticide use, and the acidification of spawning habitat (a pH of 4.8 suppresses egg development). To date in California neither of these introduced species have spread as widely and affected native anurans as has their big cousin, the American Bullfrog (*Rana catesbeiana*).

CONSERVATION QUESTION: The CDFG presents us with quite a challenge here. It lists *R. pipiens* as a CSSC but does not designate the historical areas where it may have once occurred in this state. All eight currently known populations have been introduced, and species that have been introduced from large populations elsewhere traditionally are not accorded any protective designation. Possibly to offset this problem, the CDFG also allows all fishing license holders to take four Northern Leopard Frogs and four Southern Leopard Frogs from the wild. Certainly an intensive search for native *R. pipiens* populations is needed, so that molecular analysis may be used to see if any found are indeed different from recent introductions. With respect to the latter, the authors believe that whenever possible, the removal of any introduced amphibians should be attempted, especially when such populations are small and localized. Supporting this approach to introduced Northern Leopard Frog populations is that in its current native range dead frogs occasionally turn up with a condition known as "red-leg." It is caused by various bacteria, particularly *Aeromonas hydrophila*, and it is not known whether it may spread to California native ranids. However, the risk that it might would seem to justify attempted elimination of introduced populations in California.

RIO GRANDE LEOPARD FROG
INTRODUCED
Rana berlandieri

IDENTIFICATION: ADULT total body length is 5.7 to 11.4 cm (2.25 to 4.5 in.). It resembles the Northern Leopard Frog but is generally much paler, with dorsal color varying from grayish brown to green. The dorsal spots are lighter and less clearly edged with light color than those of *R. pipiens*. There are usually dark thigh reticulations, and its dorsolateral folds are segmented in front of the groin and deflected inward. Males have a swollen and darkened thumb base and paired external vocal sacs. Its **VOICE** is a short, guttural trill or rattle, given singly or rapidly repeated. **EGGS** of this species are similar to those of the Northern Leopard Frog, but the clutch size is smaller (500 to 1,200). **LARVAE** closely resemble those of the Northern Leopard Frog.

RANGE: A disjunct population of this frog exists in the Colorado Desert in the Imperial Valley and the adjacent southwest corner of Arizona.

NATURAL HISTORY: This is a frog of both the grassland and woodland. In its primary population area of southwestern Texas and northeastern Mexico it frequents streams and their side pools, springs, arroyo pools, and more recently, stock watering tanks. Its expansion through the Imperial Valley was made possible through the extensive irrigation canal development there.

In these areas breeding is similar to *R. pipiens*, but in arid areas it spawns whenever the heavy annual rainfall occurs. Adult food consists of a great variety of wetland-associated insects.

LOWLAND LEOPARD FROG
Rana yavapaiensis

IDENTIFICATION: The snout-to-vent range is 4.6 to 8.6 cm (1.8 to 3.4 in.). It is similar to the Northern Leopard Frog but stockier with a more rounded head, shorter limbs, and slightly upturned eyes. The dorsolateral folds are usually broken into short segments in the posterior region and angled inward. The dorsum is tan, gray-brown, gray-green, or green, and the vent is yellow. There is a dark network on the upper rear legs. Its **VOICE** is a series of three to four guttural, chucklelike notes, sometimes described as a series of short, fast kissing sounds. A call may last three to eight seconds. **LARVAE** are similar to those of the Northern Leopard Frog.

RANGE: In California this species has been located only in the Colorado Desert in the Imperial Valley and San Felipe Creek. However, at this writing most populations may be greatly depleted or extinct.

NATURAL HISTORY: This frog occupies riparian areas in a variety of habitats ranging from deserts and grasslands to oak and oak–pine woodland. In these it usually stays close to a water source, which may include permanent, densely vegetated creek pools, overflow ponds, side channels of major rivers, and in drier areas, permanent stock tanks.

CONSERVATION NOTE: Within its small range in the southeast corner of California, populations of this frog have been seriously depleted or are now considered extinct, and the prospects for the survival of isolated populations in Imperial County are not good. Unfavorable agricultural practices, reservoir construction, and the introduction of predatory fishes, crayfish, and the American Bullfrog (*Rana catesbeiana*) are associated with the decline of this species throughout the desert Southwest.

AMERICAN BULLFROG
Rana catesbeiana

INTRODUCED

IDENTIFICATION: This is the largest frog in North America, with an **ADULT** body length range of 8.7 to 20 cm (3.5 to 8 in.). There are no dorsolateral folds, and the skin is usually relatively smooth in young frogs but sometimes roughens in older ones. The eardrums are as large as or larger than the diameter of the eyes in males but usually not greater than the eye in females. The dorsal body color varies from greenish to olive to almost black, often with black spots or mottlings of dull brown. The green base color is especially pronounced on the hindlimbs, which have dusky crossbars, and the throat is yellow. Juveniles often have more well-defined dark markings. The **VOICE** is a deep-pitched, hoarse, and bellowing "jug-o-rum," or "more rum" and is given only by males. When jumping into the water from shore some adults and most juveniles emit a loud squawk or chirp. **EGGS** are deposited in a disk-shaped layer at the surface, measuring up to 6.5 cm (2.5 in.) thick to 50 by 150 cm (3 by 5 in.) wide. The average number of eggs per egg mass is about 7,000, but it may contain over 20,000. The average jelly envelope diameter is 7.5 mm (0.3 in.), and that of the ova is 1.3 mm (0.05 in.). Egg color is black above and white or cream below. **LARVAL** total length is about 16.2 cm (6.5 in.). The dorsal body color is olive green with dark speckles on the body and fins. The ventral body is white or pale yellow but without a pinkish iridescence as in Red-legged Frog larvae (*R. aurora* and *R. draytonii*). American Bullfrog larvae also lack dorsolateral fold glands, and the snout is more rounded than that of *R. aurora*. The labial tooth rows vary from 2/3 to 3/3.

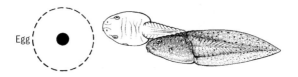

RANGE: The American Bullfrog is not native west of the Rockies, but due to widespread introductions it is now found in all but one of the lower 48 states. In California it now occurs in all of the geographic subdivisions but is restricted to widely scattered wetland sites in the three desert regions.

NATURAL HISTORY: This is a permanent water species and prefers sites with abundant submerged and emergent vegetation. Both adults and larvae usually overwinter buried in the bottom mud of such habitats. It rarely uses temporary aquatic sites except when juveniles are dispersing. In their native range American Bullfrogs are found along the shoreline of most lakes with dense inshore vegetation that apparently affords protection from large predatory fishes. The larvae are reported to have a bad taste that also protects against fish predation. Adults also have protective measures. One is a screamlike call that it often gives when initially seized and may startle a predator enough to let go. When grabbed it may also go into tonic immobility, a state in which it becomes limp and motionless as if dead. Predators that feed only on live prey may lose interest when a newly captured item suddenly becomes "dead." The squawk given by juveniles as they leap from the shore to water may also be

Plate 56. American Bullfrog in a California farm pond.

defensive. On several occasions the junior author has set off chain reactions of retreating, squawking juvenile bullfrogs along a pond shoreline by disturbing just a few in one small area. Juvenile red-legged frogs do not squawk when leaping, and this difference can help you determine which species has just splashed into the water when visibility is poor.

During the breeding season males typically move out into shallow water among emergent vegetation at night where they take up positions and give their advertisement calls. They assemble in groups, and the choruses that are formed attract sexually active individuals of both sexes. The females, however, stay close to shore until ready to lay eggs, whereupon they make their way to the areas occupied by the males, which outnumber females at breeding sites. Amplexus is usually initiated when a female approaches a male and touches any part of him with either her head or forelimb. Older females may produce two clutches in a single breeding season. In shallow ponds in Santa Barbara and Ventura counties larvae have been observed to require only six to eight months to metamorphose. However, in many Bay Area locations and northern counties larvae often overwinter. Metamorphs feed on the usual assortment of invertebrates found along a water course. However, as they grow to adult size, they attempt to capture and eat any prey that can be successfully swallowed. The countrywide list of food items that have been retrieved from American Bullfrog stomachs includes voles (*Microtis*), young Muskrats (*Ondatra zibethicus*), ducklings, sparrows, hatchling turtles, a young American Alligator (*Alligator mississippiensis*), fish, and a variety of snakes including the Western Diamondbacked Rattlesnake (*Crotalus atrox*) and a Eastern Coral Snake (*Micrurus fulvius*). Most adult bullfrog feeding is done in shoreline areas, and once a prey item is grasped, it jumps into the water to swallow it.

Given this broad-spectrum menu, the American Bullfrog heads the list of North American amphibian feeding generalists. Its ability to easily adapt to a wide range of permanent water habitats, including the many created by humans in the past two centuries, also qualifies is as a top habitat generalist. When these two features are part of a species' makeup, it can usually thrive within most regions of a country, and the American Bullfrog certainly has. Along with the "laboratory frog" (the Northern Leopard Frog), the American Bullfrog is perhaps the most recognized amphibian in the United States. For many folks in states east of the Rocky Mountains, the male's familiar "jug-o-rum" call note is the "sound of summer." The junior author, a transplanted midwesterner, has pleasant memories of being lulled to sleep on the screened porch of a Wisconsin lake cottage by male American Bullfrogs announcing their territorial position along the shore. This frog also has the dubious distinction of being considered a game animal in most states, including California, which until recently restricted the number that a fishing license holder could take.

Frogging, as this form of hunting is called, relies on the bullfrog's reclusive daytime behavior giving way to open exposure along shorelines edge areas at night. It is also easily transfixed by a flashlight beam and in such a state easily approached and speared with a three- or four-prong "frog gig" on a long, light pole. Besides the "sport" of the hunt, the other reward are the legs, by far the largest among North American frogs. Made famous in French cuisine, frog (hind) legs became a standard wild game food in eastern and midwestern states. The hind legs of an adult male American Bullfrog may be as large as those of a frying chicken, and amphibian muscle is also "white meat." Thus, this is the only amphibian that is legally hunted and raised for food in the United States today. As for its taste, it is reminiscent of chicken but relatively bland. The most common method of preparation is simply to coat the legs with flour and sauté in a light layer of butter or olive oil. Large legs tend to be stringy, however, and often marinating in a lemon juice, pepper, and salt mixture for several hours in the refrigerator is helpful. This process also takes the kick out of the legs—literally. The anaerobic nature of amphibian muscle makes possible contractions for hours after connections with the body's central system have been severed. Initial warming can briefly increase muscle enzyme activity, and occasionally a frogger and his or her family suffer a rapid appetite loss after viewing their potential meal kicking its way around the frying pan!

It was this promise of bigger and better frog legs that prompted the first introduction of the American Bullfrogs into California in 1905. As noted earlier, the beginning of the twentieth century was a period of mass introduction of eastern U.S. freshwater fish species into this state by the California Fish Commission, supposedly to enhance what was viewed as a minimal fish fauna, and the bullfrog was apparently swept along with this tide. At first it seemed to be a fortunate situation for our own native large-pond species, the California Red-legged Frog, whose numbers were already declining due to unregulated frogging for commercial markets. Because the hind legs of adult American Bullfrogs are nearly twice as large as those of *R. aurora*, froggers began to concentrate on the expanding populations of these big frogs throughout the state. Their quest for quantity was soon dampened by the lesser taste quality of bullfrog flesh compared with that of the red-legged frog, and both species continued to be hunted throughout most of the past century.

Whatever brief good the American Bullfrog's role as a buffer species may have performed, it was greatly offset by the severe and ultimately successful competition in nearly every red-legged and Oregon Spotted Frog permanent water habitat it invaded (see "Conservation Note" in California and Northern Red-legged Frogs section). Because many such sites are creek pools and farm ponds along water courses, the invasion and eventual complete takeover of such sites by Bullfrogs often occurs like a chain reaction through a valley system.

The introduction of the American Bullfrog into native amphibian aquatic habitats has not been confined to western North America. A 2007 study revealed that this frog has been brought to Europe on at least 25 separate occasions, with two such introductions occurring since the 1997 European Union banned its importation. Only five of these introductions were to establish a local supply of frog legs for food. The rest are attributed to the release of pets and other non-food-related reasons. As a result, the American Bullfrog now occupies five countries in continental Europe and has been directly linked to the spread of the aforementioned deadly chytrid fungus (see the "Introduction: The Ongoing Loss of the Habitats and Numbers of California's Amphibians and Reptiles" section), which has in several locations been devastating to local amphibian populations. The American Bullfrog's role in spreading this fungus may also be another key factor in the demise of our native California Red-legged Frog and other western ranid species when bullfrogs invade their aquatic habitats.

However, this foreign invader has a weak spot in its otherwise formidable biological armor: it cannot cope with seasonal wetland habitats that dry completely by late summer. Red-legged frogs can, and thus they still persist in such habitats. This inability of the American Bullfrog to seek estivation sites and survive annual drought periods can be used to eradicate it from red-legged frog ponds that it has invaded. It will usually remain in the moist drying pond basin along with its late-developing larvae, where both will soon desiccate in the hot late-summer weather, while resident red-legged frog adults and metamorphs seek rodent burrow estivation sites until the fall rains refill the now American Bullfrog-free wetland.

Tongueless Frogs (Family Pipidae)

These rather flattened, smooth-skinned frogs are almost completely aquatic. The head is small and tongueless, and the eyes lack movable lids. The hind feet are fully webbed, and the front foot digits are long and thin. Adults of some species may reach 25.4 cm (10 in.). Some 30 species occur in sub-Saharan Africa, Panama, and northern South America. One of the best-known members of this family is the Surinam Toad (*Pipa pipa*), in which the young develop in capped pits in soft skin on the mother's back and emerge as tiny froglets.

AFRICAN CLAWED FROG   *Xenophus laevis*
INTRODUCED

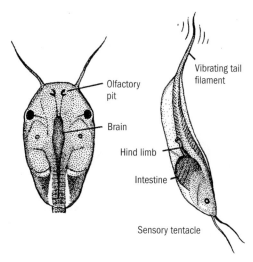

Figure 26. Larva of the African Clawed Frog. Dorsal head view (left) shows a tentacle on each side of the toothless mouth; full body view (right) reveals its finless, filament-like tail.

IDENTIFICATION: ADULT total body length is 5.1 to 14.3 cm (2 to 5.6 in.). The head and body are flattened, and the skin is smooth, except where ridges of the lateral line system have a "stitched" appearance. The head and eyes are small, and there are no eyelids or tongue. The hind feet are large and fully webbed with sharp black claws on the inner toes. The hind legs are much larger than the front, whose long digits are not webbed. The body color ranges from ochre to brown above with dark spots, blotches, or mottling. The whitish ventral side may or may not have dark spots. Males are smaller than females and lack protruding cloacal claspers. They have small dark tubercles on the forelimb and digits during the breeding season. Its **VOICE** is a two-parted short, oscillating trill, sometime given over 100 times per minute. The **EGGS** are encased in three jelly envelopes and are attached singly or in pairs on aquatic plant stems. A single clutch may number several thousand ova. The larvae are unlike those of any other anuran found in North America. There are no labial teeth, but instead a tentacle extends from each side of the soft mouth. The tail is a thin, finless filament, and the transparent body allows a view of the internal organs (figure 26). The total length ranges up to 3.7 cm (1.5 in.).

RANGE: Currently populations occur in Santa Barbara, San Diego, Orange, Riverside, Los Angeles, San Bernardino, and Kern counties as a result of numerous careless unsanctioned introductions.

Plate 57. African Clawed Frog.

NATURAL HISTORY: In its natural habitat, the high plains of East Africa, it inhabits permanent lakes, ponds, marshes, and more recently reservoirs in arid and semiarid areas. It is the most aquatic anuran now occurring in North America and seldom ventures out of water.

It breeds nearly all year in California, where April and May appear to be the peak months. Males assume a pelvic amplexus position while females deposit eggs on submerged stems. The larvae are filter feeders that forage on microorganisms suspended in the water. They hang head down in the water column and move slowly, propelled by the rapidly beating tip of their tail. They are capable of removing suspended particles equal in size to the smallest that human-designed mechanical sieves can filter. High concentrations of larvae can virtually sterilize a confined segment of water. Adults feed on a wide variety of aquatic invertebrates plus small frogs and fish. At some locations the latter group includes the Unarmored Threespine Stickleback (*Gasterosteus aculeatus williamsoni*; FE, SE, FPS), the Arroyo Chub (*Gila orcutti*; CSSC), and in Orange and San Diego counties, the Tidewater Goby (*Eucyclogobius newberryi*; FE).

CONSERVATION NOTE: The African Clawed Frog was imported to the United States by various medical laboratories in the 1940s after the discovery that the ovaries of the female will begin producing eggs when a small sample of urine from a newly pregnant human is injected into its body. This test has long since been replaced with a litmus paper–type test that detects the presence of a pituitary gland hormone that is secreted upon the onset of pregnancy. However, the African Clawed Frog proved to be an easy "lab frog" to maintain, and thus it was used in other laboratory work, which prompted further imports. Added to this use was its widespread appearance in pet stores, and thus the stage was set for the careless and currently illegal release of excess laboratory stock and unwanted pets into local aquatic habitats. Fortunately, these have to date occurred only in habitats that are not part of a large aquatic system such as the Sacramento–San Joaquin Delta. A close call along these lines occurred in the 1960s when it was introduced into ponds on the UC Davis Campus, only a short distance from a drainage that leads to the Sacramento–San Joaquin Delta. Fortunately, enlightened individuals at that institution eradicated these populations in the late 1970s. The fully aquatic nature of this species is a fortunate one, because it essentially negates the chance of upland wandering to new sites as seen in ranid frogs. However, it has been observed to retreat from sites when pond-drying is used to attempt eradication of this species.

Unfortunately, the story of this introduction does not end on the preceding encouraging note. In 1998 the chytrid fungus was identified by an Australian herpetologist, Dr. Gerry Marantelli, as the source of the disease that destroys the ability of frog skin to function as a respiratory or water absorption organ. This is affecting anurans on all continents where frogs occur except Africa. Then in 2002 he and colleagues discovered how this killer fungus had been spread literally around the world. It is believed to have been carried by imported African Clawed Frogs. These were originally all females for pregnancy testing and thus were the amphibian counterparts of the famous human female disease transmitter, "Typhoid Mary." Given these recent findings, all sales and imports of this species should be immediately banned in this country, and every possible means to eradicate existing wild populations should be undertaken.

REPTILES

Taxonomy, Anatomy, Physiology, and Behavior of Reptiles

In our introduction to the amphibians we highlighted the landmark stage in vertebrate evolution some 400 million years ago, when a group of shallow-water fishes acquired anatomical and behavioral features that enabled them to explore shoreline habitats and ultimately give rise to the first tetrapod vertebrates. In contrast to this scenario, the evolution of reptiles from what was most likely a common ancestral form for both reptiles and amphibians entailed no such dynamic transformation but instead progressed as a gradual accumulation of features that allowed many members of this new group of terrestrial vertebrates to no longer be dependent on aquatic habitats for any phase of their life history.

One cannot begin a discussion of the class Reptilia without recalling that this was the first truly dominant group of terrestrial vertebrates. This dominance is most evident in the paleo–natural history of the Archosauria. Except for the descendants of one order (Crocodilia), all other major groups within this group are extinct. Even so, many species of the group collectively referred to as dinosaurs are well known because of their starring roles in films and books, most of which feature large carnivorous species. However, it was the many, less popularized herbivorous forms that provided a wide, primary consumer base for the flesh-eating secondary consumers, thus establishing an intraclass food web that has been duplicated only once since by the mammals. The "Age of Reptiles" lasted for some 183 million years, whereas the "Age of Mammals" is only about 65 million years young.

Of the approximately 20 currently recognized orders of reptiles, only four are represented by living species, which at this writing number about 8,240. The smallest of these is the Rhynchocephalia, which contains only two surviving species, the tuataras (genus *Sphenodon*). These lizardlike reptiles have a maximum total length of about 75 cm (2.5 ft) and exist on 20 or more small islands off the New Zealand coast. They have many primitive features found in 150-million-year-old fossil reptiles from which the Tuataras evolved, and like all "living fossils" they afford a brief glimpse at reptilian life forms of the late Jurassic Period.

Another ancient order that is represented by about 23 species today is Crocodilia. These are the only reptile group of the great dinosaur subclass Archosauria that have survived the massive habitat changes that have occurred on this planet during the past 180 or so million years. The present-day alligators, caimans, and crocodiles have remained quite similar to the ancestral form that successfully filled the large, shallow-water, carnivorous feeding niche early on and never relinquished it to other reptilian or mammalian groups. Crocodilians do not currently occur naturally in California, and therefore further discussion of this group is not appropriate for this natural history guide. However, this order possesses some of the most advanced behavioral and physiological features of Reptilia, and we encourage you to explore it further in other texts.

The other two living orders, Squamata (lizards and snakes) and Testudines (turtles), do occur in this state and its boundary waters and therefore form the basis of this introduction. We present first a brief review of several features upon which the success of reptiles was and still is based. This is followed by a review of those features that are unique to our two California groups.

Some Major Anatomical and Physiological Advances of Reptiles

Reproduction

The evolution of the amniote egg by reptiles must be placed at the head of this list. Given the one prerequisite that it requires a somewhat moist substrate on or within which to develop, this egg negates the necessity of standing water for embryonic development. Instead, the reptile embryo, like those of the birds and mammals that inherited this reproductive mode, develops within a fluid-filled sac, the amnion, that substitutes for the pond, lake, or stream habitat of most amphibian embryos and larvae. The reptilian egg's large yolk store allows for complete development of the embryo to a fully functional miniature adult, ready to function on its own with no parental care once it hatches. With this advance in terrestrial reproductive biology, the larval life stage was no longer a part of reptilian, avian, and mammalian natural history. And except for the plethodontid salamanders, no terrestrial vertebrate group produces totally precocial young, which at hatching or birth are capable to completely fend for themselves with no parental care or protection.

Reptile eggs differ from the familiar bird egg in that the shells of most species' eggs are leathery instead of hard. Both types of shells are porous to allow for gas and water molecule exchange, but the flexible-shelled reptile egg permits compaction within the reptile body until it is laid on a moist substrate. Water is then absorbed to attain the final turgid egg form. The junior author discovered the Bunchgrass Lizard (*Sceloporus scalaris*) (not a California species) pictured here just as she was completing her clutch. In their fully expanded state, this group of eggs could never have fit inside the diminutive female's body. The reproductive mode seen here is the classic oviparous type, where the egg is laid and then develops under selected ambient conditions. It is nearly universal throughout the amphibians and most bony fish groups.

However, in several lizard and snake species the amniote egg minus its shell is retained in the oviduct, where the embryo undergoes complete development, and fully precocial young are born, not hatched. This is the ovoviviparous reproductive mode. A further extension of this occurs

Plate 58. Hatching kingsnake eggs.

Plate 59. Bunchgrass Lizard laying eggs in moist soil.

where the large yolk store is replaced with a system in which nutrients pass from maternal capillaries to embryonic capillaries that are separated only by a thin membrane. This is the viviparous reproductive mode found in most mammals and a few lizard, snake, and fish groups. These latter two developmental methods are usually seen in reptiles that live in northern latitudes or at high elevations where the habitat may not provide enough warm days to permit complete embryonic development in oviparous species before the next cold season begins. Although reptiles cannot produce significant body heat for internal egg incubation, daily warming through basking and other behavioral means can provide sufficient warmth for full embryonic development.

The laying of a shelled egg or its retention requires that fertilization occurs within the oviduct, and thus external fertilization, the norm in many amphibians, no longer exists in reptiles, birds, and mammals. Instead, reptiles copulate, after which spermatozoa either soon fertilize the newly ovulated ova or are stored in seminal receptacles in the oviducts. In some species sperm may be stored alive for periods of many months or several years.

Respiration

The complexity of the lung in different reptilian groups varies, but it is generally more efficient and structurally complex than that of amphibians. Of even greater importance is the presence of an expandable rib cage in all but the turtles. Crocodilians have even added a diaphragm-like structure that, along with the flexible rib cage, closely approximates the lung ventilation system of mammals. More efficient blood oxygenation in reptiles permits greater sustained activity than is seen in most amphibians.

Because the ribs of turtles are incorporated into the shell, expansion and contraction of a rib cage for ventilation is not possible. Instead specialized muscular membranes alternately expand and contract the body cavity area within the rigid confines of the carapace and plastron, and the changes in internal body pressure that this creates causes an expansion and contraction of the lungs. Aquatic turtles can also obtain oxygen from water drawn in and out of the mouth and cloaca. Here highly vascularized papillae in the pharyngeal cavity and accessory bladders on the dorsal cloacal wall allow for oxygen and carbon dioxide exchange as water is pumped in and out of these cavities. Soft-shelled turtles, some of which have been introduced into California, may acquire as much as 70 percent of their oxygen through the tough skin that covers their carapace and plastron. These alternate oxygen supply routes can support some turtles during long periods of submergence, including months of winter hibernation in the substrates of ponds, streams, and lakes.

Skin

One major result of the improved reptile respiratory system over that of amphibians is that except in the case of some turtles, the skin no longer must conform to the requirements of a respiratory organ. Thus, we see for the first time in land vertebrate evolution the appearance of epidermal structures such as scales and plates. Besides the obvious protection against abrasion these provide, especially in species that regularly burrow into soil or sand, the epidermis has given rise to structures such as the covering of the "horns" of horned lizards and the rattlesnake rattle. The avian feather and mammalian hair and fur as well as scales in both of these groups represent a continuation of this expansion of the role of the epidermis in the evolution of reptiles, birds, and mammals.

As in amphibians, the outer layers of the epithelium are periodically replaced by new underlying layers. Lymph eventually diffuses into the space between the old and new layers. As the two layers further separate, body coloration, particularly in snakes, becomes less vivid, and the normally clear scale over the eye becomes cloudy before shedding. In many lizards the old skin is shed in small fragments, and in most water turtles the covering of individual shell scutes is shed annually. However, in some lizards and nearly all snakes the old skin is shed in one complete piece that contains the embossed shape and pattern of all the scales along with subtle shading that represents the species color pattern. The recovery of such a shed skin is a real prize for the field herpetologist, because correct identification provides proof of a species' presence in a given area. Many times skins are shed when a snake or lizard is under a ground-cover object (old board, fallen log, loose rock), because it and the ground help hold the old loose skin as the animal crawls out of it. These are good sites to search for shed skins when conducting a reptile survey, but always remember to place all cover objects back in their original position.

Sensory Input
Vision
Vision is a dominant sensory input in reptiles. Many lizards have exceptionally good eyesight, and diurnal species can detect a moving ant at a distance of 2.5 m (8 ft) and will react to a human's approach at over 15 m (50 ft). However, movement of an object appears necessary for detection, and objects that are stationary before and while they are being viewed usually go undetected.

Most reptiles except snakes have a "lens-squeezing" mechanism for accommodation for near and far distance viewing that should be the envy of those persons who, with increasing age, are forced to wear glasses or contact lenses to compensate for the loss of lens elasticity. Whereas in the human eye contraction of the ciliary muscle releases tension on the small ligaments that hold the lens and allow it to assume a more rounded shape for close viewing, many reptiles increase the thickness of the lens by the direct squeezing action of this muscle on it. The snake eye employs the same focusing mechanism found in cameras, in which the lens is moved closer or further away from the retina to accommodate for near or far vision, respectively. This is accomplished by the contraction of the circular muscles of the iris that increases pressure on the vitreous body behind the lens, which then pushes it forward when near-field focus is needed. The eyes of most snakes and a few lizard groups are covered with a large, clear, convex scale called the spectacle, which prevents abrasion of the cornea below. Like human spectacles, these specialized scales gradually become scratched but are replaced with each shedding.

Many diurnal lizards and snakes have color-sensing cone cells in their retinas, while nocturnal species have primarily rod cells for dim-light viewing. One indication as to whether a snake or lizard is well adapted for seeing at night is the presence of a vertical pupil instead of the usual round one. In many gecko species there are several small pinholes present along the vertical slit when it is completely closed. Light passing through each forms complete images that are superimposed on the retina, giving these reptiles acute vision even in very dim light.

The ability to see colors probably aids diurnal insectivorous lizards in recognizing preferred insect prey. It also is important in social interactions, where the flashing of a colored belly or throat patch is a major part of a male's territorial display. Snakes such as the racers (*Masticophis*; *Coluber*) appear to locate their prey primarily by sight and often at some distance. They have very large eyes that are positioned well to the front of the head, presumably to permit more precise depth-of-field discrimination. Turtles also possess good vision. In addition to eyelids they and most lizards have a clear nictitating membrane that in turtles covers the eye when submerged. In lizards it cleans the eye surface as it sweeps laterally across the corneal surface in windshield wiper fashion. Given that many lizard species occasionally burrow into soil or sand, this is an important feature, as all who at one time may have gotten sand or sawdust in their eyes will agree.

Olfaction and Taste

These two senses are discussed together, because in many reptile species it is difficult to determine to what extent each is being used for a chemical analysis of the habitat at any given time. This is because in addition to the nasal olfactory epithelium, snakes, lizards, and turtles have paired Jacobson's organs. As previously discussed, they are also present in amphibians, where the two packets of chemoreceptors that compose this organ are at the confluence of the nasal passages and pharynx where they detect molecules in the inhaled air and from food in the mouth. This is also the state in which they exist in turtles, where they presumably allow these reptiles to taste food items before swallowing them. Hence in turtles the separation of smell and taste appears to be similar to that which we experience.

However, in snakes and many lizards the Jacobson's organs are no longer associated with the nasal passages but instead exist as paired cavities on the roof of the mouth well forward of the internal nasal openings. The forked tongue tip of those species depends heavily on this organ to pick up molecules, either from the air or substrate, and transfer them to the chemoreceptors that line these two depressions and have neural connections to the olfactory lobes of the brain. Thus, when a snake is vibrating its extruded tongue and then bringing it into contact with the Jacobson's organs while at the same time chemically sampling the same airborne molecules as it passes over its nasal epithelium, both its taste and olfactory senses are providing very similar information. The great advantage of the Jacobson's organs–forked tongue complex is that it appears to permit far better detection of non-airborne molecules. Snakes can track prey by the molecules left by their feet on the substrate, and lizards, like the whiptails, can locate buried invertebrates by "tasting" the soil surface above them. Male snakes also use this organ to track females, and both sexes employ it in the location of winter aggregation sites. Some species also identify the presence of predators with it. The characteristic defense posture of a rattlesnake to predatory kingsnakes is abolished if the transmission route between tongue and Jacobson's organ is negated, even though the kingsnake can still be seen.

As for the sense of taste as we humans experience it, it appears to be relatively poor in reptiles, except for those without the Jacobson's organs like alligators and crocodiles. Taste buds in reptiles are found mainly in the lining of the pharynx but are also present on the tongue of some lizard groups such as the iguanids. This suggests that the Jacobson's organ method of "tasting" is superior to that provided by tongue taste buds, and therefore the evolutionary selection for the latter has been greatly relaxed. Unfortunately, we humans do not have a definitive way of judging which is better, since these organs are not functional in adult mammals.

Hearing

This reptilian sensory input is by far the most difficult to analyze. Snakes, for instance, have no tympanic membranes (eardrums), and thus an initial assumption might be that they can't hear. However, tests with several species have shown that they are sensitive to both air- and substrate-transmitted sound in the low-frequency range between 100 to 700 kHz (cycles per second). They are able to do this because the columella, the single bone that in most amphibians, lizards, and turtles, transmits vibrations from the eardrum to the inner ear, is still present and attached instead to the snake's quadrate bone, which suspends the lower jaw. Insomuch as a snake's lower jaw is in contact with the ground much of the time, this appears to be a very efficient way to perceive substrate-borne vibrations, ranging from the subtle slaps of a hopping kangaroo rat's hind feet or the thundering hoofbeats of an approaching ungulate herd. In a true sense, snakes literally "have their ear to the ground." It is also significant that snakes swallow their food whole instead of chewing it first, otherwise mealtime would be a rather deafening affair.

Other living reptiles do have eardrums from which the columella transmits airborne and waterborne sounds to the inner ear. Turtles appear to have very good hearing that, as in snakes, is more acute at lower frequencies. On numerous occasions the junior author has attempted to move his canoe just a bit closer to a basking turtle whose eyes were shut tightly, only to have the slightest scrape of the paddle against the boat send it into the water. Geckos use chirps, clicks, and other sounds in breeding and territorial behavior, and adult crocodilians are also vocal and respond to the cries of their young. However, in general most lizards seem oblivious to sound. The junior author vividly recalls his first trip to the Mojave Desert with the senior author, on which they were both "packing pistols" loaded with dust shot to collect selected specimens for the Museum of Vertebrate Zoology at UC Berkeley. They happened to come upon a Common Side-blotched Lizard (*Uta stansburiana*), which allowed them to approach to within less than a meter (a few feet). This inspired an ad hoc experiment in which one of the guns was fired upward but at that short distance away from the lizard. Both experimenters flinched slightly at the sound of the discharge, but the lizard remained completely still. Did this lizard actually hear the loud sound, or was it simply not part of the spectrum of auditory stimuli that normally trigger a response?

The frequency of the shot sound and its overtones was certainly within the auditory sensitivity range of several desert lizards (900 to 3,500 mHz). This is also the range for sound made by rustling leaves, small insects, snakes crawling over a gravel substrate, and so on, and perhaps it is these sounds to which a lizard reacts. Whiptail lizards, for instance, have very large, prominent eardrums and may detect the presence of buried insects upon which they often feed by their digging sounds. The eardrum position on the head of whiptails and several other lizards is interesting to examine closely if you happen to have one in hand. The eardrums are positioned well below the main brain area, and if you view a lizard's head from one side with a bright sky or flashlight behind it, its tympanic membrane seems to glow. Then, if you take one hand and shade the opposite side of the head, the glow disappears, indicating that light was passing through the head from one membrane to the other. This little demonstration never ceases to amaze the novice herpetologist and usually brings to mind the old uncomplimentary remark that "he (or she) has nothing between his (or her) ears!"

Thermoregulation and Metabolism

Like amphibians, reptiles are ectotherms, deriving their body heat from sources outside of their bodies. Depending on their primary temperature sources and methods of warming, they may be classed as heliotherms or thigmotherms. The former obtain body heat by sun-basking, whereas the latter acquire it through contact with the substratum, air, water, or any of these. Diurnal desert lizards are primarily heliothermic, while burrowing forms and most nocturnal species are thigmothermic. Species that have a wide range of body temperatures through which they are able to conduct coordinated neuromuscular activities are referred to as eurytherms, while those that have a narrow range of operating temperature are called stenotherms. Aquatic turtles, most snakes, and some lizards, like the alligator

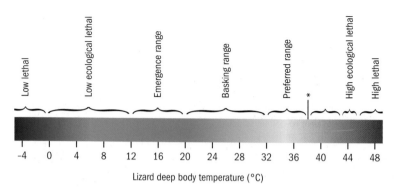

Figure 27. A thermoprofile of a typical temperate-zone heliothermic reptile, the Western Fence Lizard (Sceloporus occidentalis). The color bar denotes the rise in lizard body temperature from cool (blues) through preferred range (green–yellow) to hot (orange). ∗ = human deep body temperature.

lizards (*Elgaria*), are eurytherms, while diurnal heliothermic lizards and many tortoises are stenotherms. A stenotherm such as the Western Fence Lizard (*Sceloporus occidentalis*) has a body temperature range for normal activity from about 28 to 38 degrees C (82 to 100 degrees F), spanning 10 degrees C (18 degrees F), and on most sunny days maintains its body temperature within a few degrees Celsius of 35 degrees C (95 degrees F). In contrast, a Southern Alligator Lizard (*Elgaria multicarinata*) has a thermal activity range from around 11 degrees C (52 degrees F) to 36 degrees C (97 degrees F), an activity range of 25 degrees C (45 degrees F).

The ability to attain and maintain a body temperature in the range of 35 to 40 degrees C (95 to 104 degrees F) during sunlight hours is a major advance achieved by many reptiles. This is also the thermal range in which both bird and mammal species conduct their life processes and appears to be the optimal thermal spectrum for highly coordinated neuromuscular activity for animal life on planet Earth. As a graphic example of this "evolutionary thermal progression," we present in figure 27 the "thermal profile" of the Western Fence Lizard, compiled from field and laboratory thermal data obtained by the junior author over a five-year period. It is a companion graphic to figure 14, which presents similar body temperature data obtained in the field from the plethodontid salamander *Ensatina* by the senior author. Key features common to both profiles are the low lethal range, the temperature at which ice crystals form within cell membranes, and the low ecological lethal range, the thermal range in which enzymatic activity is slowed to such a degree that an ectotherm cannot perform coordinated body movements such as walking or righting itself if turned over and is therefore vulnerable to predation if discovered out of a protected retreat. Cellular enzymes are not destroyed, however, and normal function returns as body temperature increases.

Plate 60. Panting Western Fence Lizard.

Two added criteria for a heliotherm are the emergence range and basking range. The former represents the thermal span through which a reptile will emerge from its night retreat site and begin the basking process. Note that there is noticeable variation here, which seems to be influenced by a reptile's thermal experience over the past days or even weeks. Once emergence is initiated, basking begins, during which deep body temperature may rise as fast as a degree Celsius per minute. The next category, the preferred thermal range, is the key region for this salamander–lizard comparison. Of special note is that the salamander's high lethal temperature (35 degrees C or 95 degrees F) is the mean preferred temperature of the fence lizard.

As the fence lizard's body temperature is elevated above the preferred range, panting begins as an apparent effort to halt this rise. This is not the very fast, shallow respiration seen in dogs. However, the mouth does suddenly open as the lower panting range temperature is exceeded, and copious saliva is secreted over the buccal area to promote evaporative cooling. Panting is often accompanied by a posture seen in Plate 60, in which the body and toes are raised off the hot substrate. Other forms of emergency cooling in reptiles include the production by tortoises of frothy saliva, which flows over the thin neck skin and evaporates, and the wetting of the hindquarters with urine in many lizards. If limited evaporative cooling is unable to deter the body temperature rise of a reptile that for some reason is trapped in a thermally threatening situation, it will soon pass into the high ecological lethal range, in which, as in amphibians, enzymes begin to denature (break down), resulting in a loss of coordination similar to that seen in the low ecological range. However, here the enzymes that control cellular function are not just temporarily inactivated but instead permanently destroyed, and this dire state quickly leads to the high lethal, at which all cellular activity ceases and death ensues.

We are a classic example of a stenothermic endotherm who, like all other mammals and birds, continuously produce an excess of metabolic heat that is then precisely regulated for us by our autonomic nervous system. Therefore, we often find it hard to comprehend just what the thermal day of a stenothermic ectotherm is like. Thermoregulation is the dominant activity of these animals and is accomplished by a variety of means. We have already mentioned basking, but even that entails more than just lying in the sun as we might do at the beach. For instance, the posture assumed by a basking lizard will accelerate or retard the rate of heat gain. A cool, warming lizard flattens its body and either tilts it so that the dorsal surface is perpendicular to the sun's rays or chooses a basking site whose surface is at such an angle. Many species also darken their basic dorsal color by dispersing cutaneous melanic pigment in the melanophores. The resulting dark body absorbs more solar radiation and may greatly facilitate heat gain. Later, when the body temperature is well within the species' preferred range, the dorsal color pales through a reversal of this process, and the rate of solar heat absorption is retarded.

Other reptile groups also undergo morning basking but sometimes in a different manner. Turtles and tortoises greatly extend the neck and the foreleg that is facing the sun. This provides maximum exposure to the capillary network in the relatively thin neck and shoulder skin, and hence a faster rise in deep-body temperature. On the other hand, field experiments with the Desert Tortoise (*Gopherus agassizii*) by the junior author and his former graduate student Bill Voigt revealed that heat absorption through the carapace was relatively slow in raising body temperature. Employing head and neck "basking" to eventually warm the entire body is also utilized by substrate-burrowing species such as the horned lizards (*Phrynosoma* spp.). By warming and then circulating blood in large head blood sinuses, the preferred temperature range may be attained without exposing the entire body while at a suboptimal state for effective escape from predators.

Diurnal snakes also face this problem, because full-body basking usually entails movement into open, sunlit areas, where one of their chief groups of predators, soaring hawks, may spot them. Some years ago both authors and their then graduate student colleague, Karen Swaim, used implanted thermal tracking transmitters to document this behavior in the Alameda Striped Racer (*Masticophis lateralis euryxanthus*) in Tilden Regional Park in the East Bay hills. The Alameda Striped Racer engages in head–neck basking before emerging from a rodent burrow retreat with a body temperature within its preferred mid 30 degrees C (86 degrees F) thermal range. Another former student, Geoff Hammerson, and the junior author documented how the head and neck of this species initially warm, and the sequestered body slowly follows by implanting a miniature temperature-sensitive transmitter in the upper neck region of this species and feeding it a second unit to record its deep-body temperature changes.

Once a heliothermic reptile achieves its preferred thermal range, it maintains it there by shuttling within the thermal mosaic usually produced by vegetation shade and intervening sunlit areas. If a cool wind or excessive cloudiness should occur, it may revert to basking to again bring its temperature back into the preferred range. If midday temperatures are very high, many reptiles will abandon all attempts at active thermoregulation and retreat to burrows or cool, deeply shaded sites. The Desert Tortoise normally begins the day with a large circular foraging walk after an initial head–neck basking period near its burrow entrance. Usually by late morning when the desert surface is becoming quite hot, this movement path returns it to its burrow, where it retreats until late afternoon when a shorter circular foraging trip is taken. On hot days most basking lizards undergo a similar midday retreat. This is an important bit of reptilian natural history for the reptile watcher and photographer. "Prime time" for viewing heliothermic reptiles is during the morning basking and foraging period. However, on hot days your subjects usually pass the thermally stressful midday hours in a cool retreat, and therefore you might as well do the same.

There are a few examples of marginal endothermy in the reptile world. The junior author recorded a 2 degree C (4 degree F) increase in deep-body temperature above that of Gulf of California water in free-swimming Green Sea Turtles (*Chelonia mydas*). This is a rather significant rise, because water takes away body heat several times faster than air. A similar example of endothermy has also been found in the Leatherback (*Dermochelys coriacea*). Incubating Rock Pythons (*Python molurus*) (an Asian species) can elevate their temperature 4 to 5 degrees C (39 to 43 degrees F) through spasmodic contraction of body muscles, and actively foraging whiptail lizards (*Aspidoscelis* spp.) are capable of a small body temperature increase above ambient. However, most reptiles do not engage in sustained muscular activity, and without epidermal insulation such as fur or feathers, small increments of body heat cannot be retained.

The primary reason for the absence of endothermy in reptiles is the lack of any significant metabolic heat production (thermogenesis). The amount of metabolic heat per square unit of body surface in a reptile is normally only about one fifth to one seventh of an average mammalian and avian endotherm, and as low as one tenth that of high-metabolism forms like hummingbirds and shrews. Upon first hearing these facts, one might be tempted to view the ectothermic life mode as vastly inferior to the endothermic world of which we mammals are a part. However, recalling that reptiles have been around for the past 180 million years or so, there must be some advantages to it.

One of the most important of these is energy conservation. Most endotherms produce metabolic heat 24 hours each day, usually well in excess of that which they need to maintain a high body temperature. This excess production is most apparent during the period when an endotherm sleeps. Energy in the form of heat is constantly being lost, while during the same period no energy sources are being taken into the body to replace it, and of course the high degree of neuromuscular activity that is made possible by high-temperature stenothermy is not being utilized. Conversely, when a heliothermic reptile retreats for the night, its metabolic rate declines with its body temperature, and both remain low until the next day's activity period is at hand. The cost of endothermy is also very high with respect to amount of food needed to continuously supply this high-temperature metabolic machinery. Many endotherms, particularly herbivores and insectivores, spend most of their waking hours feeding. Indeed, shrews, which are similar in size to many smaller lizards, spend most of each 24-hour period foraging. Even endothermic carnivores like hawks, owls, foxes, and bobcats must capture and consume at least one moderate-sized rodent, bird, or reptile each day or night to stay in good physiological trim, whereas a snake can do well on one or two such meals a month. And of course when cold weather sets in, all reptilian feeding is suspended during the winter season. The ability to survive long periods of food scarcity is perhaps one of the greatest advantages of the reptilian ectothermic system, and in the evolutionary "long haul," it may be this factor more than any other that could ensure Reptilia another 180 million years or more on Earth.

Osmoregulation

One other major reptilian adaptation to a fully terrestrial existence has been the ability to greatly reduce body water loss in those habitats where standing water is scarce or absent for much of the year. Reptiles obtain water in a variety of ways. Species that live in or are closely associated with freshwater simply drink, and some can even absorb water through the skin in the manner of amphibians. Partially dehydrated Florida Worm Lizards (*Rhineura floridana*) were shown to rehydrate when placed in wet sand where free-standing water was not available for drinking but water molecules on the sand grain surfaces could be taken up through the skin. A unique role of the skin in obtaining water had been documented in the Australian lizard, the Moloch (*Moloch horridus*), which is often seen lying belly-deep in ground-surface water. It was originally thought that it also absorbed water through its skin until tests showed that the latter was impervious to water passage. However, further research revealed that the water was moving up through the stratum corneum layer of the skin by capillary action to the mouth, where it was then ingested by movements of the jaws. This remarkable phenomenon has more recently been reported in the horned lizards (*Phrynosoma* spp.) of North America.

Species not associated with standing water often lap dewdrops or drink from small catchments in the hollows of leaves, rocks, or other irregularities. However, many species that inhabit arid habitats probably never drink and instead depend entirely on the preformed water in their food and the metabolic water that is produced in their cells by the oxidative reduction of the molecules that compose that food. Some lizards can also store water in the bladder, lymph spaces, and various body tissues, and these sources can be drawn upon during periods of food scarcity.

The skin of most reptiles is far more resistant to water passage than that of amphibians. The general lack of skin glands and the thicker outer layer are important factors, because a thin, moist epithelium is no longer required for respiration. However, the presence of scales may contribute little to resisting water loss. Some years ago a herpetological colleague of both authors, Dr. Paul Licht, surprised them with his new laboratory "pet," a young scaleless Gopher Snake (*Pituophis catenifer*), a genetic anomaly that he happened to come by. He conducted cutaneous water loss tests with it and a scaled Gopher Snake of the same size and found no significant difference. Nevertheless, water is lost in varying degrees through reptilian skin, and there are marked differences depending upon the species. In general, those from arid environments lose water far less rapidly than those from humid or aquatic habitats. Water loss per surface area is 19 times higher in the aquatic Spectacled Caiman (*Caiman crocodilus*) (not a native California species) than in a California species of desert lizard, the Common Chuckwalla (*Sauromalus ater*). There can be up to a 100-fold difference in rate of water loss between tropical and desert snake species, with the tropical forms losing water at a rate equivalent to that of amphibians.

Reptiles have also made a major evolutionary advance in body water conservation by greatly limiting the amount lost in the excretion of nitrogenous wastes. Amphibians excrete their nitrogenous wastes in the form of urea, a water-soluble molecule that requires relatively large amounts of water for elimination. This method is adequate for most amphibian species, given their requirement of a moist or standing-water habitat that readily permits body water replacement. Mammals also employ this form of nitrogenous waste excretion. We are quite familiar with the need to drink water throughout the day, much of which is used to dissolve and transport urea from our kidneys to outside the body.

In contrast to this system, reptiles excrete their nitrogenous wastes chiefly in the form of uric acid, a process known as uricotelism. Unlike urea, which is a water-demanding solute, uric acid is a precipitate and needs only enough water to keep it in suspension as a thin paste for elimination. Your most familiar view of this excretory product is probably through your car windshield, because birds also are uricotelic. In lizards the uric acid is usually excreted as a small cylinder of chalky substance attached to the end of a fecal pellet. It is often deposited near the close of the basking period and can provide a helpful clue to the field herpetologist as to what sites are commonly used by lizards in a given area.

One other means of significant body water expenditure in amphibians and mammals is in the excretion of another solute, salt. The occasional salty meal or snack that we eat usually brings forth the desire for a second glass of water (or other refreshment), so that the excess salt load can be eliminated in the urine. However, in those reptiles that inhabit water-scarce environments, salt excretion is handled primarily by special salt-secreting glands that produce a highly concentrated fluid that is then secreted directly to the outside, thus eliminating water expenditure at the kidney. In lizards these glands are located in nasal passages, and the secreted salt concentrate is snorted out of the external nares. Close examination of the nasal area of most California desert lizards will reveal dried salt residue from this process.

There is one other group of "desert reptiles" within the territorial limits of this state that also depend heavily on salt-secreting glands. These are the six species of marine turtles that occasionally enter California's inshore waters. The sea is actually an "aquatic desert," where the only freshwater occurs in the form of occasional rain, and that might only be utilized by marine vertebrates if they could possibly position their open mouths upward at the surface for extended periods during a storm. This to the best of our knowledge does not happen, and so marine reptiles like sea turtles and sea snakes obtain all of their water from their food. To negate the loss of this limited source, marine turtles have salt-secreting lacrimal glands located in the head with ducts that carry the secretions to the corner of the eyes, where it is shed like tears. This is the origin of the term "turtle tears," since these secretions continue when a turtle is out of water and the salty tears can be seen running down the dry snout area.

The combination of reptilian thermoregulation, a highly flexible metabolism, and the body water conservation measures just reviewed make desert reptiles the best-adapted vertebrates for arid environments. A small select number of rodents like the kangaroo rats and pocket mice, several bird species such as the Roadrunner and Gambel's Quail, and the previously discussed spadefoot toads are also well adapted to the deserts of the American Southwest. However, it is the approximately 38 lizard species, 28 snake species, and one tortoise that compose the dominant group of vertebrates that inhabit our California desert areas and whose intriguing life histories along with those of all other reptiles in this state we will briefly explore in the following pages.

Lizards and Snakes (Order Squamata)

Most lizards and snakes are quite different from one another, and it may be surprising to some people that they are grouped in the same order along with one other small legless suborder, the worm lizards (Amphisbaenia). However, upon close inspection, it is hard to find even one sound feature that separates these two major groups. The one that may first come to mind, the presence or absence of legs, is not satisfactory, because although no snakes have fully formed legs, there are several legless lizard species, including one in California, the California Legless Lizard (*Anniella pulchra*). Features shared in common include a scaled skin covering, paired copulatory organs (hemipenes), and a jaw that is suspended from the skull by a moveable element, the quadrate bone. Additional features that at first may seem to be unique to snakes, such as a fixed, clear, lidless eye covering and the absence of an eardrum, are also found in some lizards. How, then, do you determine if that limbless reptile you may encounter in the field is actually a snake or a lizard? The best way that we can suggest is to look it in the eye. None of the North American legless lizards have acquired the clear-scale eye covering of snakes but instead have retained the moveable eyelid found in most other lizards. Thus, if your legless discovery winks at you, it's a lizard!

Several major anatomical and physiological features shared by both groups have already been reviewed in the introductory comments for reptiles. One of these that merits further discussion is this suborder's unique copulatory organs, the hemipenes. These are paired tubular structures that lie under the ventral tail skin just posterior to the cloacal opening. They are erected by being turned inside out and everted, like the fingers of a rubber glove when it is pulled off the hand. This is accomplished by the filling of blood sinuses in the hemipenes. During copulation the male mounts the female and positions his tail base on either side of hers, after which the appropriate hemipenis is inserted into the female's cloaca through her vent. A groove on the dorsal surface of the hemipenis guides seminal fluid containing the spermatozoa that has been emitted into the male's cloaca on into that of the female. In California lizards the hemipenes are usually straight tubes, but those of many snakes have spines and fingerlike projections that help hold it in the female's cloaca or oviduct during what may be a lengthy copulation. Upon completion, the tip end of the hemipenis is withdrawn back into its pocket recess by a long retractor muscle. Either hemipenis may be used in copulation, apparently depending on which side the female is approached. A more definitive investigation as to whether male lizards and snakes are "left or right hemipenised" has yet to be done.

In most snakes and many lizards, the determination of a specimen's sex by features such as coloration or size may often be difficult if not impossible. However, all male saurians have hemipenes, and by using the appropriate-sized small, blunt object such as a bobby pin, one can attempt to probe posteriorly along the edge of the vent on either side of the midline. If it easily slips into a recess for about 10 mm (0.4 in.) or so in an average-sized adult snake or lizard, your probe is in a hemipenial pocket, and a male snake or lizard is in your hand. When employing this sex-determination technique, great care must be taken so that the soft, delicate tissues in this area are not bruised or torn. One should observe this method being employed by an experienced naturalist before attempting it oneself.

Lizards (Suborder Sauria)

Eyelid Geckos (Family Eublepharidae)	247
Western Banded Gecko (*Coleonyx variegatus*)	247
Barefoot Gecko (Switak's Banded Gecko) (*Coleonyx switaki*)	250
Gecko (Family Gekkonidae)	251
Leaf-toed Gecko (*Phyllodactylus nocticolus*)	251
Mediterranean House Gecko (*Hemidactylus turcicus*)	252
Moorish Wall Gecko (*Tarantola mauritanica*)	253
Iguanids (Family Iguanidae)	253
Desert Iguana (*Dipsosaurus dorsalis*)	254
Common Chuckwalla (*Sauromalus ater*)	257
Collared and Leopard Lizards (Family Crotaphytidae)	260
Collared Lizards (Genus *Crotaphytus*)	260
Great Basin Collared Lizard (*Crotaphytus bicinctores*)	261
Baja California Collared Lizard (*Crotaphytus vestigium*)	263
Leopard Lizards (Genus *Gambelia*)	263
Long-nosed Leopard Lizard (*Gambelia wislizenii*)	264
Blunt-nosed Leopard Lizard (*Gambelia sila*)	267
Cope's Leopard Lizard (*Gambelia copeii*)	269
Family Phrynosomatidae	270
Zebra-tailed Lizard (*Callisaurus draconoides*)	270
Fringe-toed Lizards (Genus *Uma*)	272
Coachella Valley Fringe-toed Lizard (*Uma inornata*)	274
Colorado Desert Fringe-toed Lizard (*Uma notata*)	276
Mojave Fringe-toed Lizard (*Uma scoparia*)	277
Spiny Lizards (Genus *Sceloporus*)	278
Western Fence Lizard (*Sceloporus occidentalis*)	279
Island Fence Lizard (*Sceloporus becki*)	279
Sagebrush Lizard (*Sceloporus graciosus*)	282
Southern Sagebrush Lizard (*Sceloporus vandenburgianus*)	282
Desert Spiny Lizard (*Sceloporus magister*)	284
Granite Spiny Lizard (*Sceloporus orcutti*)	286
Side-blotched Lizards (Genus *Uta*)	288
Common Side-blotched Lizard (*Uta stansburiana*)	288
Brush and Tree Lizards (Genus *Urosaurus*)	290
Long-tailed Brush Lizard (*Urosaurus graciosus*)	290
Black-tailed Brush Lizard (*Urosaurus nigricaudus*)	292
Ornate Tree Lizard (*Urosaurus ornatus*)	294
Rock Lizards (Genus *Petrosaurus*)	295
Banded Rock Lizard (*Petrosaurus mearnsi*)	296
Horned Lizards (Genus *Phrynosoma*)	298
Coast Horned Lizard (*Phrynosoma blainvillii*)	300
Desert Horned Lizard (*Phrynosoma platyrhinos*)	302
Flat-tailed Horned Lizard (*Phrynosoma mcallii*)	304
Pigmy Short-horned Lizard (*Phrynosoma douglasii*)	305

Night Lizards (Family Xantusiidae)	307
Desert Night Lizard (*Xantusia vigilis*)	307
Wiggins' Night Lizard (*Xantusia wigginsi*)	307
Granite Night Lizard (*Xantusia henshawi*)	310
Sandstone Night Lizard (*Xantusia gracilis*)	310
Island Night Lizard (*Xantusia riversiana*)	312
Skinks (Family Scincidae)	313
Northern Skinks (Genus *Plestiodon*)	314
Western Skink (*Plestiodon skiltonianus*)	314
Gilbert's Skink (*Plestiodon gilberti*)	316
Teiids (Family Teiidae)	318
Whiptails (Genus *Aspidoscelis*)	319
Western Whiptail (*Aspidoscelis tigris*)	320
Orange-throated Whiptail (*Aspidoscelis hyperythra*)	323
Alligator Lizards and Relatives (Family Anguidae)	325
Alligator Lizards (Genus *Elgaria*)	325
Southern Alligator Lizard (*Elgaria multicarinata*)	326
Northern Alligator Lizard (*Elgaria coerulea*)	329
Panamint Alligator Lizard (*Elgaria panamintina*)	330
North American Legless Lizards (Family Anniellidae)	331
California Legless Lizard (*Anniella pulchra*)	332
Venomous Lizards (Family Helodermatidae)	334
Gila Monster (*Heloderma suspectum*)	335

Key to the Families and Genera of Lizards

All lizards have a scaled skin, legged forms have clawed toes, and most species have moveable eyelids. These are present in our one legless species, a feature that distinguishes it from snakes, which don't have eyelids. Figure 28 presents many of the scalation and anatomical features used in identifying lizards.

1a	Eyes have a transparent covering but no eyelids	2a
1b	Moveable eyelids present	3a
2a	Toe tips very broad with a pair of large, flat scales . . . . Leaf-toed, Mediterranean House, and Moorish Wall Geckos	
2b	Toe tips not broadened	Night Lizards
3a	Entire body with fine, granular scales, pliable skin, large eyes, vertical pupils	Western Banded and Barefoot Geckos
3b	Medium to large scales on part or all of body, eyes not unusually large; pupils round or not easily observed	4a
4a	Legless, snakelike body; tiny eyes with moveable eyelids that blink	California Legless Lizard
4b	Limbs present	5a
5a	Body (except for head) covered with smooth, shiny, cycloid scales	Skinks

5b	Scales not cycloid over entire body	6a
6a	Projecting structures on back of head, ranging from spikes to small nubbins; usually one or two rows of enlarged, pointed scales along sides of body	Horned Lizards
6b	No "horns" or enlarged, pointed lateral scale rows	7a
7a	Fold area containing small granular scales on sides of body that separate large, squarish dorsal and ventral scales	Alligator Lizards
7b	No lateral folds separating large dorsal and ventral scales	8a
8a	Dorsal and lateral body surfaces covered with large, beadlike scales; bold black and orange-yellow reticulate color pattern; tail sausage-shaped and much shorter than body	Gila Monster
8b	Tail as long or longer than body and not swollen in appearance	9a
9a	Very, small, fine granular dorsal scales; large ventral scales, often arranged in rows	Whiptails
9b	No great size difference between the size of dorsal and ventral scales	10a
10a	Single row of enlarged scales down middle of back	Desert Iguana
10b	No row of enlarged scales down middle of back	11a
11a	Large, stocky lizard; no large rostral scale at tip of snout	Common Chuckwalla
11b	Large rostral scale present at tip of snout	12a
12a	All dorsal scales keeled and pointed; incomplete gular fold on throat skin	Spiny Lizards
12b	Some or all dorsal scales granular; if keeled scales present they are never pointed; complete gular fold on throat skin	13a
13a	Underside of tail usually with black crossbands or spots; upper lip scales separated by diagonal furrows	Zebra-tailed and Fringe-toed Lizards
13b	Underside of tail without black crossbands; upper lip scales separated by vertical furrows	14a
14a	Very small scales on top of head and behind eyes	Leopard and Collared Lizards
14b	Scales on top of head and behind eyes relatively large	Side-blotched, Brush, Tree, and Rock Lizards

To date (2011) about 4,770 species of lizards have been described. This suborder is worldwide in distribution, and its members have adapted to many different environments. Most are terrestrial, but many are arboreal and several are semimarine. A few species even engage in gliding flight,

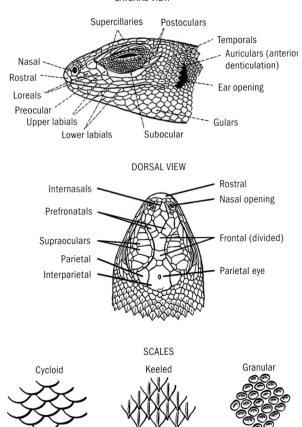

Figure 28. Head anatomy, scale names, and basic scale types of lizards.

using "wings" formed from a membrane stretched between elongate free, extensible ribs. There are also forms that have adapted to a burrowing life and in the evolutionary process lost all four limbs. However, the great majority of lizards have two pairs of well-developed limbs, with five digits bearing claws on each foot. Many species are adept at climbing and have no difficulty in clinging to vertical or even inverted surfaces. In cursorial forms like the Zebra-tailed Lizard (*Callisaurus draconoides*) the limbs are long and slender but quite muscular. It and other similar species run bipedally and may equal or exceed the average human in speed. Some of these high-speed runners and slower forms as well regularly burrow headfirst into soft sand or soil, a behavior that could clog straight nasal passages. Such lizards have a "kitchen–sink trap" conformation to the nasal passageways that captures soil particles before they can enter the lungs; particles are then snorted out upon emerging. Most burying lizards also have erectile tissue in the anterior portion of the vicinity of the nostrils that can be enlarged and thus reduce the size of passage while burrowing.

Many lizards have preanal and femoral glands. The latter occur as a series of pores along the posterior edge of the upper part of the hind legs. The dermal pits under the pores accumulate waxy cellular debris, and this substance may extend from the pores in the form of small columns, somewhat like small, thick hairs. This material wears off on rocks and other surfaces over which a lizard climbs. Femoral pores are usually larger in males than females, and territorial males are occasionally seen dragging the hind legs along as if injured, with the pore secretions in contact with the substrate. The waxy material is evidently odoriferous and is possibly used to mark territories. It also may be important in species, sex, and even individual recognition.

The tail of many species breaks easily when pulled or injured, or in some geckos it may be shed or "cast" even without being touched. Those species adapted for tail loss have a breakage zone in the tail vertebrae that allows for easy separation (figure 29). The contraction of segmental muscles attached on each side of these fracture zones produces the rup-

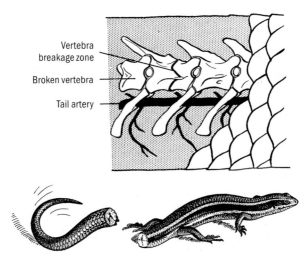

Figure 29. Lizard tail breakage results from a weak zone in the center of the tail vertebrae. Muscular valves in the tail artery below the vertebrae constrict to stop blood loss after a break.

ture. Sphincter muscles on the caudal artery and valves in the caudal vein anterior to the break prevent excess bleeding. Like tail breakage in plethodontid salamanders, the severed segment wiggles violently for a minute or more, as a result of the anaerobic nature of reptilian muscle. This often attracts the attention of a predator as the lizard escapes.

Unlike the nearly perfectly duplicated tail that is regrown by salamanders, the replacement in lizards is more of a pseudo tail. Instead of new vertebrae, a fibrocartilage rod is grown for support and the new muscular structure is not segmental as in the original. The new scalation appears to be embossed in the epidermis and differs from the original, and the coloration is more subdued. A regrown tail may break again, but this usually occurs anterior to the original break where the tail vertebrae with their fracture zone and segmental muscle attachments are still present. It is therefore the fortunate lizard whose first tail breakage occurs quite distally, because it has several more tail breakages left if needed. The frequency of regenerated tails in a population has been used by some as an indication of predation pressure. However, tails are also often lost in head-to-tail fighting among males, and the frequency of tail breakage in that sex has been shown to be higher than in females in some species. As in many mammals, the tail of lizards appears to be important for balance. Climbing species will continuously reposition it to offset the momentary tilt of their body, and in lizards that run in a bipedal manner it is usually elevated to act as a counterbalance.

Many lizards have a parietal or "third eye" located in the middle of the large interparietal scale on top of the head. The term "eye" is basically a correct one here, because it has a cornea, a lens, a retina containing photosensory elements, and a nerve that connects the retina to the pineal apparatus of the brain through a hole in the skull (figure 30). What is lacking is an iris diaphragm, and thus light intensity reaching the eye cannot be controlled. Thus, the parietal eye functions more like a radiation dosimeter than an image-producing camera.

After numerous experiments, the senior author and his former graduate student Dan Wilhoft propose the following theory of function for this unique structure. The parietal eye seems to register a lizard's exposure to light and heat, and adjusts this exposure to daily and seasonal changes, ensuring proper timing of photothermally controlled events in the reproductive cycle. Thus, during periods of unusually bright warm weather when overacceleration of the cycle might occur, the third eye tends to inhibit light-seeking behavior. On the other hand, unusually cool, overcast weather results in a reduction of parietal eye stimulation. This, in turn, removes any restraint on light-seeking behavior, and the cycle is freed to proceed driven by its endogenous cycling mechanism.

The inhibitory role of the third eye has been further explored by the senior author through experiments with lizards that had their parietal eyes surgically removed (parietalectomy). When compared to a control group,

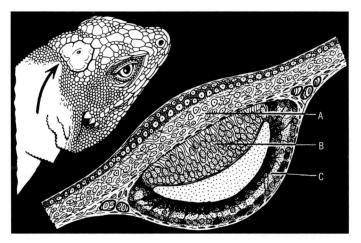

Figure 30. The parietal eye of an iguanid lizard, appearing as a small dot in the center of the large interparietal head scale (ahead of arrow). It is composed of a cornea (A), lens (B), and a retina (C).

the parietalectomized lizards exhibited an acceleration of the gonadal cycle, which implies the inhibitory role of this structure. The parietal eye may be producing its effect by providing sensory input to the pineal gland, whose hormone melatonin has been implicated in the control of reproduction. Experiments by both authors have indicated that the loss of third-eye function is followed by increased photothermal exposure as a result of an increase of light-seeking behavior and prolonged time spent in the open above ground.

This latter result was vividly displayed to the junior author one early fall day while completing a series of hourly observations of a small population of Western Fence Lizards (*Sceloporus occidentalis*) in an outdoor enclosure area. One group had been parietalectomized, and the other served as a control. Individuals were identified by color-coded dots on their back and scored as to their exposure to sunlight or bright hazy sky. On this particular sunny day clouds appeared in early afternoon, and in less than an hour a light rain began to fall. Upon seeing raindrops on his laboratory windows, he initially decided to terminate his observations for that day, because temperate-zone heliotherms seek retreats with the onset of rain. On second thought, however, he felt it would be scientifically prudent to complete the day's observation sequence and so once more went out to the enclosure. His diligence was rewarded by the sight of two adult lizards sitting out in the rain, much like two frogs on a lilypad. He quickly read their color codes, after which a check of his color key revealed that these were two of his three parietalectomized males, which literally no longer knew enough to come in out of the rain!

The function of this built-in radiation dosimeter may be far more important to a heliothermic lizard's well-being than originally suspected. Many desert species have the opportunity to remain active throughout the long, hot midsummer days when they could extend their photothermal exposure well beyond what they experience in spring. However, most do not and instead greatly reduce their time of exposure above ground during the period of long, hot days. Some years ago the junior author decided to see what changes might occur in a Western Fence Lizard's physiology if it were given the "opportunity" to live the thermal life of a mammal for several weeks. Adults were maintained in an experimental chamber with continuously available food and water under a normal day–night light cycle. However, the temperature in the chamber was held constant at 35.5 degrees C (96 degrees F), the mean preferred body temperature of this species. After only a few weeks under this imposed "endothermic regimen," the experimental group had lost weight and appeared to be in a highly stressed state. Thyroid gland activity was measured with radioactive iodine and was significantly higher than that of a control group, as was their resting metabolic rate as measured by oxygen consumption. Exploratory surgery also revealed that the fat storage bodies and the testes had all but completely disappeared. It was quite apparent that these lizards were in the process of metabolically "burning out," and that the heliothermic lifestyle, though similar to our own in its optimal thermal range, is vastly different from the mammal–bird endothermic system.

Unlike snakes, which are primarily carnivorous, lizards as a group display a wide variety of feeding habits. This spectrum includes carnivorous, insectivorous, herbivorous, and omnivorous species, many of which also often scavenge. Herbivorous species like the Common Chuckwalla (*Sauromalus ater*) and the Desert Iguana (*Dipsosaurus dorsalis*) tend to be large, because considerable digestive space is required to process plant food. The vast majority of California lizards are insectivorous, a category that also includes the eating of other small invertebrates such as spiders and scorpions. Our most carnivorous species is the Gila Monster (*Heloderma suspectum*), but even it won't pass up a ground nest of bird eggs or a large, plump insect. Many lizards have excellent eyesight and detect their prey by its movement, which seems to be the stimulus for attacking it. Others, such as skinks, appear to identify prey suitability by chemoreception as well as by vision. The sense of taste as perceived by the Jacobson's organ is also a major means of locating prey in whiptails (*Aspidoscelis* spp.) and alligator lizards (*Elgaria* spp.). The teeth of most saurians are short conical structures of nearly one shape (homodont) and attached to the inner surface of the jawbones. In general there is little mastication of food. Instead most species seize their prey, work it about in the jaw, bite it a few times, and then swallow it.

Many lizard species are territorial. They defend favorite sections of their habitat and living area, particularly during breeding season. Defense of territory helps to ensure a good distribution of a lizard population throughout areas containing food and retreat shelter, and ensures that the more vigorous individuals will perpetuate their kind. Females are usually less territorial than males, although they may become aggressive and defend territories, especially at the time of egg-laying when they may compete for prime nest sites.

Territorial defense is by display and occasional fighting. There is visual recognition of conspecifics by size, shape, pattern, behavior, and odor, and the combination of such cues and their relative importance varies among species. In diurnal lizards visual displays are especially important. A typical territorial male assumes an erect posture, well up on its legs, and flattens its body from side to side while lowering its dewlap to reveal any bright ventral signal colors it may have. Once this posture is assumed, it often arches its back and bobs. Bobbing involves flexing and extending the limbs, particularly the forelimbs, and sometimes includes nodding movements of the head. This action is often compared to the human "pushup" exercise. The displays of a territorial spiny lizard involves a series of stereotyped movements called the display action pattern (DAP). Displays may be low key, such as "assertion displays" that serve to announce the presence of a territorial male. Here the movements are of low amplitude and often contain only a portion of the complete species' DAP, and there is little posturing. If an approaching male is not deterred by this initial action, the "assertion display," a performance of high intensity with strong orientation toward the intruder, high-amplitude movements, and full posturing, is given.

The research of Dr. Charles Carpenter has revealed that the DAP is innate and species specific, with each species having a characteristic pattern of body movement. Differences occur in the amplitude and configuration of bobs and their order of appearance, cadence, and duration. The rhythm of the movements is the prime information-yielding part of the display. This behavior is as distinctive as the vocalizations of anurans or birds, and serves a comparable function in these nonvocal vertebrates. There is also some evidence that in some species each individual has, in addition to bobbing elements of that species, movements that are distinctively its own, perhaps making possible individual recognition.

As with some male-based human encounters, if all the display efforts of a territorial male fail to keep an intruder off his turf, fighting ensues. In lizards, the combatants commonly stand high off the ground and broadside, often facing in opposite directions with sides flattened and dewlaps lowered. They may circle each other in this position, and then with a sudden lunge the more aggressive male tries to seize the neck, leg, or tail of the other, or he may flex his body and slap with his tail. If a hold is secured the combatants may roll over and over until the one that is held is able to break free. Such fights are generally ritualistic and seldom draw blood. However, it is during these fights that tails are often broken. When an opponent gives way, the now-dominant male often pursues him. Males of some species are able to terminate aggression by another by lying flat with the body, head, and tail against the substrate in an apparent complete reversal of the posturing display. Some may even roll over on their back and wave a forelimb in the air. It is often tempting when trying to understand the behavior of other animals to ally it to portions of our own, and here the similarity of such a lizard and the loser of a human fight who is on his back and waving off his opponent is tempting. However, we should not "read into" such behavior various human attributes. Instead a more "scientific" explanation might be that all of the sign stimuli that initiated the encounter are now completely eliminated.

Male lizards' recognition of a female of their species is primarily by visual cues, although in some species scent or taste may also be important. In most species female display color is less intense than that of the male, or absent. However, in a few California species such as the leopard lizards (*Gambelia* spp.), the collared lizards (*Crotaphytus* spp.), and the Desert Spiny Lizard (*Sceloporus magister*), breeding females take on their own bright distinctive coloration. In most species male recognition of females is usually based on the absence of any male display behavior as they approach his territory while being displayed to by the owner. In the absence of a male-type response, he approaches her with head lowered and vibrating in a series of fast, short bobs, after which he may nudge and lick her. A receptive female holds ground upon the approach or makes only what appears to be a halfhearted attempt to escape. The male then secures a grip with his jaws on her head, neck, or shoulder region, the exact site varying with genus or family. In most horned lizards (*Phrynosoma* spp.) a head spine or "horn" is grabbed and held. He then places one hindlimb over her sacral region and brings his vent region in contact with hers to achieve insertion of one of his hemipenes into her cloaca.

After copulation female lizards commonly lose interest in mating. In heliothermic lizards they typically assume a rejection pose resembling the stance described for a displaying territorial male. If the male is not deterred and attempts copulation, the female may shake her head free of his grasp and roll over on her back to assume the submissive pose that a subordinate male takes when being attacked.

Oviparous females are very adept at selecting good nest sites. These must be hidden from predators and usually direct sunlight, and they must retain enough soil or organic debris moisture to keep the eggs from desiccating during the one to two or more months of incubation. Finally, the nest site must be one that warms sufficiently each day to ensure that hatching occurs while foraging and thermal conditions in the habitat are still favorable for the hatchlings. The thermal requirements of the developing reptile embryo are quite different from those of birds. In the latter case, once the hen begins to incubate her clutch, the developing embryos must be maintained at approximately the same, constant high temperature as the body of the adult bird, and any significant decrease from this high stenothermic state for even a short time may terminate development. In contrast, the embryos of lizards and other reptiles are eurythermic like the adults and are able to withstand wide daily thermal swings, during which cell division proceeds more rapidly during the warm daytime hours. Thus, instead of a standard, relatively short developmental period such as seen in small birds (about 21 days), lizard eggs take from one to two or even more months to hatch. The prehatching lizard develops a small, sharp epidermal ridge or egg tooth on its snout, and through upward thrusts of the head it slices a slit in the leathery shell through which it then emerges, completely developed and ready to fend for itself.

Some lizards retain the eggs once embryonic development begins. Laying may occur when the embryo is in an early stage of development, or the eggs may be retained in the ovoviviparous state in which the embryo develops fully within the female's oviduct, nurtured by the egg yolk, and is then "born." True viviparous development occurs in California night lizard species (*Xantusia* spp.), in which the young are born after being nourished through a placenta-like connection with the maternal blood supply. Clutch size in lizards varies from one or two eggs in geckos to more than 20 in some heliotherms, and this may be repeated several times a year. However, in multiple-clutch species early clutches tend to be larger than later ones, because ovarian function wanes as the summer season progresses.

The time from hatching or birth to sexual maturity varies greatly. Some Central American species of *Anolis* may be able to reproduce at the tender age of two months, whereas a slow-growing lizard such as the Common Chuckwalla (*Sauromalus ater*) may require five to six years. Lizards can be grouped into two reproductive categories: one contains those early-maturing species that produce two or more broods per year and have a relatively short life span, and the other are those late-maturing species that produce a single brood each year and have a long life span.

Geckos (Families Eublepharidae and Gekkonidae)

Geckos are a worldwide group that occurs in warmer areas between latitudes 50 degrees north and south. They are also widespread on oceanic islands where they have often been introduced. There are about 88 genera, and three native and two introduced species are found in California. Adult geckos vary in size from about 2.5 to 17.5 cm (1 to 7 in.) in snout-to-vent length. Their heads are relatively large, and their bodies are usually more or less flattened dorsoventrally. In most species the undersides of the toes have pads composed of subdigital plates covered with microscopic and often branched hairlike bristles (setae) with spatulate tips. This enables many species to move with ease on the walls and ceilings of houses and public buildings in the tropics. Most species have a soft skin that readily moves over the underlying tissue and in some species like the Leaf-toed Gecko (*Phyllodactylus nocticolus*) may slip off when grasped by a predator. Rapid tail breakage is also employed as a predator escape mechanism. The scales are usually granular, which gives the body a soft, silky feel, and the eyes are large, usually with vertically elliptical pupils. Most geckos are nocturnal or crepuscular.

The habits of geckos vary greatly. Many are arboreal, others are ground or rock dwellers, and some are sand dwellers. Habitats range from tropical forests to extreme deserts. Many species have adapted to human habitation and are common occupants of homes, hotels, and other buildings in the tropics where they contribute to the control of insect pests. The best gecko collecting ever experienced by the junior author was in the old village church of San Blas, Nayarit, Mexico. With the coming of night to the California deserts, insectivorous bird species are replaced by a wide spectrum of bats and a few specialized avian species such as night hawks, all of which feed on the wealth of flying nocturnal insects that inhabit these areas. Numerous species of lizards also feed on ground-dwelling insects during daylight hours, but at night it is only one or two gecko species that come forth to occupy the night insect-feeding niche. It may be that the characteristically cool desert night climate selects for thigmothermic lizard species that are proficient at acquiring body heat from warm objects such as large rocks in their habitat. Geckos appear to do this most efficiently.

Many geckos produce vocal sounds, a rare capability among lizards. These are squeaks, chirps, and barks that they use in defense of territory and sometimes when feeding and during courtship. The use of voice for communication seems to have been prompted by the nocturnal activity of many species, because diurnal geckos appear insensitive to sound and are generally nonvocal. Some geckos have no definite reproductive cycle and mate throughout the year, while others have cyclic breeding that is restricted to a short period. Females of some species are able to store sperm for a considerable period. Geckos commonly lay two (occasionally one) round or oval white eggs, often depositing more than one clutch a year. Depending upon the species, the eggs are either hard- or soft-shelled. Three New Zealand and New Caledonian genera are live-bearing, and one species is apparently parthenogenic.

Eyelid Geckos (Family Eublepharidae)

WESTERN BANDED GECKO   *Coleonyx variegatus*
DESERT BANDED GECKO   *C. v. variegatus*
SAN DIEGO BANDED GECKO   *C. v. abbotti*

Desert Banded Gecko *C. v. variegatus*

San Diego Banded Gecko *C. v. abbotti*

IDENTIFICATION: The snout-to-vent length in the Western Banded Geckos ranges from 5.1 to 7.6 cm (2 to 3 in.). The body and tail are rounded, and the tail is constricted at the base. Its scales are granular, the skin soft and delicate, and the toes slender with no pads for adhesion to surfaces. The eyes have vertical pupils and moveable lids. The dorsal body is flesh-colored, cream, or yellowish with brown crossbands on body and tail. These bands tend to break up with age or in some localities into a blotched, mottled, or spotted pattern. The ventral side is whitish. Some individuals become decidedly spotted. Males have a spur on each side of tail base and preanal pores. The corresponding scales in the female are usually enlarged and sometimes pitted, but without pores, and spurs are absent.

RANGE: In California it occurs from northern Inyo County south through the Great Basin, Mojave, and Colorado deserts, and southwest through the Kern River Canyon in the southern Sierra Nevada and on through the Tehachapi, Transverse, and Peninsular ranges. It is absent from some southern coastal areas.

NATURAL HISTORY: The Western Banded Gecko occurs in arid and semiarid environments of broken chaparral, creosote brush desert, sagebrush flats, desert grassland, and subtropical scrub. The terrain in which it is found varies greatly from rocky canyons, hillsides, and flats to alluvium, washes, and sand deposits. It is chiefly nocturnal, and during the day seeks shelter in rock crevices, beneath rocks, thin exfoliating rock flakes, plant debris (yucca stalks, logs), boards, and other refuse including dried cowpies and dried carcasses of animals where it is perhaps attracted by insect food. It is less completely a rock dweller than the Leaf-toed Gecko (*Phyllodactylus nocticolus*), which rarely leaves its rocky abode.

Plate 61. Western Banded Gecko on a vertical wall.

It is often encountered on highways at night, where it can be mistaken for a small, pale twig. Its attraction to road surfaces, especially dark ones, stems from the fact that it is one of the few nocturnal lizards of California. As such it is a true thigmotherm, deriving its body heat from warm objects and from air. When populations occur near hard-surfaced roads, the nightly foraging brings some individuals in contact with these areas, which stay warm well after sundown and where they then linger in much the way that most nocturnal snakes do. When disturbed this gecko may run with its tail curled scorpion-like over its back or at one side. When molested, it often gives a mouselike squeak, stands stiff-legged with head and tail elevated and writhing, apparently attracting attention to this expendable part. The frequency of tail loss is accordingly high, and in some instances may actually be shed or cast through the contraction of the segmented tail muscles as opposed to being grabbed by a predator.

The Western Banded Gecko is abroad from late March to October or November, with some activity, particularly of immatures, occurring in winter in the warmer parts of its range. Most mating appears to occur at night in May and June. The male approaches the female with nose and body low, tail waving. He may poke her with his snout or lick her, then bite her tail, leg, flank, shoulder, or neck. Upon obtaining a hold he moves her jerkily forward. Often the tail is seized first, then he shifts his grip to the skin of her neck. He strokes her tail base with his swollen cloacal region and the projecting hardened secretion of his preanal pores. With his forelimbs astride her body and a hindlimb over her lower back, he curls his tail under hers, drawing one of his cloacal spurs a number of times conclusively across her cloaca. The spur is used to open her vent. As the hemipenis touches the female's lower cloacal lip, her cloaca opens to allow intromission.

This species' eggs have flexible shells. Two is the usual number per female, and probably at least two clutches are laid a year. The incubation period is about one and one half months. Many hatchlings reach adult size when less than a year old, but some do not mature until their second year. An individual in captivity lived over 14 years. It feeds on a wide variety of primarily nocturnal insects and their larvae plus spiders and solpugids. Excellent night vision permits prey location at night.

BAREFOOT GECKO
(SWITAK'S BANDED GECKO)

Coleonyx switaki

PROTECTIVE STATUS: ST

IDENTIFICATION: It has a rounded body and tail, and a snout-to-vent length range of 5.1 to 8.4 cm (2 to 3.3 in.). The tail is constricted at the base, and the scales are granular except for small, evenly spaced spines extending about halfway down the upper sides from the back of the neck to the base of the tail. The skin is soft and delicate, and the toes are slender with no pads. Like the Western Banded Gecko, it has vertical pupils and eyelids, but it differs from this species in that it has small, sooty tubercles on scales on its upper sides, back of the neck, and the base of the tail. The dorsal color varies from pale beige or yellowish cream to yellowish olive. The head and body are spotted with round to oval dark spots and generally larger round to oval light-yellow spots that tend to be arranged in transverse rows. The tail is sometimes strikingly black-and-white banded. One unique coloration feature of this species is that males often turn bright yellow during the May-to-July breeding season.

RANGE: Barefoot Geckos occur from near sea level to around 700 m (2,297 ft). They are well known from the vicinity of Borrego Springs, San Diego County, southward along the eastern edge of the Peninsular Ranges and into Baja California, Mexico. There are unconfirmed reports as far north as the Palms to Pines Highway area in Riverside County.

NATURAL HISTORY: This is a rock-oriented gecko. It inhabits arid hillsides and canyons with large boulders and massive rock outcrops with deep crevices. Both granite and volcanic rocks are frequented. Its nocturnal nature, along with the use of deep rock crevices for daytime retreat, make this species hard to locate. This gecko has a relatively strong tendency to lose water for a lizard that prefers very arid habitats. Its lower metabolic rate compared with other *Coleonyx* species, plus its preference for cool ambient temperatures (16 to 20 degrees C or 61 to 68 degrees F) and relative humidities above 30 percent may serve to offset this apparent osmotic handicap. Vegetation at these sites consists of desert and thorn scrub.

The Barefoot Gecko squeaks like *C. variegatus* when disturbed and may walk with its tail curled, waving, and elevated. Its use of deep rock crevices and nocturnal habits make field observations of this species difficult. Its reproductive behavior is presumed to be similar to that of the Western Banded Gecko.

Geckos (Family Gekkonidae)

LEAF-TOED GECKO *Phyllodactylus nocticolus*

IDENTIFICATION: Its snout-to-vent length is 4.1 to 6.3 cm (1.3 to 2.5 in.), and the head, body, and tail (especially at base) are flattened dorsoventrally, an adaptation for crevice-dwelling. Two large plates separated by a claw occur at the tip of each toe. Dorsal body scales are granular with a scattering of larger keeled scales in irregular rows. The ventral surfaces are covered with somewhat larger flat scales. The eyes are large, without moveable lids but with vertical pupils. Its dorsal color varies from pale brown, gray, pink, or almost cream, depending upon color phase, and marked with irregular bars, crossbanding, or spotting of brown. The underside is yellowish to white, often flecked in varying amounts with brown.

RANGE: It is found from sea level to around 610 m (2,000 ft) in the desert slopes of the Peninsular Ranges in southern California from south of San Gorgonio Pass to the tip of Baja California, Mexico.

NATURAL HISTORY: In California the Leaf-toed Gecko occurs in canyons with massive rocks and populated by thorn scrub desert and chaparral. As far as is known this lizard is strictly nocturnal. In southern California it is active from March to October. It is an excellent climber capable of running with ease over the sides and undersurfaces of rocks. Although delicate in appearance, it survives in hot, dry places. Nocturnality and crevice-dwelling habits make this possible. This gecko spends the day in crevices of boulders, beneath the bark of trees such as mesquite and palo verde, beneath fallen prickly pear pads, and in fissures in the large cardon cactus. Crevices just wide enough to accommodate the flattened head are favored.

A clutch of two whitish, brittle-shelled eggs is laid in late May and June, and there may be more than one clutch a year. In southern California the young first appear in June and are found through August. Like other California gecko species it feeds on nocturnal insects and spiders that frequent its upland rock habitat.

MEDITERRANEAN HOUSE GECKO *Hemidactylus turcicus*
INTRODUCED

IDENTIFICATION: It has a snout-to-vent length of 4.4 to 6 cm (1.75 to 2.3 in.), large eyes with no eyelids, vertical pupils, and broad toe pads. Its dorsal surface is covered with knobby and often keeled tubercles. This gecko is capable of marked color change. Its light phase is a light-pink, pale-yellow, or whitish hue with brown or gray blotches. With the expansion of its melanophore pigments it changes to a gray or brown, and the dark blotches become less distinct. The tail usually has a banded pattern, and the ventral body skin is somewhat translucent.

RANGE: In California this species is currently found at localities in Riverside, Imperial, San Bernardino, Inyo, and San Diego counties. The species is native to India, Somalia, the Middle East, and the Mediterranean basin.

NATURAL HISTORY: The Mediterranean House Gecko is one of several species of reptile that has been introduced into California and established breeding populations. Unlike the two introduced freshwater turtle species that now pose a serious competitive threat to our native Western Pond Turtle (*Actinemys marmorata*) at sites where both occur, it appears that this foreign lizard may not do the same to our native geckos. The reason is that it is an "urbanized" species that in its homeland lives in or near human dwellings, where it feeds mainly on nocturnal insects that are attracted by artificial lighting. However, it has also been found in rock crevices and under tree bark, so a possibility that it may invade some native gecko habitat still remains.

The advertisement call of the male is a series of clicks, and squeaks are also emitted when fighting with other males. Females produce one to three clutches of one or two hard-shelled eggs per year. Eggs are often laid in communal nests, and such aggregations concealed in plant nursery stock shipped to southern California from the Mediterranean or from Ensenada, Mexico, where it also has become established, may account for its arrival in California during the late twentieth century. Geckos are also among the most popular reptile pets, and thus the release of unwanted pets must also be considered as a possible source of their presence in this state.

MOORISH WALL GECKO *Tarentola mauritanica*
INTRODUCED

IDENTIFICATION: It is very similar in appearance to the Mediterranean House Gecko, but with more pronounced keeled tubercles along the dorsal midline of the body and tail and also on the ventral–lateral aspect of the tail, and its snout is slightly more pointed. However, the Moorish Wall Gecko grows larger than the Mediterranean (8 cm snout-vent length versus 6 cm).

RANGE: Small established populations occur in San Diego County. It is native to Mediterranean Europe and Africa.

NATURAL HISTORY: This is another foreign gecko species that appears to be established in urban areas in the southwest corner of California, where its life history parallels that of the closely related Mediterranean House Gecko. Given the continuing interest in keeping snakes and lizards—particularly geckos—as pets, we can expect to see the appearance of more such introductions as a result of escapes from home terrariums or purposeful releases. At this writing, there is a report of a new breeding population of the Ringed Wall Gecko (*Tarentola annularis*) in one Southern California city. In such areas, foreign gecko species find a year-round mild climate and numerous human dwellings with night lighting that attracts nocturnal insect prey, just like "back home," and they settle down to raise a family. Fortunately, that process is quite slow compared to most other lizard species, because the average gecko clutch size is only one or two eggs.

Iguanids (Family Iguanidae)

Until recently this family contained 24 California species, but with the revision of this formerly large taxon, that number has been reduced to two. This is a family of mostly large, herbivorous lizards. Two of its members, the Marine and Land Iguanas (*Amblyrhynchus cristatus* and *Conolophus* spp., respectively) of the Galapagos Islands, are familiar to many through television wildlife documentaries, and the Green Iguana (*Iguana iguana*) of Central and South America has achieved pet status in U.S. households. The two North American species, the Desert Iguana (*Dipsosaurus dorsalis*) and the Common Chuckwalla (*Sauromalus ater*), occur in California and several other southwestern states.

It is interesting to note that of the 91 reptile species currently found in mainland California, only these two lizards and the Desert Tortoise (*Gopherus agassizii*) are primarily herbivorous. This is most likely because the present feeding niche for small vertebrate herbivores is almost completely dominated by rodents and lagomorphs. However, these three reptiles have successfully competed with these mammals for a portion of the desert plant food resource.

DESERT IGUANA *Dipsosaurus dorsalis*

IDENTIFICATION: This is our third largest California lizard, with a snout-to-vent range of 10.1 to 14.6 cm (4 to 5.75 in.). Its head is relatively small in comparison to its large, round body. The tail is long, almost twice the length of the body, and ringed with keeled scales. The body has small, granular scales on the sides, weakly keeled scales on the dorsal surfaces, and a row of slightly enlarged keeled scales down the middle of its back, from head out on to the tail. The dorsal color is gray with brown to reddish brown lateral barring, and the tail usually has distinct brown rings. The male lacks large postanal scales, but its head is usually more angular and broader than that of the female. Male femoral pore secretions are also more conspicuous than those of the female, and the tail base of the male is flattened on the ventral side. During the breeding season both sexes develop a pinkish, buff, brownish red, or brownish orange ventrolateral patch on each side of the belly.

RANGE: It occurs from the northern Mojave Desert near Bishop, Inyo County, south through this desert and the Colorado Desert.

Plate 62. Desert Iguana.

NATURAL HISTORY: The Desert Iguana is usually found on sandy plains, often where sand has accumulated about plants to form hummocks. It also may be found in rocky areas and on firm clay soil near the lower Colorado River but is less common there. It favors areas having some firm soil suitable for burrow construction and where rodent burrows also provide places of retreat, although it is able to construct its own. Over most of its range it frequents creosote bush desert, but in the southern part as in Sonora and Baja California, Mexico, it is found in arid subtropical scrub.

This lizard is primarily diurnal but is occasionally abroad after dark. It is a late riser, probably because of its size and high thermal preference. The junior author has recorded deep-body temperatures by radiotelemetry from free-ranging Desert Iguanas on the Colorado Desert near Palm Springs. On several occasions these exceeded 47 degrees C (116.6 degrees F), the highest body temperature ever obtained from a free-living terrestrial vertebrate. He was also able to document thermoregulatory behavior within their retreat burrows as they moved to warmer upper segments and elevated their body temperatures above 40 degrees C (104 degrees F). After warming in the safety of their burrows, they then emerged to forage at a body temperature range that permits optimal neuromuscular coordination. This ability is apparent when it runs bipedally at speeds up to 28 km per hour (18 miles per hour), a pace that most readers and certainly the authors would have difficulty matching. This high thermal tolerance also allows the Desert Iguana to be abroad when most large, diurnal snakes and other predators retreat during the hot midday desert hours.

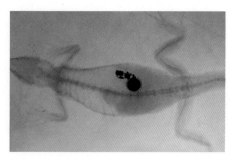

Plate 63. Thermal transmitter in a Desert Iguana.

Although these lizards may at times form congenial aggregations where food such as the luxuriant growth of blooming annuals is abundant, adult males usually defend territories and fight, even to the point of drawing blood. Male challenges and courtship resemble that described for heliothermic lizards. When fighting, males often approach each other facing opposite directions and lash with their tails. When mating, the male approaches in the same direction that the female is facing and grasps her shoulder skin in his jaws. He sometimes turns her on her back and rubs his abdomen against hers before assuming the back-straddling copulatory position.

In southern California breeding occurs from mid-April until near the end of July, and the sexes may be observed together during this period. The presence of males with testes fully active in July (the hottest month) shows the great heat tolerance of spermatozoa of these lizards. Females go underground for egg-laying during the latter part of June and early July, after which they resume surface activity. Only one clutch of three to eight eggs is laid per season, and a given female may not lay every year. About two and one half months are required for incubation. Hatchlings appear from mid-August to October and measure approximately 44 mm (1.7 in.) from snout to vent. Three years are probably required to reach minimum adult size, and some individuals apparently do not reach breeding capability until their fifth year.

The Desert Iguana feeds extensively on vegetation and occasionally may climb to procure food, sometimes to a height of over 1.5 m (5 ft). It eats the buds, flowers, and leaves of desert plants, and in spring depends heavily on annuals and the flowers of the creosote bush. Insects, other arthropods, and carrion also form a portion of its diet, particularly when plant foods are scarce. Animal foods include spiders, termites, cockroaches, scale insects, beetles, ants, and caterpillars. Fecal pellets are regularly eaten, particularly during spring months. This habit also occurs in rabbits, where a second passage of vegetation-based food through the digestive tract evidently releases nutrients not extracted the first time.

COMMON CHUCKWALLA *Sauromalus ater (=obseus)*

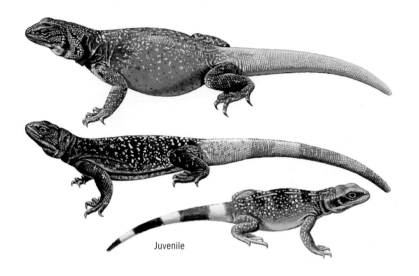

Juvenile

IDENTIFICATION: With a snout-to-vent range of 12.7 to 22.8 cm (5 to 9 in.), this is the second largest lizard native to western North America, surpassed only by the Gila Monster (*Heloderma suspectum*). It is a robust lizard with somewhat flattened body and tail, and loose folds of skin on the neck and sides of the body, except when it inflates its lungs. The scales on the sides and back are small and rounded, and the tail is thick at its base and tapers to a blunt tip. Unlike the tail of most other California lizards, it is not easily lost but can be detached by excessive pressure by predators attempting to pull the lizard out of crevice retreats. It also differs from all other western lizards in that it lacks a rostral scale. The color varies considerably with age, sex, and locality. In general males have a black head, neck, shoulders, chest, and limbs, and are sometimes spotted and flecked with gray and whitish. Individuals from the southern Mojave and Colorado deserts may have a torso suffused with red (Plate 64), while those in dark-substrate habitats such as Pisgah Crater are black. The tail is usually grayish or straw-colored and not banded. The female is gray or brownish gray with a dark head and limbs with the juvenile pattern of crossbands on the body and tail. Juveniles are distinctly crossbanded, most conspicuously so on tail where bands are black on olive gray on a yellow background.

Plate 64. Chuckwalla.

RANGE: It occurs throughout the Mojave and Colorado deserts.

NATURAL HISTORY: The Common Chuckwalla is a classic example of a habitat specialist and is closely restricted to rocky areas. Hillsides, canyons, and even isolated outcrops are frequented. A typical habitat consists of a group of massive angular, fractured rocks that provide elevated outlooks, deep crevices, and platforms for basking.

Because chuckwallas are almost exclusively herbivorous and presumably rarely drink, their activity is controlled to a considerable extent by plant growth. In the Mojave Desert they are active five or six months a year, emerging in late February or March and reaching peak activity in April, May, and June. Activity declines in late July and August, and they return to hibernation by mid-August or September. In dry years when plant growth is scant, activity is largely confined to late-winter and spring months, and the summer is spent in estivation. In the morning individuals emerge in a dark body color mode from rock crevices and sun themselves at prominent basking stations. Their color then lightens as they warm. Basking sites are usually strewn with their cylindrical leaf and seed-filled droppings, which provide the field herpetologist with an excellent indicator of chuckwalla presence in an area when thermal conditions at a given site are not appropriate for its emergence.

In addition to behavioral adjustments to aridity, this lizard is able to store water in large lymph sacs beneath their lateral body folds and to excrete excess salts, obtained from their plant foods, via nasal salt glands. The sole defense of the Common Chuckwalla is to retreat deep into crevices in or under large rocks and boulders. They appear to know exactly where these are and rarely stray from them. Head thickness determines the minimum-size crevice that can be entered. When attempts are made to probe a chuckwalla from its retreat, it wedges itself tightly in place by inflating its lungs. The remarkable capacity for distention of the body and the basis of its previously used Latin name *obesus* may be seen by patting and scratching the back and sides of a hand-held individual.

The larger more aggressive males establish territories, which they defend with bobbing displays, chases, and occasional fights. They may have harems of two to five females within their territories. The bobbing display, involving head nodding and "pushups," conserves energy, because often it alone is sufficient to prevent territorial intrusion and makes a chase unnecessary. Each adult appears to have a distinctive set of body movements used in display, a "signature" that identifies it to other resident chuckwallas in the area. Given this lizard's large size and the habit of giving displays on high, prominent sites, males may be able to recognize others up to about 100 m (328 ft) away. Territories are evidently marked with the exudate from the femoral pores, which are well developed in males. Odor (pheromones) seems to be very important in the life activities of this species. Both sexes frequently lick the ground, crevice entrances, and each other, rub the pelvic region against rock surfaces, and rub each other and the substratum with the side of the head.

Males court members of their harems almost daily from March to May. Mating occurs in late May and early June. Large adult females tolerate little premating contact by males, but young and immature females engage in extensive contacts. This consists of licking, crawling under the body of the male, onto his back, leaning against him, embracing him with the forelimbs around his neck while astride his back, and rubbing with the side of the head. Males also lick and nudge the female, and courting pairs often circle one another before copulation. Clutch size ranges from five to 16 eggs, which are laid from June to August. Females may skip a year or two between clutches. Hatchlings size is around 45 to 60 mm (1.7 to 2.4 in.) snout to vent. They appear in late summer and fall when plant food is scarce and apparently soon go into hibernation.

Growth of the young is relatively slow. Males may not reproduce until six or seven years old, although they might attain sexual maturity in two or three years. Females begin reproduction at five or six years of age. Large adults may be very old, some probably over 25 years.

Chuckwallas are almost completely herbivorous, although the diet of very young animals is poorly known. They feed on the flowers, seeds, and less often, leaves of the creosote bush (*Larrea tridentata*). Little feeding occurs in summer because of the scarcity of good-quality annual plants. At that time chuckwallas frequently must subsist primarily on perennial plants, which are often considerably desiccated. Although creosote bush is usually widely available, they make little use of its leaves. Annuals commonly predominate in the diet in spring, especially during good wildflower years. Large rock and boulder mounds where Chuckwallas occur often have a relatively lush growth of annual grasses and forbs immediately adjacent to their base. This is because much of the limited rainfall that strikes such areas runs off the rocks and waters the basal soil strips, thus promoting plant growth far beyond that seen in other areas. These seasonal "gardens" are therefore conveniently available within a short distance from a favorite basking and retreat site.

CONSERVATION NOTE: Neither the Desert Iguana nor the Common Chuckwalla has been yet accorded any protective status in California. Instead, they continue to be listed each year by the CDFG as two of 64 reptile species that may be legally taken from the natural habitat by anyone with a current fishing license or who is under 16 years of age. These species have two of the lowest maturation times and reproductive turnover rates of any California lizard, and concentrated collecting in unprotected areas could eliminate entire populations. Given the current attractiveness of large, herbivorous reptiles, such listings are unwarranted.

Collared and Leopard Lizards (Family Crotaphytidae)

Although members of this family feed on a wide variety of invertebrates, their relatively large size, strong jaws, robust build, and pugnacious behavior make this small taxon of 12 species one of the most predatory of all North American lizard groups. The five species that occur in California are widely distributed in this state's arid and semiarid habitats. Within these regions the two genera occupy two different physical habitats. Collared lizards (*Crotaphytus* spp.) are mainly rock dwellers and usually have preferred basking and lookout sites within their upland rocky realms. In contrast, the leopard lizards (*Gambelia* spp.) are primarily ground dwellers in desert, grassland, and chaparral areas and appear to be at home on a wide range of substrates as long as there are some scattered clumps of vegetation present. It is in the mottled shade of such vegetation that they often lie in wait for prey to pass by.

One curious and as yet not fully understood aspect of this family's reproductive biology is the development in females of bright-orange or red markings on the sides and neck and, in *Gambelia*, on the underside of the tail. Speculation as to the purpose of such bright color in the female rather than the male ranges from its possible function as a male attractant to that of deterring males once a female has been bred. It is most likely hormonally directed, because it is associated with breeding, but the lack of a definitive answer at this writing is yet another example of the many intriguing questions yet to be answered in reptilian natural history.

Collared Lizards (Genus *Crotaphytus*)
These are large, robust lizards with stout bodies, large chunky heads, blunt snouts, moderately developed dewlaps, large hindlimbs, and, in California

species, long more or less laterally compressed tails that are not easily broken. The dorsal scales are granular, and the ear openings are conspicuous. A gular fold and femoral pores are present. As their name indicates, all have bold dark collar markings. Males have large postanal scales, larger femoral pores, and a broader tail base than females. During the breeding season, females develop bright-orange or red markings on the sides of the body and neck. These lizards differ from *Gambelia* in having a better developed dewlap in males, relatively larger hindlimbs, a less fragile tail, and fewer teeth. They are predatory, diurnal lizards, attracted to rocky habitats where they perch on rocks to bask, defend territory, and detect prey. Locomotion is bipedal at high speed.

GREAT BASIN COLLARED LIZARD *Crotaphytus bicinctores*

IDENTIFICATION: This is a relatively large lizard with a snout-to-vent range of 8.6 to 11.2 cm (3.3 to 4.3 in.). The tail is often twice as long as the head–body length and sometimes somewhat compressed laterally in the male. There are two dark collar markings, separated by a light area in the shoulder region. The anterior collar completely encircles the throat area. The dorsal body is brown to grayish-brown and has numerous white dots and dashes. In males these markings are underlaid by vertical orange or pinkish crossbars on the back and sides. When in full breeding color, this is considered by some people to be the most strikingly colored lizard in California.

RANGE: In this state it occurs in most of the Great Basin, Mojave, and Colorado deserts from the California–Nevada border to the eastern base of the Sierra Nevada and northern base of the Transverse Mountains. However, there are several localities beyond the Mojave Desert in the North Fork of Lytle Creek, San Bernardino County, in San Antonio Canyon, and the headwaters of the East Fork of the San Gabriel River in Los Angeles County.

Plate 65. Male (left) and female Great Basin Collared Lizards.

NATURAL HISTORY: Like other collared lizards this species inhabits rock and boulder areas in the California desert areas. It is agile, swift-footed, and an excellent jumper, leaping nimbly from rock to rock. At high speeds it is bipedal. It is most commonly seen on the tops and gently sloping surfaces of rocks. When alarmed it usually retreats to a self-made burrow beneath a boulder or to a rock crevice or rodent burrow.

Most breeding occurs from April to June. The male approaches the female with a slithering, loose-jointed gait, rapidly nodding his head. He rubs his hindlimbs and tail base against her. His comblike femoral pores may provide tactile stimulation to both sexes. He often nips her or flicks her with his tongue, and often the female may reciprocate and sometimes initiate courtship by nudging the male. In mating, the male seizes the female's neck or shoulder skin in his jaws. A nonreceptive female flattens her body dorsoventrally, puffs out her throat, elevates her tail, presents her hindquarters, and then stalks away in a stiff-legged fashion from the courting male. (Similar behavior may occur in a subordinate male approached by a dominant one.) If the male seizes her, she breaks free by rapidly shaking her head from side to side as if to say "NO!," turning on her side, or rolling over on her back, clawing and kicking. Head shaking may occur even before contact.

Eggs are usually laid in June or early July. Nests have been found in well-illuminated areas under large flat limestone rocks with the burrow opening plugged with earth. Clutch size varies from one to 12, with most clutches around four to seven eggs. Older females tend to produce the larger complements. There are one or two clutches laid in a season, and incubation lasts approximately seven to 10 weeks. Hatchlings appear from August to October and measure approximately 34 to 38 mm (1.3 to 1.5 in.) in snout-to-vent length. They usually reach sexual maturity by the following spring.

This lizard is primarily carnivorous and insectivorous, but occasionally takes some plant food. It usually watches for prey from an elevated location and captures it with a rush. It eats other lizards (*Sceloporus*, *Phrynosoma*, *Uta*, *Plestiodon*, *Xantusia*, and *Aspidoscelis*), small snakes, snails, spiders, centipedes, and a wide complement of large insects. Grasshoppers and beetles are particularly common in the diet. Among plant foods taken are the fleshy red berries of desert thorn (*Lycium andersonii*). Individuals have been seen to jump upward some 18 in. to catch a flying insect. This habit can be annoying when trying to catch this lizard with a noose made of copper wire or other highly reflective material on the end of a long pole. Instead of sitting still or retreating upon the approach of the noose, it will often leap at it when it is still some distance away. The glitter of sunlight on the noose strand apparently simulates reflections of an insect's shiny exoskeleton and initiates the leap-and-strike response.

BAJA CALIFORNIA COLLARED LIZARD *Crotaphytus vestigium*

IDENTIFICATION: This species has a maximum snout-to-vent length of 12.7 cm (5 in.). The dorsal color is brown or grayish with well-spaced slender crossbars of white or aligned white spots and intervening areas of white dots and dashes. The black collars have wide dorsal gaps between the right and left segments on the dorsal side but meet ventrally.

RANGE: It occurs on the desert side of the Peninsular Ranges from the north slope of the San Jacinto Mountains southward through Baja California, Mexico.

NATURAL HISTORY: This is a lizard of rocky desert hillsides, canyons, alluvial fans, and lava flows. Its basic natural history follows closely that described for this genus, especially that of the Great Basin Collared Lizard (*Crotaphytus bicinctores*).

Leopard Lizards (Genus *Gambelia*)

These are large, robust lizards with an elongate body, well-developed limbs, a weakly developed or absent dewlap, and a long, rounded, and rather fragile tail. The dorsal scales are granular, femoral pores are present, and they have well-developed palatine teeth. There are no collar markings.

Sexual differences in color are slight except during breeding season, when females bear orange or reddish spots and blotches on the sides of the neck and body and orange or reddish color on the underside of the tail. Males of one species, the Blunt-nosed Leopard Lizard, also produce a salmon coloration over the body during the breeding season. Leopard lizards are predatory diurnal lizards, found primarily in arid foothills and flatlands.

LONG-NOSED LEOPARD LIZARD *Gambelia wislizenii*

IDENTIFICATION: This is a large, robust lizard with a snout-to-vent range of 8.2 to 14.6 cm (3.25 to 5.75 in.). It has a long tail, relatively large head, and long snout. The head width usually is less than the distance from the nostril to the anterior border of the ear. The dorsal body is gray or brown with dusky "leopard" spots on body, limbs, and tail. The throat has dusky parallel longitudinal streaks, sometimes fragmented into short lines or spots, or fused laterally. Males have large postanal scales, a broad tail base, large femoral pores, and an average size less than that of females. When in breeding condition females develop a red-orange color on the underside of the tail and red-orange bars and spots on the face, neck, and body.

RANGE: In California it occurs east of the Cascade–Sierran Mountain system north of the Transverse Ranges and in the Peninsular Ranges and the Mojave and Colorado deserts. It is also found in the upper part of the Kern River drainage and in the areas such as Kelso Valley in the southern Sierra Nevada outside of this drainage.

NATURAL HISTORY: The Long-nosed Leopard Lizard inhabits arid and semiarid hills, washes, and flatlands, usually where there are scattered bushes and/or grass clumps that provide shelter and places of concealment from which to ambush prey. When retreating for the day, they usually

Plate 66. Female (above) and male Long-nosed Leopard Lizards.

enter burrows and plug the entrance with soil. The ground surface may be of hardpan, alluvium, gravel, or sand. Scattered rocks or rock outcrops are often present, upon which these lizards may bask and watch for prey. In dunes they are sometimes found well within areas of nearly barren windswept sand.

This is an active, diurnal lizard. In some areas they are wary, and careful stalking is necessary to get close to them. At other sites they appear little afraid and often merely flatten themselves on the ground and remain immobile when approached, or may dash to the shade pattern of a bush and remain motionless. When taken in hand they often attempt to bite and will hold the mouth open in a threatening manner, revealing a black throat lining. Territories are large and widely dispersed. Head-bobbing displays by territorial males can last four to eight seconds, considerably longer than in the Blunt-nosed Leopard Lizard (*G. sila*). In some areas males are encountered more frequently than females. Perhaps stresses associated with reproduction cause a somewhat higher mortality in females, or there may be seasonal differences in activity.

The development of the brilliant red-orange of breeding females is primed by estrogen and under the direct influence of progesterone. Females develop the color at approximately the time of ovulation. There is evidence that copulation occurs before or at about the time the nuptial coloration appears. The color remains until the eggs are laid and then fades to a darker red and eventually disappears. Nuptial colors may reappear if another clutch is laid. Egg-laying occurs from May to June. There is some evidence that communal laying occurs where there are limited sites suitable for egg deposition. Usually only one clutch is laid each year. However, under favorable conditions some individuals lay two clutches, but following droughts there may be no reproduction at all. The incubation period is about one and one half to two months. Hatchlings measure approximately 38 to 50 mm (1.5 to 2 in.) snout to vent. They appear from July to September and may be active until mid-October. Sexual maturity usually does not occur until around two years of age.

This is an opportunistic forager, feeding chiefly on large, active arthropods and lizards, but it will also eat some plant food, mainly blossoms and fruit. Lizards eaten include *Uta*, *Callisaurus*, *Sceloporus*, *Phrynosoma*, and *Aspidoscelis* spp., and they are also sometimes cannibalistic. Lizards are occasionally captured by ambush. A leopard lizard may position itself at the edge of the shade cast by a bush where its spotted and barred patterns blend, and it then dashes out to capture passing prey. Once seized, the prey is usually held tenaciously. This behavior was vividly displayed to the junior author and his herpetology class on one of his annual field trips to the Mojave Desert. These always included a stop at the massive Kelso Dunes to observe the Mojave Fringe-toed Lizard (*Uma scoparia*). However, the fringe-toes were not cooperating that day and instead were dashing across the sand surface at the first sign of people. Eventually one student spied a large juvenile sitting still with front legs fully extended. The cautious approach of the rest of the class failed to startle it, and soon the lizard was completely encircled by camera-clicking humans. What they all (including the professor) failed to notice was that there was a second lizard within their circle, its dappled body pattern blending in beautifully with the mottled shade of the small shrub under which it crouched. The only individual that apparently did know of the second lizard's presence was the fringe-toed lizard, and hence its behavior of remaining motionless. Eventually excessive movement of one overzealous photographer elicited a slight shift in head position by the fringe-toed, and that was all the stimulus the second lizard required. Bursting forth with a shower of sand from under the brush came an adult female Long-nosed Leopard Lizard. It grabbed the young fringe-toed lizard at mid-body in full stride, then rose up on the long, muscular hind legs and ran bipedally with her prize through the circle of amazed onlookers. During discussions of this unique observation on the way back to camp, the analogies to *Tyrannosaurus rex* were numerous.

CONSERVATION NOTE: Scattered populations of the Long-nosed Leopard Lizard in the Cuyama Badlands in northwestern Ventura County have characteristics that suggest past interbreeding with its close relative, the Blunt-nosed Leopard Lizard (*G. sila*). Unfortunately, the habitat of these presumed hybrid (or intergrading) populations have been degraded by off-road vehicle driving and agricultural development.

BLUNT-NOSED LEOPARD LIZARD — *Gambelia sila*

PROTECTIVE STATUS: FE, SE

IDENTIFICATION: The species is slightly smaller than *G. wislizenii*, with a snout-to-vent length range of 7.5 to 12.5 cm (3 to 5 in.). It has a short snout and more truncate body and head than the Long-nosed Leopard Lizard. Its head width is equal to or greater than the distance from the nostril to the anterior border of the ear opening. The dark spots on the dorsum are less numerous than in *G. wislizenii*, and they tend to be arranged in rows on either side of the midline. The throat color is blotched with the blotches sometimes merging to form a network. Juveniles have red spots on the body and lemon yellow on the thighs and underside of the tail. Young females are blotched laterally with yellow. The male is larger than the female, unlike our other two leopard lizards, and has a broader head and larger femoral pores than the female, enlarged postanal scales, and a swollen tail base. When in breeding color he may have a pink or salmon suffusion on the chest and sides, which may sometimes extend over the entire dorsum except for the head and over the entire undersurfaces of the body and limbs, including the throat. A breeding female has crimson spots or blotches on the sides and sometimes considerable red color on the head, around the eyes, on the thighs, and on the underside of tail. This lizard tends to match the soil color of its habitat, and on alkali flats some individuals may become nearly white when in the light phase.

RANGE: It is now confined to a small portion of the western San Joaquin Valley and fringing foothills, the adjacent Carrizo Plain, and south over the Temblor and Caliente ranges into the middle Cuyama Valley. It was formerly widespread throughout the San Joaquin Valley and its surrounding foothills.

NATURAL HISTORY: The Blunt-nosed Leopard Lizard inhabits flatlands, foothills, canyon floors, and washes where the soil is sandy, gravelly, loamy, or in varying degrees consolidated, even forming hardpan. In some areas hard alkali flats are frequented. The arid grasslands of the middle Cuyama Valley appear to be optimal habitat. Where burrows of kangaroo rats, ground squirrels, gophers, and badgers are scarce, these lizards construct burrows of their own under rocks or in earthen banks. Population densities tend to be correlated with the number of mammal burrows present, and when frightened it often seeks refuge in these retreats.

This is a wary species, running and crouching to escape detection. It is occasionally bipedal at high speed. It often bites, and commonly hisses and waves the tail when captured. Males defend territories by bobbing displays and fights, facing each other broadside with sides inflated, throat distended, and body lifted posteriorly. Attempts are made to bite the rival's flanks, and battles are often brief but vicious. Bobbing sequences in this species last only about two seconds.

The mating season is from late April through early June. The male licks the area above the groin or on the lower back of the female, crawls over her with a slight rubbing action, and secures a grip on her neck or upper back with his jaws. He straddles her back with one hind leg to achieve copulation. A nonreceptive female arches her back and jerks her head from side to side, sometimes violently shaking her entire body. If the male holds on, she may kick free with her hind legs or roll over several times. Breeding color in females may appear about one week after copulation and becomes vivid within two weeks, suggesting that in this species it may serve as a deterrent to further male advances. From two to five (usually three) eggs are laid in early June and July. Females may have up to three clutches per season. Larger females tend to have the larger clutches. Eggs are laid in burrows that are then plugged with soil. Hatchlings appear from early July to September and measure 42 to 47 mm (1.7 to 1.9 in.) snout to vent. Females become sexually mature following their second hibernation. Males may not mature until after their third hibernation.

This lizard feeds on grasshoppers, crickets, cicadas, beetles, lepidopterous larvae, other insects, mice, lizards such as *Uta*, *Sceloporus*, *Aspidoscelis*, and *Phrynosoma*, and is occasionally cannibalistic. Like the Long-nosed Leopard Lizard, they are also capable of leaping high in the air to catch insects.

CONSERVATION NOTE: Historically the Blunt-nosed Leopard Lizard probably occurred from Stanislaus County south through the San Joaquin Valley to the Tehachapi Mountains in Kern County. Today a drive along the approximately 200-mile stretch of Interstate 5 from northern Stanislaus County to Bakersfield presents a vivid picture of what has become of the native valley sink scrub, valley saltbush scrub, and valley needle grassland plant communities, the natural habitat of this lizard that once occupied nearly this entire area. More than 95 percent of this land has been converted to large-scale agriculture, intensive livestock feedlots, and urban and petroleum development. Even the remnants of these native plant communities that can still be glimpsed here and there along this route are frequently compromised by off-road vehicles. Clearly this lizard's "double-endangered" status has been a direct result of essentially uncontrolled and unregulated land conversion. It is currently protected in the Pixley Wildlife Refuge and a few other localities in the San Joaquin Valley and the extensive Carrizo Plain National Monument west of the San Joaquin Valley. However, biologically effective conservation easements and reserves in remaining prime habitat areas where this endangered lizard still persists are urgently needed to prevent it from becoming a relic species in the very near future.

COPE'S LEOPARD LIZARD *Gambelia copeii*

IDENTIFICATION: Its maximum snout-to-vent length is 12.7 cm (5 in.). Cope's Leopard Lizard resembles the Long-nosed Leopard Lizard but has much fainter dorsal spotting, and spots may be absent entirely on the head and greater neck region. The spots on the tail are large and unite to form dark crossbands. Females are usually larger than males and acquire a red or orange color pattern during the breeding season similar to that of females of the other two California leopard lizard species.

RANGE: This is primarily a Baja California (Mexico) species that enters just the extreme southwest corner of California, where it occurs at Cameron Corners, Campo, and Potrero Grade in the Peninsular Range.

NATURAL HISTORY: In California it can be found in chaparral, coastal sage scrub, and oak woodland. It is most often seen when basking on rocks or soil mound areas along small roads. Its reproductive biology probably follows closely that of *G. wislizenii*, with egg-laying occurring from April to late July. It feeds on a variety of large insects and other lizards.

Family Phrynosomatidae

This family contains the Zebra-tailed Lizard, fringe-toed lizards, rock lizards, spiny lizards, tree lizards, horned lizards, earless lizards, and the side-blotched lizards. Representatives of all except the earless lizards occur in California. This widely varied collection of lizard body forms is what was left after the former large family Iguanidae was split up, placing California species such as the Desert Iguana and Common Chuckwalla under the original family name and another small group, the collared and leopard lizards, in the family Crotaphytidae. Given this situation, it is rather difficult to present some finite features that members of the family Phrynosomatidae have in common. One of the very few but consistent characteristics is their specialized nasal passage, which has a "U"-shaped bend that traps fine particles so they do not reach the lungs. The majority of species in this family either occasionally bury themselves in sand or fine soil, or live in areas where windblown dust is common, so this one unifying feature is indeed an important part of their existence.

ZEBRA-TAILED LIZARD　　　　　　　　*Callisaurus draconoides*

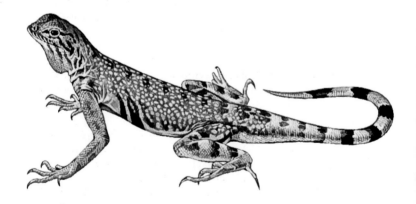

IDENTIFICATION: This is a very trim lizard with a snout-to-vent range of 6.3 to 10.1 cm (2.5 to 4 in.) and a considerably flattened body and tail. The

Plate 67. Female (left) and male Zebra-tailed Lizards.

dorsal scales are granular and unkeeled, and the ventral scales are larger, flat, and smooth. The upper labial scales are separated by oblique sutures. Ear openings are present, but without projecting scales along the anterior border. The dorsal color is pale gray to brownish gray dotted with white to yellow spots. The underside of the tail has black bars ("zebra" markings) on a white background that wrap around dorsally. There are two black bars on either side of the ventral midline that extend slightly up the sides and in the adult male are flanked by a vivid blue-green color. In the female ventrolateral markings are vague or absent. Both sexes have lemon yellow on the flanks, which is more extensive and conspicuous in females.

RANGE: It occurs throughout the southwest portion of the Great Basin Desert, Mojave Desert, and Colorado Desert from the Nevada and Arizona borders to these deserts' western edges. It can still be found on the coastal side of the mountains in southern California along Cajon and San Jacinto washes, although increasing human disturbances may be diminishing its presence.

NATURAL HISTORY: The Zebra-tailed Lizard occurs in sandy and gravelly areas, desert flats with scattered small rocks, and occasionally in predominantly rocky areas. It appears to favor the sand of desert washes, open areas of alluvium, hardpan desert, and the margins of dunes, but it is less strictly a sand dweller than its close relatives, the fringe-toed lizards (*Uma*). In the Great Basin and northern Mojave Desert it occurs in open areas in the desert thorn–hop sage plant community.

The male's side display colors are among the most vivid in California desert lizards. He flashes them by rising up on the front legs and expanding (lowering) the rib cage. Females lay from one to as many as five clutches ranging from two to 15 eggs each between June and August.

The major food of this lizard consists of larger insects (grasshoppers, beetles, robberflies, spiders) and small lizards. Like a number of other desert insectivorous species, it occasionally eats some plant material.

Flat open areas of rather firm soil seem especially suited to this fast lizard, and this plus their abundance in many areas make this species a favorite subject for desert lizard watchers. It is usually seen as it waits motionless in the open or in mottled shade for prey to appear, which it then runs after. When about to run, it curls and wags its long tail. When running bipedally at full speed (about 28 km or 18 miles per hour), the tail remains curled and is brought forward over the body. At such times bold black-and-white stripes on the underside of the tail may divert hawks and other avian predators to this appendage, which can be regenerated if lost.

Fringe-toed Lizards (Genus *Uma*)

The body and tail of these lizards are noticeably flattened. The scales are smooth and granular, and have a velvet texture dorsally. There is a fringe of elongate pointed scales along the rear side of each toe, which provides better traction when running and burrowing into soft sand. The lower jaw is countersunk, which gives the head a wedge-shaped profile for easy

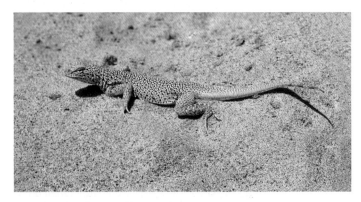

Plate 68. Mojave Fringe-toed Lizard in loose sand habitat.

penetration into sand. The eyelids have protruding scales, and when fully closed, the upper lid broadly overlaps the lower. There are also well-developed ear scales. The dorsal color is whitish or buff with numerous black spots and reticulations. The venter is white with dusky lines in the gular area and black bars toward the tip of the tail. There is a large black mark on each side of the abdomen in the Colorado and Mojave species. The Coachella Valley species lacks these conspicuous black markings.

Fringe-toed lizards are restricted to dune areas of fine windblown sand, desert flats with sand hummocks about bushes, and sand accumulations associated with playas and washes. Sand grain size is important, because they spend the night buried in the sand and retreat to it during the day to escape predators, aggressive rivals, and extremes in temperature. If sand grains are much larger than 0.375 mm (0.04 in.), quick penetration and burial is not possible and such areas are avoided. Burial is shallow, usually at depths of 0.6 to 3 cm (0.25 to 1.5 in.), but may be a depth of 15 cm (6 in.) or more as on the slip faces of dunes. Refuge is also sought in the burrows of round-tailed ground squirrels, kangaroo rats, and other desert vertebrates. Submergence may be so rapid and complete that unless the lizard is watched closely, its burial site is easily overlooked. When fleeing, fringe-toed lizards often turn sharply at the end of the run and bury themselves at nearly a right angle to their course. This occasionally happens after the lizard is momentarily out of sight after running behind a clump of vegetation. A human pursuer who is unaware of this behavior is often left standing in complete bewilderment upon seeing the lizard's tracks stop in the middle of a large expanse of sand.

Fringe-toed lizards rival *Callisaurus* and *Dipsosaurus* in speed, despite their unstable substratum, and are bipedal at high speed, touching down only occasionally with a forefoot to maintain balance. Because they overstep the small forefoot marks with their hind feet, which leave a much larger dent in the sand—especially where the heels strike the surface as opposed to the toes, the direction of travel can easily be determined.

The habitat of *Uma* is characterized by annual plant growth (following rains) and scattered perennial shrubs, within and especially around the periphery of the sand deposits. These lizards feed mainly on arthropods attracted to these plants or that are blown into the sand deposits. Insects that occur on the perennial shrubs are their mainstay, and they may at times climb to a height of 0.6 m (about 2 ft) or more to catch them. Buried *Uma* are able to detect prey on the surface, and surface-foraging individuals may stop, dig, back off, and then dart forward to catch an insect they have uncovered in the sand. They perhaps hear the sounds of such buried prey. Fringe-toed lizards occasionally eat other lizards and have been observed to be cannibals in captivity. They may also eat their own shed skins, and leaves and blossoms are taken in considerable quantity at times.

Fringe-toed lizards are strongly territorial, and in captivity they form dominance hierarchies. The male display pattern is quite distinctive.

They flatten the body and tilt it toward the opponent, drawing the lateral abdominal skin upward on the side closest to their rival and thereby displaying the ventrolateral markings. This method of ventral marking display may be related to the flattened body form, an adaptation for sand burrowing. During displays bobbing may occur, and sometimes the hind leg on the far side is lifted off the ground when tilting. The tail may be twitched from side to side and the dewlap lowered. The rivals, usually facing in opposite directions, circle and occasionally lunge at and bite one another. In the presence of a dominant individual, a subordinate may assume a submissive pose, lying flat on the sand with its head down, or it may bury itself. In captivity, dominants sometimes were observed digging out such buried individuals to pursue them further.

During courtship the male approaches the female with head lowered and rapidly nodding. He alternately moves his forefeet in a rapid waving motion. A hold is secured on the female's neck or shoulder with his jaws, and he scratches her sides with his feet, evidently in an effort to arrange her body so he can achieve union. Refractory females present their hindquarters to the male, stand high with body flattened and back slightly arched, and elevate and wave their banded tail. Breeding begins in April and extends to late July, mid-August, or mid-September, depending on the species. Eggs are laid in the late spring and summer. In dry years when food is scarce, reproduction may not occur.

COACHELLA VALLEY FRINGE-TOED LIZARD *Uma inornata*

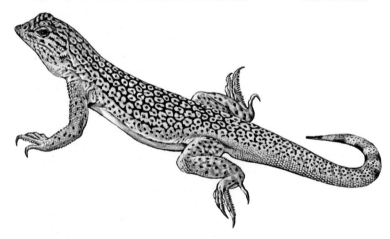

PROTECTIVE STATUS: FE, ST

IDENTIFICATION: It is similar in size to the Colorado Desert Fringe-toed Lizard (*U. notata*) with a snout-to-vent range of 7 to 12.4 cm (2.75 to 4.9 in.), but it lacks conspicuous black ventrolateral markings, and males lack orange on the sides of their abdomen. Ventrolateral black markings are either absent or evident only as a minute cluster of black dots, a single dot, or a small narrow black bar on each side of the belly. The dorsal background color is ash-white in contrast to the back color of other fringe-toed species, which is buff. During the breeding season the lower side area of adults is suffused with pink, and in some females this color may intensify to bright orange. Orange also occurs on the festoons of the eye.

RANGE: This lizard originally occurred in soft-sand areas of the Colorado Desert throughout the Coachella Valley and the San Gorgonio Pass area at the Valley's northwest end, from Snow Creek wash southeast to the vicinity of La Quinta and Indio, Riverside County. At present its greatly reduced numbers are mainly restricted to three small reserves.

NATURAL HISTORY: Its habitat consists of wind-sorted sand dunes, hummocks, and flats with scattered desert bushes such as creosote bush, burrobush, indigo bush, and mesquite clumps cresting the sand hummocks. Desert willow is often present in areas of favorable water table. Sand verbena and other annuals are extensive in some seasons. The sand color in its range is ash-white or pale gray in contrast to the buff sands of the Mojave Desert to the north and the Colorado Desert to the south.

Emergence from hibernation begins in late February and March, although young may be abroad on warm days in winter. The active period lasts until October. When two individuals approach, one or the other may engage in "foot waving." This is done with the forelimbs, first one and then the other, elevated and rapidly vibrated in a vertical plane.

Breeding occurs from April to mid-August and during this period adults become more colorful. Two to four eggs form a clutch, and probably more than one clutch is produced during favorable seasons.

Its food consists of insects such as beetles, grasshoppers, ants, and other arthropods that inhabit soft-sand areas. Blossoms, buds, and leaves are also occasionally eaten. In dry years these lizards have been found climbing to a height of several feet in perennial bushes, evidently in search of insect prey.

CONSERVATION NOTE: This lizard and the Coachella Valley ecosystem of which it is a part continue to be seriously compromised by human developments and activities. These include housing and golf course development, road construction, mining of sand, agriculture, windbreaks that impede sand movements, and off-road vehicles. All this has resulted in a current estimated loss of 75 to 90 percent of this lizard's original habitat. This unfortunate situation is a classic example of what can happen to a species when the economic value of its geographic range elevates to such an extent that concerns for the local fauna are consistently overridden at all levels of government. Fortunately, a few small reserves have been established at Thousand Palms, Willow Hole, and Whitewater River. However, they collectively protect only a small portion of this lizard's original range, and few *U. inornata* now exist outside their boundaries.

Even within these small preserves this species now experiences a new threat, Asian Mustard (*Brassica tournefortii*) that has invaded the desert and sensitive dune areas in recent years. Management of this invasive exotic is required.

COLORADO DESERT FRINGE-TOED LIZARD *Uma notata*

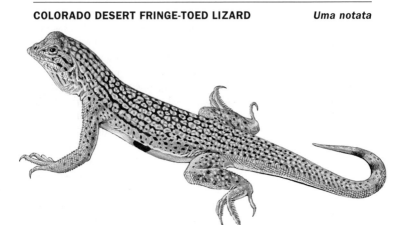

PROTECTIVE STATUS: CSSC

IDENTIFICATION: This fringe-toed lizard has a snout-to-vent range of 7 to 12.2 cm (2.75 to 4.8 in.). Its dorsal ground color closely matches the buff sand color of its habitat. The dorsal pattern consists of small black flecks and ocelli, which are suggestive of dark particles of sand and shadows cast by sand grains. The venter is white with dark diagonal lines on the throat, black crossbars on the underside of the tail, and a large black blotch on each side of the belly. In adults the sides of the belly are often pinkish or orange, which is especially vivid in breeding individuals.

RANGE: It occurs on the Colorado Desert in the vicinity of the Salton Sea and the Imperial Sand Hills in California, and south across the Colorado River delta to Tepoca Bay, Sonora, and northeastern Baja California, Mexico.

NATURAL HISTORY: The Colorado Desert Fringe-toed Lizard is restricted to fine, loose sand deposits with grain size no coarser that 0.375 mm (0.04 in.). Sand grains larger than this size may hinder speedy burrowing, and thus sites where it occurs may be avoided. Predominant plant species present are creosote bush, burrobush, mesquite, palo verde, saltbush, buckwheat, croton, and the desert sunflower, but barren sand dunes, completely devoid of plant life, are also entered.

Breeding occurs from April to mid-September. The nuptial colors, present throughout the breeding season, usually do not reach maximal development until May. Not all adults become colorful, and probably not all breed every year. When mating, the male bobs rapidly in front of the female, then approaches her from behind and seizes her neck skin in his jaws. He occasionally lifts and alternately vibrates his front legs. He then scratches her side with his adjacent front and hind feet, shifting from one side of her body to the other while maintaining the neck hold, and positioning his tail beneath hers, so as to insert a hemipenis. Oviductal eggs have been found from mid-May to late August, indicating an extended laying period with most commonly two to three eggs per clutch. Hatchlings are found from mid-July to early September, the time varying with climate and other factors.

CONSERVATION NOTE: The Colorado Desert Fringe-toed Lizard and its companion species, Coachella Valley Fringe-toed Lizard (*U. inornata*) and the Mojave Fringe-toed Lizard (*U. scoparia*) are three examples of extreme habitat specialists that can thrive only in the widely scattered soft, small-grain sand areas of the Colorado and Mojave deserts, respectively. Given the ongoing impacts to these habitats by off-road vehicles, sand mining, and ongoing and expanding agriculture in the Imperial Valley, they are well deserving of their listing as a California Species of Special Concern by the CDFG. It is therefore especially disturbing that this same agency lists them as two of the many California reptiles that fishing licenses holders can legally take two of each from their unique natural habitat. One would think that the current plight of the Coachella Valley Fringe-toed Lizard would be warning enough that unregulated, legalized collecting of these special-status species has no place here.

MOJAVE FRINGE-TOED LIZARD *Uma scoparia*

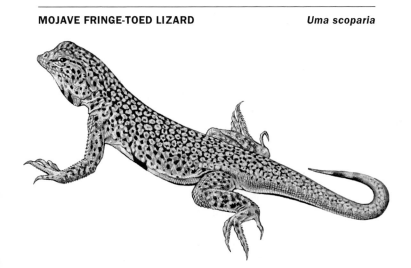

PROTECTIVE STATUS: CSSC

IDENTIFICATION: It is similar in size to other western *Uma* species, with snout-to-vent length of 7 to 11.6 cm (2.75 to 4.5 in.), but differs in usually having five internasal scales instead of three. The dorsal ocelli are scattered rather than arranged in rows, and there are well-defined black gular crescents. Like the Colorado Desert Fringe-toed Lizard, it also has large black blotches on each side of the belly. The breeding color in both sexes includes pinkish to orange suffusion along the ventral edge of the lateral fold, but in the breeding females it is more pronounced.

RANGE: This species occurs in the Mojave Desert from the Dumont Dunes in northeastern San Bernardino County north to the southern end of Death Valley National Park, and west to Lovejoy and Piute Buttes, Los Angeles County.

NATURAL HISTORY: Its habits and habitat are similar to those of the Colorado Desert Fringe-toed Lizard. Breeding usually occurs from April to late July. Nuptial colors usually do not reach maximum development until May, and not all adults develop breeding colors. Successful breeding depends on winter rainfall, which makes possible the plant growth upon which dune-oriented insects feed. The food of the Mojave Desert Fringe-toed Lizard includes dune-dwelling ant species, beetles (including ladybird beetles), butterfly and moth larvae, ant lions, spiders, and plant seeds and flowers. The conservation note for this species is the same as that for *U. notata*.

Spiny Lizards (Genus *Sceloporus*)

Spiny lizards have relatively large heads well set off from the neck; the body may be either rounded or somewhat flattened, the limbs are well developed with long, slender digits without expanded tips, and the tail is longer than the snout-to-vent length. All members of the genus have pointed, keeled, and strongly overlapping scales, which are the most defining characteristic of this group. Males of most species have a blue patch on the throat and an elongate patch of blue in the ventrolateral area on each side of the belly. Males have larger postanal scales, larger femoral pores, and a more swollen tail base than females.

Across North and Central America members of this genus occupy a wide variety of habitats including deserts, grasslands, chaparral, and woodland; deciduous, coniferous, and tropical forests; and alpine meadows. Within these habitats species are segregated by their preferences for rock outcrops, sand dunes, beaches, crevices, tree trunks, tree canopies, bunchgrass, shrubs, salt flats, and gravel plains.

Spiny lizards are often called "blue-bellies" or "swifts." The blue or sometimes pink markings, best developed in males, are generally not noticeable until the lizard displays by flattening its sides and lowering its dewlap. These displays of color are coupled with pushup and/or head-bob sequences for territorial defense and the courtship of females. These lizards are also classic heliotherms and begin each day by basking in an open elevated place. Such basking sites are often preferred male display sites and are usually situated near the center of territories. Adult breeding males will often remain at a display post all day and thermoregulate by moving back and forth between the sunlit and shaded parts of the habitat. These behaviors are easy to observe, and along with the large geographic range of two species, the Western Fence Lizard (*S. occidentalis*) and the Sagebrush Lizard (*S. graciosus*), make this group the best known among California lizards.

All California spiny lizard species lay eggs, but elsewhere some are live-bearing. Many species have a well-defined breeding season, whereas others breed over much of their season of activity and may produce several clutches or broods. In those oviparous species studied, embryos in different stages of development have been found at the time of oviposition, suggesting that these lizards retain their eggs for varying lengths of time and lay them when rainfall or other factors are advantageous for embryo development, hatching, and juvenile foraging.

WESTERN FENCE LIZARD	*Sceloporus occidentalis*
NORTHWESTERN FENCE LIZARD	*S. o. occidentalis*
SIERRA FENCE LIZARD	*S. o. taylori*
ISLAND FENCE LIZARD	***Sceloporus becki***

Western Fence Lizard

TAXONOMIC NOTE: Morphological evidence was recently presented to recognize the Island Fence Lizard at the species level, as it was on its original description in 1905.

IDENTIFICATION: The Western Fence Lizard has the classic lizard body form with a snout-to-vent range of 5.7 to 8.9 cm (2.25 to 3.5 in.). Two species characteristics are the separation of the anterior border of the supraocular scales from the large head plates by a row of small scales and the pointed keeled scales on the rear of the thigh that extend nearly to the femoral pore row. The dorsal color ranges from nearly black in cold individuals to brown or gray and is usually marked with blotches or scallops of dark brown or black arranged in two longitudinal rows down the back. The venter is whitish, often with some dark spotting, and the posterior of the limbs, especially the thighs, is usually yellow to orange. Males have large postanal scales and a broader tail base and larger femoral pores than females. They also have a prominent blue or blue-green patch on each side of the belly and usually a blue throat with grayish or black chin. In the female blue areas are less intense than in the male or may be absent. The Island Fence Lizard species is similar to its mainland counterpart, but males have a black throat with black lines radiating forward over a blue chin and blue speckling on the face. All ages of both sexes have thin, vertical black lines across the mouth. The body size of the Sierra Fence Lizard is consistently at the upper end of that for the Western Fence Lizard, and the males usually have solid blue venters.

In some areas of the state *S. occidentalis* coexists with two very similar-appearing species, the Sagebrush Lizard (*Sceloporus graciosus*) and the Southern Sagebrush Lizard (*S. vandenburgianus*). Both of these lizards usually have a rust-colored patch on the side of the neck and body (seldom seen in *S. occidentalis*) and a black bar on the shoulder but no yellow color on the rear of the limbs. The dorsal scales are smaller, and the blue throat patch of males is flecked with white or pink.

RANGE: The Western Fence Lizard occurs in all 12 California geographic subdivisions and spans elevations from sea level to around 1,800 m (7,000 ft). The Sierra Fence Lizard occurs above 1,800 m (7,000 ft) to around 3,353 m (11,000 ft) from the Tuolumne River drainage to Sequoia National Park. The Island Fence Lizard occurs on three of the Channel Islands.

NATURAL HISTORY: *S. occidentalis* is perhaps the best example of a habitat generalist among California lizards. Its varied habitats include grassland, broken chaparral, high-desert brushland, woodland, streamside growth, and open coniferous forest to timberline. Within such habitats it frequents talus and rock outcrops on hillsides in canyons. In areas of human occupancy included in the section "Amphibian and Reptile Distribution throughout California's Habitat Complex," it is attracted to old buildings, woodpiles, fences, and stone walls. The two large redwood barns, tank house, and various woodpiles on the junior author's farm probably house more Western Fence Lizards today than existed in that small area when it was a part of the upper flood plain of the nearby Stanislaus River.

The Western Fence Lizard is perhaps the best-known lizard species in California. Its extensive geographic range in this state includes many areas of high human population density, where it may regularly be seen in parks and outlying suburban yards and gardens. The basking and display behavior described in the genus introduction make it an ideal subject for the beginning lizard watcher. On a warm spring morning you may observe it at relatively close range (4 to 5 m or 13 to 16 ft) without causing significant disturbance. The first activity you will see is the initial morning basking behavior, which may be brief or lengthy, depending on air temperature. At this time a fence lizard will be much darker than it will appear for the rest of the day, because of the expansion of the dark pigment in its melanophores to facilitate heat absorption. It also orients its body so that the dorsal surface is approximately at a right angle to the sunlight path. During this basking period, which may be as brief as five to 10 minutes, the lizard's deep-body temperature rises from that of the subsoil area to within a few degrees of our own. Once the preferred body temperature range is attained, it reorients its body axis so that it is parallel to the sunlight path but may revert to a full or partial basking position several more times, especially if the day is cool.

Once at full "operating temperature," males normally spend the rest of a spring day displaying and feeding from one or more high perches, which are usually in the heart of its territory. The colorful male in Plate 69 was photographed while performing pushup displays on lichen-covered boulders in the Merced River Canyon.

Like other *Sceloporus* lizards, the Western Fence Lizard usually captures insect prey by first locating it visually, often at a distance of a meter (a little more than three feet) or more, and then dashes out to grab it and return quickly to its lookout site. The intervals between feeding are used for the pushup

Plate 69. Male Western Fence Lizard on lichen-covered rock.

and blue-belly flashing display for which this genus is known. These are usually elicited by the appearance of another fence lizard within the male's field of vision, whether it be a male or a female. The latter will usually not return the display, an indication that it is a female. She will then be courted and perhaps bred if she is receptive. A nonreceptive female signals her unwillingness by laterally flattening her body, humping her back, and moving with short stiff-legged hops. If instead the intruder is another male that also displays, it will receive more pushups from the territorial male, and if these do not repel it, body jousting and tail biting will ensue. The resident male, especially a large one, almost always wins such bouts and retains his real estate.

If you happen to approach a territorial male too closely, you may also receive a series of pushups. This is an example of displacement behavior, a situation in which a normal activity, in this case territorial display, is presented out of context. It often occurs when an animal is in a highly energetic state, of which aggressive or defensive behavior is usually the most common.

Eggs are laid from early April to mid-July, depending on locality, in a hole dug by the female. Clutch size ranges from three to 16 eggs, with seven to nine most common. From one to three clutches are produced by a female per year. Embryo development time is about 10 weeks but will vary with nest temperature. Hatchlings appear from mid-July to early October and measure 22 to 26 mm (about 1 in.). Some individuals reach sexual maturity by June following their first hibernation, but others do not breed until after their second hibernation.

A large variety of insects, particularly flies, beetles, ants, and grasshoppers, are present in the diet. Spiders, centipedes, isopods, scorpions, and mites are also taken. Some cannibalism of the young occurs. Termites are captured chiefly during periods of emergence of winged sexual stages.

SAGEBRUSH LIZARD — *Sceloporus graciosus*
SOUTHERN SAGEBRUSH LIZARD — *Sceloporus vandenburgianus*

Sagebrush Lizard

PROTECTIVE STATUS OF S. VANDENBURGIANUS: CSSC

TAXONOMIC NOTE: Based on a monograph by Wiens and Reeder (1997), many now recognize a former subspecies, the Southern Sagebrush Lizard, as a separate species, and for the sake of consistency with several other recent field guides, we have followed suit. Because the morphology and natural history of these two species are nearly identical, we discuss both together, pointing out differences where they occur.

IDENTIFICATION: In both body form and color they closely resemble the Western Fence Lizard (*S. occidentalis*), which occurs together with them in many places in California. However, there is often a black bar on the shoulder of sagebrush lizards, and the axilla areas and sides of the neck are usually rust-colored. The scales on the back of the thighs of the two sagebrush lizards are normally granular, whereas these are mostly keeled in *S. occidentalis*. A main distinction from *S. occidentalis* is that the dorsal scales are much smaller and more numerous, resulting in a much less spiny appearance in both sagebrush lizards.

The differences between these two sagebrush lizards are far more subtle. The snout-to-vent range for the Sagebrush Lizard is 4.7 to 8.9 cm (1.9 to 3.5 in.), and that for the Southern Sagebrush Lizard is 4.8 to 7.3 cm (1.9 to 2.9 in.). In *S. graciosus* there are 50 to 68 (average 61) scales between the interparietal scale and the point at which the posterior aspect of the thigh joins the body, while in *S. vandenburgianus* this scale number ranges between 48 and 66 (average 55). In the male Sagebrush Lizard the blue throat and belly patches are separated by whitish areas, and the female is whitish below. In male Southern Sagebrush Lizards the blue or black belly color often joins the blue throat patch with no separating light area. The ventral surfaces of both the tail and thighs are frequently blue, and females are often dusky rather than whitish below. Breeding females of both species exhibit intensified orange-red coloration along the sides of the neck and body.

RANGE: The Sagebrush Lizard occurs throughout the North Coast Range, the Cascade Range, the Sierra Nevada, the Tehachapi Range, and the northern part of the South Coast Range. It is absent from the Great Valley except in the Sutter Buttes. Populations also exist on Mount Diablo in the Bay Area, the San Benito Mountains, and on Telescope Peak in the Panamint Mountains in the Mojave Desert. It also has an extensive range throughout the sagebrush regions of the west. The Southern Sagebrush Lizard occupies two separated regions in the Transverse Range and the Peninsular Range. Both species occur from around 150 to 3,180 m (500 to 10,500 ft) above sea level.

NATURAL HISTORY: The Sagebrush Lizard is most prevalent in large sagebrush stands, but also occurs in chaparral, piñon–juniper woodland, pine and fir forests, and even in riparian situations in coastal redwood forests. The Southern Sagebrush Lizard is found primarily in chaparral and montane plant associations. The one common characteristic of these varied plant associations with respect to both species is the presence of areas with good sunlight penetration for basking and scattered low brush, into which they often retreat and climb.

These two sagebrush lizards are similar in structure, behavior, and food habits to the Western Fence Lizard. However, a few subtle differences may serve to reduce competition between them. Sagebrush lizards are on the average smaller and in general appear to be better able to function efficiently at low temperatures. On Mount Diablo, Contra Costa County, Steve Ruth, a colleague of the authors, noted that *S. graciosus* preferred areas of massive rock (including rock faces), whereas *S. occidentalis* occupied the less steep talus and open grassy areas. When *S. occidentalis* entered areas of massive rock, it appeared to be less agile in climbing than *S. graciosus*.

Sagebrush lizards are also less fecund than the Western Fence Lizard, with an average clutch size of three to four eggs. They feed on a wide variety of small ground-dwelling insects and other arthropods including spiders, mites, ticks, and scorpions.

DESERT SPINY LIZARD *Sceloporus magister*

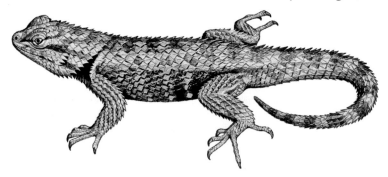

IDENTIFICATION: This is a large, stocky, rough-scaled lizard with a snout-to-vent range of 8.2 to 14.2 cm (3.25 to 5.7 in.). Adult males are extremely robust and well deserving of the "majestic" designation. The body is straw-colored, gray, or brown with vague to well-defined dusky crossbands or spots. The sides are often rust-colored, and there are two dark lines on the face, one extending backward from the eye and the other from below it. A black wedge-shaped mark is present on each shoulder and edged posteriorly with pale yellow or white. These marks are usually not connected at dorsal midline. Juveniles have numerous small blotches of irregular size and shape that tend to be arranged in four longitudinal rows. Males are on average slightly larger than females and have a blue-green throat patch and ventrolateral areas edged medially with black. The head of a breeding female is orange or reddish.

Plate 70. Boldly marked underside of a male Desert Spiny Lizard.

RANGE: This species occurs throughout much of the Mojave and Colorado deserts and from the Panoche Hills south to Buena Vista Hills in the eastern portion of the South Coast Range. It has also been found on the floor of the southern Great Valley south of McKittrick along Highway 33.

NATURAL HISTORY: This is a spiny lizard of arid and semi-arid regions. Like most other members of its genus, it is a good climber and usually associated with trees, large shrubs, or rocks. Vegetation at occurrence sites includes Joshua trees, mesquite, arid grasslands, palo verde, ironwood, smoke tree, juniper, mesquite woodlands, and ghost (gray) pine in the Diablo Range. Prominent display sites are an important element in this lizard's habitat, and along some desert roads where there is low vegetation but no trees or large rocks, wooden electric and telephone poles are often utilized for both basking and display. Males will climb 6 m (20 ft) or more to engage in such activities and are quite visible while driving a desert road. Retreat sites include rock crevices, tree cavities, woodrat middens, and rodent burrows.

At breeding time a pair will circle one another as the male approaches with dewlap extended, sides flattened, and head moving rapidly up and down. When courting, females may seek out males, crawling over and under them and licking their noses and cloacal regions. Both sexes often lick the substratum where body contact has occurred. Egg-laying occurs in May and June. Females produce more than one clutch a season. Clutches vary from four to 19 eggs, often containing around seven to 12. Hatchlings appear from late June through October and vary in snout-to-vent length from around 29 to 41 mm (1.15 to 1.63 in.). They may reach sexual maturity before they are two years old.

GRANITE SPINY LIZARD *Sceloporus orcutti*

IDENTIFICATION: This lizard is similar in structure to *S. magister* but has a smaller snout-to-vent length range (8.2 to 11.7 cm or 3.25 to 4.6 in.). It differs in having dorsal scales with shorter points and weaker keels, shorter more rounded auriculars, and generally darker coloration. Males have blue to purple in the dorsal scales, and the ventral surface of the body is typically solid blue. There is a dark shoulder wedge on each side in front of the forelimbs, but it is often obscure because of dark coloration. Juveniles have a rust-colored head and contrasting dark and light crossbanding on the body and tail. Males are on average slightly larger than females and are bluish and yellow-green dorsally, the latter especially during breeding season. They also have a broad purple stripe down the middle of their back. The ventral blue color is reduced or absent in females, and females have dorsal barring, giving a banded appearance.

RANGE: *S. orcutti* occurs from the north side of San Gorgonio Pass in the Transverse Range southward on each side on the Peninsular Ranges to near the tip of Baja California, Mexico.

Plate 71. Basking Granite Spiny Lizard.

NATURAL HISTORY: On the desert side of the mountains it is found in rocky canyons and on the rocky upper portions of alluvial fans where there is sufficient moisture for growth of chaparral, palms, or mesquite. On the coastal side it occurs in areas of coastal sage and chaparral where there are large boulders and rock outcrops. It also ranges into the ponderosa pine–oak zone of the Peninsular Ranges.

This is a rock-dwelling lizard, often found basking on rock prominences. Because of its size and dark coloration, it is conspicuous on light-colored surfaces. Shelter is often sought beneath rocks or in crevices, occasionally by climbing trees.

Most adults apparently remain in hibernation between mid-November and early March, no matter what the weather. They may aggregate under rock overhangs, where as many as 37 individuals have been found together. A "den" site near Hemet, Riverside County, was used over a period of several years. The den was a horizontal crevice in a large granite monolith about 3 m (10 ft) high. *S. orcutti* converged on the site from surrounding areas in the fall and departed over a week's period in the spring. This behavior is reminiscent of that seen in communal denning snakes and suggests that prime winter retreat sites are scarce. On warm bright days from 19 to as many as 23 individuals, mostly adults but also immatures including some very small young, were seen basking close to the crevice opening. Once the spring dispersal is over, this lizard is usually seen singly.

Eggs are laid from May to July. Clutch size ranges from six to 15, and a single clutch a year is probably the rule. As in other oviparous sceloporine lizards studied, embryonic development may have begun at the time of laying. Hatchlings appear mostly from late July to late September, but some have been seen as late as November. In one study in the Riverside area, reproductive success of this species varied greatly from year to year. This was evidently in response to the amount and distribution of the annual rainfall.

Side-blotched Lizards (Genus *Uta*)

These are small terrestrial lizards found in arid and semiarid environments. There are five to 10 species, depending on one's taxonomic viewpoint. All of them except *Uta stansburiana* are confined to islands in the Gulf of California and off the west coast of Baja California, Mexico.

COMMON SIDE-BLOTCHED LIZARD　　　　　　　　*Uta stansburiana*

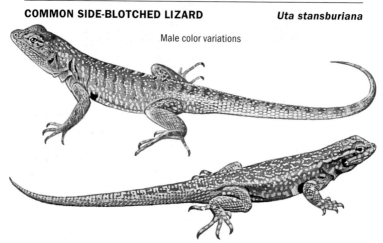

Male color variations

IDENTIFICATION: This is one of our smallest heliotherms, with a snout-to-vent range of 3.8 to 6.3 cm (1.5 to 2.5 in.). The dorsal scales are small, but some enlarged ones occur on the head, and the mid-dorsal scales are keeled. The dorsum is brown-rust to gray in California individuals and may be blotched or speckled with blue, yellow, and orange. There is an intense black to faint bluish black spot in the axilla (the basis for its common name), usually set off with an edging of lighter color. There are no ventrolateral belly color patches. Males have a more well-defined side blotch than females and are capable of marked color change, involving suffusion of the upper surfaces with pale-blue flecks. Females have brown and white blotching with occasional stripes but no blue speckling, and the "side blotch" is less defined.

RANGE: In California it ranges south from the Bay Area through the South Coast Range, Great Valley, Southern Coastal California, and the three

Plate 72. Male (left) and female Side-blotched Lizards.

desert regions, including the northern segment of the Great Basin Desert. It also occurs on Santa Cruz, Anacapa, San Nicolas, San Clemente, and Santa Catalina islands off the Pacific coast of California, throughout the mainland, and on many islands associated with Baja California, Mexico. It is absent or in scattered local pockets in the coastal fog belt from Santa Barbara County northward.

NATURAL HISTORY: This is primarily an arid to semiarid flatlands and foothills species, but it also sometimes ranges well up the slopes of mountains into upland areas. The substratum may be of coarse alluvium, small rocks, gravel, sand, or hardpan, but usually some sand is present. It frequents open shrub and grass habitats, areas of broken chaparral, and open forests. Areas of close plant growth are generally avoided. It is widespread in the creosote bush, burrobush, and basin sagebrush plant communities containing grass clumps, rock crevices, small-animal burrows, or woodrat nests, and some loose soil for nesting and burrowing. Rocks, logs, and sand hummocks that form at the base of bushes serve as lookout perches.

This is chiefly a ground-dwelling species, but it ascends rocks, logs, and other objects for display, defense of territory, and the sighting of prey, while all the time remaining near cover. At higher elevations activity ceases or is greatly reduced in winter months, but to the south individuals may be abroad in winter during warm clear weather. The small size of these lizards allows quick warming and makes it possible for them to take advantage of short periods of favorable temperatures. Males patrol their territories and engage in pushup displays. A display usually involves elevation of the foreparts by flexing and extending the forelimbs. When a display is intense, the hindlimbs are also used. In a challenge display, the dewlap is lowered and the sides flattened, revealing the axillary spot. A pushup display usually contains one to four bobs, often ascending in height, the highest one sometimes given as the lizard rises on its front toes. The number of bobs varies between and within populations.

Strong aggressive behavior between individuals of the same sex is often expressed by polygynous, orange-throated males, while other males (blue-throated) often exhibit monogamous relationships. Yellow-throated males are "sneakers" that resemble females, allowing them to slip in and copulate while other males fight and maintain their territories. However, males spend much time courting any female they encounter near or within their territory. This helps to ensure contact with the female during her short period of receptivity.

Clutch number and size varies with latitude and may be as high as eight clutches varying from one to eight eggs in the southern part of the range. These numbers may vary from year to year at a given locality, depending on rainfall and other factors that affect food supply. Incubation periods lasting from around one and a half to slightly over two months have been recorded. Hatchlings appear from late May to September depending upon locality and the timing of the clutch, and measure 20 to 25 mm (0.8 to 1.0 in.) in snout-to-vent length. Most individuals in southern populations reach sexual maturity in a year or less, and a few adults live to breed a second season. The high reproductive capacity and short life span of this lizard is reminiscent of that of microtine rodents, and like that group it provides a prey base for larger lizard and snake species. *U. stansburiana* feeds on a wide variety of small insects and other arthropods. Its apparent preference for and ability to catch relatively small invertebrate prey appears to be a key factor in feeding niche separation between it and larger insectivorous lizards.

Brush and Tree Lizards (Genus *Urosaurus*)

These lizards appear somewhat similar to the spiny lizards (genus *Sceloporus*), but they have weakly keeled scales only on the dorsal body and unlike that group have a fully developed gular (throat) fold.

LONG-TAILED BRUSH LIZARD *Urosaurus graciosus*

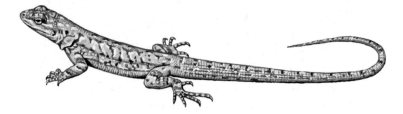

IDENTIFICATION: This is a very slim-bodied lizard with a snout-to-vent range of 4.7 to 6.6 cm (1.9 to 2.6 in.). However, its total length is much greater, because its tail is often over twice as long as its head–body length. It has a broad longitudinal band of enlarged, keeled, and overlapping scales down the middle of its back. The remaining dorsal scales, not including the head scales and keeled pointed scales of tail and upper surfaces of limbs, are small and smooth. The dorsum is dark to light ash-gray (depending on color phase), with dark to vague crossbands on the back. The vent is beige with some grayish flecking. A throat patch of reddish, orange, or lemon-yellow is present in both sexes. Males have powder-blue to greenish ventrolateral belly patches flecked with white.

RANGE: In California it is found in open brush habitats of the Mojave and Colorado deserts.

NATURAL HISTORY: This lizard follows closely the range of the creosote bush–burrobush desert scrub community. It is present over miles of flatland desert in such habitat. It also occurs in trees and bushes of wash and riparian environments, and ascends the alluvial fans and lower slopes of the mountains. Although widespread and common, it is seldom seen because of its small size, brush-dwelling and arboreal habits, and cryptic, bark-matching coloration. It descends to burrow into the soil beneath the trees and bushes it inhabits to avoid wind or unfavorable temperatures or to lay eggs. Thus, the presence of nearby loose soil, alluvium, or sand is of importance. Bushes at the crest of windblown sand hummocks are often frequented, especially those with exposed root tangles that provide

Plate 73. Long-tailed Brush Lizard on creosote bush branch.

above-ground cover. It enters dune habitats where there are mesquite, galleta grass, creosote bush, or other vegetation suitable for climbing.

This lizard spends most of its time in bushes and trees, where it usually conceals itself by lying with its body parallel and closely applied to a branch. Its coloration blends with the surface upon which it rests, and its long, slender tail resembles a twig. The long tail seems to be important as a balancing organ in climbing, and it is notable that tail loss frequency is low. It is also capable of marked and rapid color change, especially when disturbed. In creosote bushes, it is often found toward the outer part of the bush on the upper surfaces of branches, and frequently in a head-down position, which permits quick escape to the denser growth of larger branches in the interior of the bush or the exposed root tangle sometimes present at its base. When basking, especially in the morning or late afternoon, it seeks the sunny side of a bush. To find these lizards, carefully scrutinize branches, one at a time, of mesquite, creosote bushes, and other shrub growth in sandy areas. Even a careful inspection of limbs may not be successful at first, because this lizard will often inch its way to the side of a branch opposite from the viewer to avoid detection. The best way to counter this move is for a second observer to approach from the other side.

In creosote bush habitat this lizard appears to be chiefly monogamous, with a single pair to a bush or group of bushes. Typical iguanid bobbing display, fighting, and mating behavior have been observed. A displaying individual raises and lowers its entire body by fully extending and flexing both front and hind legs. In a challenge display, the dewlap is lowered and the sides are compressed. Bobbing starts with a full bob, followed by several partial pushups, then a pause at two thirds limb extension, and finally a series of four to six jerky pushups progressively decreasing in height. One or two clutches, averaging four to five eggs each, are laid between May and August. Hatchlings appear July to October and average about 23 to 24 mm (about 1 in.) in snout-to-vent length.

Foraging commonly takes place in the outer canopy of vegetation that is usually little penetrated by other climbing species with which it may share this habitat. There it lies in wait for its prey, seizing ants and other insects that pass by. It may also actively forage on limbs or on the ground.

BLACK-TAILED BRUSH LIZARD *Urosaurus nigricaudus*

Plate 74. Black-tailed Brush Lizard.

IDENTIFICATION: This small climbing lizard has a snout-to-vent range of 3.8 to 5.1 cm (1.5 to 2 in.). Its dark tail is consistently shorter than that of the Long-tailed Brush Lizard, and it has a broad area of enlarged keeled scales down the center of the back. Small granular scales cover the sides. The dorsum varies from gray to brown-black, and the tail is sooty to black. There is a row of dark crossbars on each side of the back. Males have a central orange or yellow throat pattern and blue to blue-green belly patches. Females have lighter throat coloration and no belly patches. This species also has a dark color phase.

RANGE: *U. nigricaudus* occurs throughout the length of Baja California, Mexico, and extends across the U.S. border into the southern part of Southern Coastal California and the Peninsular Ranges, where it occurs mostly on the desert side of these mountains.

NATURAL HISTORY: This lizard prefers rocky habitats near streams or desert springs with stands of oak, sycamore, desert willow, and assorted chaparral shrubs. It is an agile climber on rocks and human-made structures as well as brush and trees. It is seldom seen on the ground. Like the Long-tailed Brush Lizard (*U. graciosus*), its climbing ability permits feeding niche separation between it and the assorted ground-dwelling insectivorous lizard species that occur with it.

ORNATE TREE LIZARD *Urosaurus ornatus*

IDENTIFICATION: This slim-bodied lizard is similar in form to *U. graciosus* but has a shorter tail, usually less than twice its snout-to-vent length of 3.8 to 5.7 cm (1.5 to 2.25 in.). It has a strip of enlarged scales on the back that is broken by a central longitudinal row of small scales. The dorsum is brown, tan, dusky. Rust color is often present on sides of the tail. Males have blue, green, or greenish yellow ventrolateral patches on the belly, sometimes united and occasionally connected to the throat patch. The throat patch may be blue, green, yellow, orange, or combinations of these colors, depending on population and region. Females lack blue belly markings.

RANGE: In California this lizard occurs along the Colorado River in the extreme southeastern part of the Colorado Desert.

NATURAL HISTORY: This is a lizard of rocky areas chiefly in arid and semiarid country. It is often found in the vicinity of permanent or intermittent rivers or streams, and along boulder-strewn canyon bottoms. It is attracted to trees and larger bushes but may occur in treeless areas. When in trees it tends to use the trunk and larger branches instead of distal sites. It has been found in mesquite, oak, pine, juniper, alder, cottonwood, and nonnative trees such as tamarisk and rough-bark eucalyptus. When encountered on the ground it usually runs to a rock or tree and climbs.

U. ornatus is diurnal and an excellent climber. It is usually encountered in elevated situations, on rocks, ledges, cliffs, in bushes and trees, and on the sides of buildings or other structures where it often perches head up or down. Like *U. graciosus*, it is sometimes easily overlooked because of its habit of lying quietly against a bark or rock background that it closely resembles. During winter dormancy these lizards often aggregate in rock crevices, beneath bark, or in other retreats.

Plate 75. Ornate Tree Lizard.

Assertion, challenge, courting display, and fighting all follow the typical basking-lizard pattern. When this lizard bobs, it raises and lowers the entire body by flexing and extending both front and hind legs. Dominant males are dark, while subordinates are lighter colored. The dark color fades when a male loses his dominance.

The reproductive season in the southern part of the range in the United States extends from March to August. To restrain the female before copulation, the male sometimes seizes her by the tip of her tail, crawls forward retaining his hold and curling her tail until in a position to seize her neck skin.

Clutch size ranges from two to seven (often four to six) eggs. More than one clutch a season is the norm in the southern part of the range, and there may be three or more under good habitat conditions. Hatchlings have been reported from mid-June to late August and measure about 20 to 22 mm (0.8 to 0.9 in.) from snout to vent. Sexual maturity may be reached in less than a year.

Rock Lizards (Genus *Petrosaurus*)

These lizards of the desert slope occur from the Peninsular Ranges to the tip of Baja California, Mexico, and on islands in the Gulf of California. This genus contains four species, one of which occurs in this state. Along with geckos, these are the acrobats of the lizard world, able to run swiftly even while clinging upside down to the undersurface of boulders.

BANDED ROCK LIZARD *Petrosaurus mearnsi*

IDENTIFICATION: This is a medium-sized lizard with a snout-to-vent range of 6.6 to 10.6 cm (2.6 to 4.2 in.). Because of the black collar markings it superficially resembles a collared lizard, but its body is greatly flattened dorsoventrally, and there are keeled pointed scales on the tail and the anterior and upper surfaces of limbs. The dorsum is olive or brown-gray with wavy dusky crossbands. Bluish color tends to predominate on the tail. There are many small spots of white or bluish between, and sometimes within the bands. The black collar is often bordered behind by white. The venter is white with a suffusion of blue on belly and limbs. Males are larger and have slightly larger postanal scales, a broader tail base, larger femoral pores, and a more intense gular pattern and bluish ventral color than females. During the breeding season, females express bright orange-red coloration on the face and throat.

RANGE: In California it occurs from the south side of San Gorgonio Pass, Riverside County, south through the desert side of the Peninsular Ranges to the United States–Mexico border and on into northern Baja California, Mexico.

NATURAL HISTORY: This is a strict rock dweller, frequenting chiefly areas of massive rock found in narrow canyons, on steep hillsides, along arroyos, and among rock mounds. The vertical surfaces of piled large boulders and canyon walls are often favored, especially where there is a mosaic of light and shade. Refuge is sought beneath exfoliating rock shells and in fissures. Some cracks about 1.25 cm (0.5 in.) in width usually must be present to harbor adults. These lizards occasionally descend to canyon bottoms to forage along streams but, in general, seldom leave the areas of large rock. On the desert slopes where it occurs in southern California, the vegetation consists of creosote bush, incense bush, desert mint, Mojave yucca, and cholla, which are replaced at higher elevations by chaparral species.

Plate 76. Banded Rock Lizard.

Banded Rock Lizards usually begin basking when nearby rock surfaces receive full sunlight and remain out until the rocks become hot in late morning or midday, whereupon they retreat to crevices or shaded areas. After the rocks cool in late afternoon, activity is often resumed. In moving about they hold the body and limbs close to the surfaces over which they crawl and are able to move nearly as fast on the underside of a rock as on the sides or top. The gait is sprawling with the limbs held well out from the sides. This is an amazing feat to observe, especially because this is a moderately large lizard, unlike some much smaller and lighter gecko species that can also perform the "running upside down" maneuver.

Adult males patrol their territories performing displays consisting of a slow elevation of the foreparts by extension of the front legs, followed by a series of bouncing bobs or pushups. Challenge displays with dewlap lowered and sides compressed also occur and may be followed by a face-off, pursuit, and the nipping of the intruder. Adult males sometimes even stand their ground against human intruders. Monogamy appears to be common. Adults tend to be associated in pairs, with each defending a territory against members of its own sex.

Spiders, scorpions, centipedes, and a variety of insects constitute most of its food. Plant materials including buds and blossoms are common in the diet, as are the occasional small lizards such as Granite Night Lizard (*Xantusia henshawi*), also a crevice dweller. In California eggs are laid from June to August. Clutch size averages four eggs, and presumably only one clutch is laid a year. Hatchlings are about 35 mm (1.3 in.) snout-to-vent length, and some may reach adult size after their first winter but probably most do not mature until their second spring.

Horned Lizards (Genus *Phrynosoma*)

These lizards are familiar to most people, mainly because of their unique body form and their popular association with arid areas, often under the erroneous name of "horned toads." Their body is flattened and oval in outline when viewed from above. The head is broad and the snout blunt, and most species have sharp-pointed horns at back and sides of the head. These horns are solid bony spikes attached to the skull and covered with skin. The dorsal scales are mostly small and granular with larger, pointed keeled scales scattered among them, and there are one or two rows of projecting pointed fringe scales on the lateral edge of the body. The imbricate, smooth ventral scales are larger than the dorsal granular scales. The dorsal coloration tends to match the soil background color where they occur. Males usually have a swollen tail base, enlarged postanal scales, larger femoral pores, and an average size smaller than the females.

They are diurnal ground-dwelling lizards specialized for feeding on ants, thus they are most likely to be encountered when these insects are active, usually in early morning and late afternoon in desert areas. To escape the midday heat and avoid predators they seek the shade of bushes, enter rock piles or small-animal burrows, or bury themselves in loose soil. Like other lizards that dwell on sand or loose soil, they enter the soil by shaking the head rapidly from side to side and propel the body forward and downward by alternate movements of the hindlimbs in a manner resembling swimming. The forelimbs are held out of the way along the sides. Soil is tumbled over the tail by vibrating it rapidly from side to side after forward motion stops. They usually come to rest just under the surface.

Many of their physical and behavioral characteristics relate to feeding on ants. To procure sufficient numbers of their small prey they spend much time exposed and preoccupied with foraging near the openings of ant nests or by ant trails. Their ground-matching coloration, protective spines, and tendency toward immobility help protect them against predation at such times. Even so, they sometimes fall prey to snakes, carnivorous lizards, and birds such as the Loggerhead Shrike (*Lanius ludovicianus*) and the Greater Roadrunner (*Geococcyx californianus*). Predators occasionally meet with disaster in attempting to swallow this thorny prey. Snakes have been killed by the head spines piercing their esophagus.

To accommodate the large numbers of ants required for a meal, horned lizards have relatively larger stomachs than most other insectivorous lizards. This and their sedentary behavior may explain their stocky, short-limbed form. The blunt snout and dentition are well suited for eating ants. Ants are caught by a quick flick of the sticky tongue as the snout is brought close to the ground. The blunt, peglike teeth crush the prey. Although it might be expected that these lizards would be unaffected by the bites and stings of ants, they are, in fact, vulnerable and may show extreme discomfort when attacked. They have been seen to run and bury themselves when mobbed by ants. On the other hand, there are indications that ants sometimes recognize horned lizards as a threat and suspend activity in their presence. Although ants are important in the diet of all species of horned lizards, other insects are also eaten.

When a horned lizard is threatened by a predator, it stands high on extended legs, arches its back upward, inflates its lungs, and hisses with its mouth open, sometimes tipping its head downward to position the horns for action. It may lunge forward with mouth open, hissing as it does so, and the expanded, flattened body may be tipped in the direction of danger. If the lizard is seized it twists its head from side to side jabbing with the head spines. In some species, often as a last resort, blood is spurted from the "eyes." Dogs, kit foxes, and perhaps other animals, are repelled by the blood. The blood-spurting mechanism appears to be unique. Contraction of muscles around the internal jugular veins slows drainage of blood from the head, distending the postorbital blood sinuses (one behind each eye). Muscles in the eye region contract, increasing pressure upon the sinuses, and blood breaks through the orbital sinus wall into the membranes of the eyelids. There it forces its way into the duct of the Harderian gland, which carries its secretions to the eyelid. Here it apparently picks up a substance that provides the blood with its irritating antipredator properties. The blood sprays from the eyes between closed lids in a fine stream for a distance up to about 2 m (about 6 ft). In a large individual perhaps 0.5 to 1 ml (0.03 oz) of blood may be expelled. Blood spurting is uncommon and seems to be a last-ditch lifesaving effort, for considerable body fluid is lost. In observations by the senior author, it has always been preceded by eye protrusion and occurred more frequently in torpid individuals than active ones.

Little is known about these lizards' social interactions such as territorial and dominance behavior, fighting, and sexual displays, although some fights have been witnessed in the field. DAP behavior, common in most phrynosomatid lizards, is little developed. Usually only the head is bobbed, and there are evidently no pushups.

With the exception of the Pygmy Short-horned Lizard (*P. douglasii*), horned lizard species of California are oviparous. Litter and clutch sizes are large, but usually only one clutch is laid in a season. Hatchlings have good protective coloration, but their short head spines afford little protection against predators. Sexual maturity usually occurs after the second hibernation. Horned lizards seldom live long in captivity, usually surviving less than two years. They are difficult to keep because of their ant-eating habits.

These delightful and popular animals are now seriously threatened by human population increases in the arid southwest, as the pet trade, loss of habitats, and destruction by off-road vehicles take their toll. They are especially vulnerable to vehicles, because they are not easily seen, tend to sit still when threatened, and, when buried, they lie just under the soil surface. Arizona, New Mexico, and several other states have passed laws prohibiting the capture and sale of these lizards. However, at this writing

Californians can legally capture and possess two of the state's four species if they have a freshwater fishing license or are under the age of 16.

COAST HORNED LIZARD *Phrynosoma blainvillii*

PROTECTIVE STATUS: CSSC

IDENTIFICATION: Its snout-to-vent length range is 6.3 to 11.4 cm (2.5 to 4.5 in.). There are two elongate occipital horns, and their bases are usually separated. It also has two rows of projecting fringe scales on each side of the body and usually two or three rows of enlarged overlapping pointed gular scales on each side of the throat. The dorsal color is brown, reddish, or yellow. There are large black marks on each side of the neck, extending onto the shoulder, and undulating dark-brown to black marks on the back, edged posteriorly with white or yellow. The venter is cream to yellow with dusky spots. The male has larger back and head spines, a broader head, large postanal scales, a swollen tail base, and larger femoral pores than the female.

RANGE: It occurs in select regions of the Coast Range, the Sierra Nevada foothills, parts of the Great Valley, the South Coast, Tehachapi, Transverse, and Peninsular ranges, Coastal Southern California, and south into Baja California, Mexico. However, intensive agriculture has eliminated it from much of the Great Valley.

NATURAL HISTORY: This horned lizard inhabits valleys, foothills, and semiarid portions of mountains, often in sandy areas such as washes, flood plains, or windblown deposits. It is usually associated with grassland, broken chaparral, open coniferous forest, coastal sage scrub, and broadleaf woodland. It requires open areas for basking, bushes or grass clumps for cover, patches of loamy soil or sand for burrowing, and an abundance of ants. Despite extensive agricultural development in the floor of the Great Valley, it still persists in some grape vineyards, grassland remnants, and in the less disturbed riparian habitats.

As in other horned lizards, display behavior appears to be weakly developed. The actual mating act in the Coast Horned Lizard differs from other *Phrynosoma* species; after the male obtains a neck grip on the female, he raises himself on his forelegs, jerks his head quickly to one side while at the same time rushing forward, thus turning the female over on her back. A hemipenis is then inserted into the female's cloaca, and copulation takes place in a venter-to-venter position.

The egg-laying period is from April to June. Clutch size varies from six to 21, with the average about 12. Hatching usually occurs from July to September. Hatchlings measure about 25 to 27 mm (about 1 in.) from snout to vent and may bury themselves shortly after hatching.

Its food consists of ants, wasps, beetles, grasshoppers, flies, caterpillars, and other insects.

CONSERVATION NOTE: Unlike climbing species such as the Western Fence Lizard (*Sceloporus occidentalis*), which has adapted relatively well to some human land use practices and the artificial features these create (wood fences, rock walls, wooden barns, junk piles), the ground-dwelling Coast Horned Lizard does not fare well in the face of land development. The continuing conversion of most of the Great Valley floor to highly mechanized agriculture has been the greatest single impact to this species. Flat or moderately rolling land with the soft soil and sand conditions that this species requires has also become a prime target for housing developments

Plate 77. Panting Coast Horned Lizard.

throughout its range, especially in southern California. Its main forms of defense, sitting motionless or shallow burrowing, are not only useless but highly detrimental as an attempt to avoid off-road vehicles, which continue to increase in unprotected pockets of coastal and foothill lands. This behavior has also made it an easy target for the illegal collecting and sale of California reptiles. One further impact has been the introduction of the Argentine Ant (*Linepithema humile*) into southern California. This species is not eaten by this lizard and also displaces its native ant food. Given a set of biological features that are among the most incompatible with human activities to be found in California reptiles, the listing of Coast Horned Lizard as a Federal and State Threatened status is warranted.

DESERT HORNED LIZARD *Phrynosoma platyrhinos*
NORTHERN DESERT HORNED LIZARD *P. p. platyrhinos*
SOUTHERN DESERT HORNED LIZARD *P. p. calidiarum*

IDENTIFICATION: This species is similar to *P. blainvillii* but has a smaller snout-to-vent range of 6.7 to 9.5 cm (2.6 to 3.75 in.), shorter occipital horns, and only one row of lateral abdominal fringe scales on each side of the body. The dorsal color resembles the soil color of the habitat and may range from beige, tan, reddish, or gray to nearly black in areas of dark lava flows. There are wavy dark blotches on the back and a pair of large dark neck blotches. The venter is white and speckled with dusky spots.

RANGE: The Desert Horned Lizard occurs from below sea level (Death Valley and Salton Sink) to around 1,540 m (5,000 ft). In California it is found in the Great Basin, Mojave, and Colorado deserts. It also penetrates the South Fork of the Kern Valley and Upper Kelso Valley in the southern Sierra Nevada.

Plate 78. Desert Horned Lizard at edge of an ant nest.

NATURAL HISTORY: This horned lizard inhabits arid and semiarid flatlands, alluvial fans, foothills, and low desert ranges where vegetation is sparse and where there are areas of fine loose soil or sand. Sandy, gravelly washes with scattered rocks seem especially favored. The habitat usually contains low bushes with intervening areas of bare soil. Areas where plants are close together are avoided.

It generally emerges from hibernation in March and April, and activity often reaches a peak in May and June, when most breeding occurs. There is little evidence for territoriality, and in some habitats there may at times be as few as one lizard per hectare (100 acres). Mating occurs from April to June. An average of six to eight eggs constitutes a clutch. There may be two clutches a year, but in most years only one is laid. Hatchlings appear from mid-July to early September. They measure around 27 to 32 mm (1.1 to 1.3 in.) snout to vent. Sexual maturity is reached after the second hibernation, and females lay their first eggs at 19 to 22 months. Some individuals may reach an age of seven to eight years or more in the wild.

Although ants and occasionally beetles usually predominate among foods taken, many other arthropods, including spiders, ticks, solpugids, plus some plant material such as the bright-orange berries of the desert thorn (*Lycium andersonii*) are also eaten. The lizards evidently climb the bushes to obtain these berries and on occasion cram their stomachs full with them.

FLAT-TAILED HORNED LIZARD *Phrynosoma mcallii*

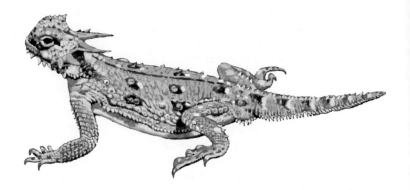

PROTECTIVE STATUS: CSSC

IDENTIFICATION: Its snout-to-vent length is 6.3 to 8.3 cm (2.5 to 3.25 in.). The head spines are longer and more slender, and its tail is flatter and longer than in other horned lizard species. As in the Coast Horned Lizard, there are two rows of lateral abdominal fringe scales. It has a distinctive narrow, dark vertebral stripe on the dorsal ash-gray, buff, or rusty-brown color. There are also dusky spots in two longitudinal rows, one row on each side of the midline. The venter is uniformly white except for a short longitudinal black or dusky stripe at the midbelly. The basic horned lizard body design has been modified for burrowing in fine sand. Stresses encountered in burrowing appear to be responsible for degeneration of the eardrum and middle ear. The former is covered with skin and encroached upon by bone, and the latter has been invaded by jaw muscle, a condition resembling that found in snakes and certain burrowing lizards. Other burrowing modifications are the flattened body and tail, reduced spininess on its back, and streamlined horns aligned with the long axis of the head.

RANGE: This species is restricted to the arid windblown sand deposits of the Coahuila Basin in Mexico and nearby desert areas in southeastern California, southwestern Arizona, and at the head of the Gulf of Mexico. The northern limit of its range is in the vicinity of Thousand Palms, Riverside County.

NATURAL HISTORY: This is a horned lizard of the sandy flats, dunes, and sand hummocks. It centers its activities near desert bushes, grasses, and other vegetation, such as creosote bush, burrobush, and galleta grass. The dorsal coloration of this lizard closely matches the prevailing sand color. When sand surfaces rise above about 41 degrees

C (106 degrees F) these lizards usually seek the shelter of bushes or bury themselves in the sand.

P. mcallii is very adept at "sand swimming." It dashes off and, while in full flight, may plunge headfirst into the sand, wriggling quickly out of sight. Some runs may be around 50 m (164 ft), but most are much shorter. At other times the lizard may simply lie flat to avoid detection. Burial often occurs after a fleeing lizard is out of sight beyond a bush or hummock. It is so skillfully executed that often only the lizard's tracks and three "thumb print" dents reveal the location. This behavior is very similar to that of the fringe-toed lizards (*Uma* spp.), and a running individual is difficult to distinguish from a half-grown *Uma*. Despite their squat form, this horned lizard can run surprisingly fast.

Ants are the chief food of *P. mcalli*, but some beetles and other insects are also taken. Little is known about reproductive habits of this species. Most mating probably occurs in April and May, and eggs are laid in May and probably June.

CONSERVATION NOTE: The Flat-tailed Horned Lizard is now seriously threatened by off-road vehicle recreation (dune buggies, motorcycles, etc.) throughout much of its restricted habitat. It is on the verge of extinction in the Coachella Valley, where human activities such as agriculture, urban, and golf course development, sand and gravel mining, and other land use changes have destroyed much of its habitat. Asian Mustard (*Brassica tournefortii*) has also invaded much of its remaining habitat in recent years, which may impact daily activities of this species. As with other species with specialized habitat requirements and relatively small and highly compromised range, the Flat-tailed Horned Lizard needs to be granted threatened or endangered status and have specific reserve areas set aside for it if it is to persist in California. While threatened or endangered status is recommended for this species, it should be noted that the Colorado Desert Fringe-toed Lizard is even more restricted in range than the Flat-tailed Horned Lizard. Both species would benefit from such listing under the Endangered Species Act.

PYGMY SHORT-HORNED LIZARD   *Phrynosoma douglasii*

IDENTIFICATION: As its common name implies, this is the smallest of the horned lizards, with adults rarely exceeding 6 cm (2.5 in.). It is readily distinguished from our other horned lizards by its short head spines and a single row of lateral abdominal fringe scales. Its dorsal color varies, and depending on local substrate, color may be slate, ash-gray, or various shades of brown, buff, red, or yellow. There are usually two rows of black or dark-brown blotches arranged on each side of the midline. The venter is white or yellow with varying amounts of dark spotting. Males are smaller than females and have enlarged postanal scales, a broader tail base, and slightly larger femoral pores.

RANGE: In California this species occurs only in the extreme northeastern part of the state to as far south as Mt. Shasta.

NATURAL HISTORY: In California the Pygmy Short-horned Lizard inhabits open-plain areas of sagebrush and bunchgrass, piñon–juniper woodland, and open pine forests. It is a cold-adapted species capable of coordinated movements at a body temperature of around 3 degrees C (37 degrees F). Territorial behavior appears to be weak. Bobbing displays usually consist of only three bobs of very low amplitude, the entire display lasting only about half a second. Defense is spirited, however, and when threatened the head is lowered, the horns are tilted forward, and the back is arched. These lizards seem to show a special aversion to dogs, perhaps a carryover of a response to wild canids such as foxes and coyotes. They rise high on their legs, puff out the body, and hiss with open mouth, revealing the orange or yellow lining that occurs at the corners of the mouth.

This is a live-bearing (ovoviviparous) species, one of only eight among California lizards. By retaining her eggs instead of laying them, a female can provide the proper temperature range for development by basking,

Plate 79. Juvenile Pygmy Short-horned Lizard.

a condition that might not be possible if they were laid in a cool climate. Births occur in July, August, and September after a gestation period of about three months. Litter size ranges from three to 15 young, the larger females tending to have the larger broods.

Newly born young often vary in size, ranging from 12 to 25 mm (0.5 to 1 in.) in snout-to-vent length. They are very active immediately after birth, often feeding on small ants and rapidly burrowing into soft sand or soil. Insects, including ants, spiders, snails, and perhaps some plant materials compose the adults' food. However, young appear to be especially partial to ants.

Night Lizards (Family Xantusiidae)

These are relatively small lizards with soft and pliable dorsal skin covered with tiny granular scales and occasionally scattered tubercles. On the venter there are large, rectangular, nonoverlapping scales. A gular fold is present as well as folds of skin on neck and lateral body. The eyes have vertically elliptical pupils and transparent spectacles but no moveable eyelids. These are secretive lizards that generally shun bright light. The retinas of *Xantusia vigilis* and *X. riversiana* are known to be adapted to conditions of low light intensity. Our five California species are viviparous live-bearers and have a placental-like connection between the female parent and the developing embryos.

DESERT NIGHT LIZARD — *Xantusia vigilis*
YUCCA NIGHT LIZARD — *X. v. vigilis*
SIERRA NIGHT LIZARD — *X. v. sierrae*
WIGGINS' NIGHT LIZARD — *Xantusia wigginsi*

Desert Night Lizard

PROTECTIVE STATUS OF *X. V. SIERRAE*: CSSC

TAXONOMIC NOTE: Until recently, the disjunct population in central southern California was assumed to be that of the Desert Night Lizard, *X. vigilis*. However, recent mitochondrial DNA studies indicate that this population is an extension of the Baja California species Wiggins' Night Lizard, *X. wigginsi*, into California.

IDENTIFICATION: These are the smallest California lizards, with a snout-to-vent range of 3.8 to 7 cm (1.5 to 2.75 in.). The ventral scale plates are arranged in 12 longitudinal rows, and the dorsal scales are granular. There is one row of supraocular scales. The dorsum is gray or shades of yellow, brown, or olive with small dark spots that vary in definition with different individuals and color phases. In some *X. vigilis*, there is a beige stripe edged with black extending from the eye to the shoulder. The venter is whitish, dull greenish, yellow, or pale gray, usually with some small spots or blotches. These lizards are capable of rapid color change from a dark to a light phase. In the extreme dark phase they may be sooty or deep olive and in the light phase pale buff or cream. Males do not have large postanal scales but do have larger femoral pores that give the thighs an angular contour in cross section. Males average smaller than females and usually have shorter, stouter tails.

X. wigginsi is similar to *X. vigilis*, but is usually patternless in California and is slightly smaller. *X. v. sierrae* is similar, but is slightly more dorsoventrally compressed due to its crevice-dwelling life style, has wider yellowish-tan stripes on its head, and has a dark reticulated pattern on a dark, grayish dorsal background.

RANGE: In California *X. vigilis* occurs throughout the Mojave Desert and portions of the Tehachapi and Transverse ranges. *X. wigginsi* occurs in the southern Peninsular Range and the southwest segment of the Colorado Desert. *X. v. sierrae* occurs in a localized area on the western slope of the Greenhorn Mountains, Kern County. Isolated populations of *X. vigilis* exist in Panoche Hills and Pinnacles National Monument, San Benito County, and Carrizo Plain, San Luis Obispo County. In the Mojave Desert it has been found at elevations as high as 2,830 m (9,300 ft) on Telescope Peak in the Panamint Mountains.

NATURAL HISTORY: These closely related species are small, soft-skinned lizards, highly susceptible to desiccation and thermal extremes, yet they live in dry, warm areas. They depend for shelter chiefly on rock crevices and plant debris that provide small openings among which the lizards can move. The recumbent rosettes of dead leaves around the bases of yuccas or fallen yucca stalks, nolina, and agaves provide such conditions, as do the fallen spiny branches of the Joshua tree (*Yucca brevifolia*). In the Inner Coast Range, *X. vigilis* is also found beneath the bark and in the heartwood of decaying gray pine and interior live oak logs (see figure 37, in the "Observing and Photographing Amphibians and Reptiles" section). They have been found under boards, pieces of metal, and dried cowpies in creosote bush desert. Details of the life history of *X. vigilis* are far better known than that of *X. wigginsi* but are believed to be very similar, and thus those documented for *X. vigilis* are presented here.

Although formerly considered rare, *X. vigilis* is now known to be one of the most abundant lizards in the Southwest. In the Mojave Desert, where they have been most studied, they are active beneath surface cover

Plate 80. Desert Night Lizard on dead Joshua tree branch.

throughout the year with a peak from March to June, when food is usually abundant and breeding occurs. In favorable habitat, as in Joshua tree woodland in the western Mojave, they reach high population densities. During the spring and summer breeding period adults are territorial and fighting occurs, dispersing the populations. Seldom are more than two to three adults found under the same shelter. Both sexes bite when fighting, and tails and toes may be lost. After the breeding season, antagonism subsides and groups of adults may be found under the same objects.

In the western Mojave Desert ovulation occurs from early June to mid-July but may be delayed to mid-June in wetter, cooler years, and in dry years there may be little or perhaps no reproduction. Both species are viviparous live-bearers, with a gestation period of about three months. A litter size of two young is most common, and there is only one brood a year, composed of one to three young. The young have a snout-to-vent range of 22 to 24 mm (0.8 to 0.9 in.) at birth. Males reach sexual maturity late in their second year and females in their third year. Growth rate is strongly influenced by rainfall, which affects plant productivity and abundance of insect food. Females must survive, on the average, at least three years for population replacement to occur. Their average life expectancy at birth is four years, but some may live as long as 10 years.

These lizards feed on a wide variety of insects that utilize the same ground-cover retreats as it does. Ants and small beetles are two of the most important foods.

GRANITE NIGHT LIZARD
SANDSTONE NIGHT LIZARD

Xantusia henshawi
Xantusia gracilis

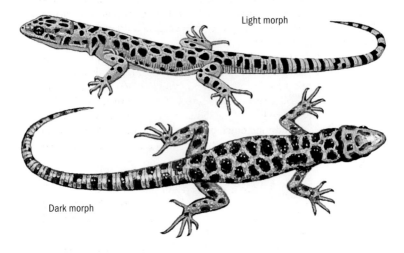

Light morph

Dark morph

PROTECTIVE STATUS OF *X. GRACILIS*: CSSC

TAXONOMIC NOTE: As with several other subspecies that are now recognized in a number of current publications as separate species based primarily on varying differences in mitochondrial DNA, we have treated the Sandstone Night Lizard as such here for the sake of consistency with these reports, combined with morphological and ecological evidence presented on its original description. However, because of its close similarity to the Granite Night Lizard and lack of natural history information for this recently elevated species, we present a combined account for both.

IDENTIFICATION: Both of these small night lizards have a snout-to-vent range of 5.0 to 7.0 cm (2 to 2.8 in.). The head and body are greatly flattened, a modification relating to their crevice-dwelling habits. They resemble the Desert Night Lizard in its scalation and general form but differ from it in having 14 longitudinal rows of scales across the midbelly rather then 12 as in that species. A structural difference between the Granite Night Lizard and the Sandstone Night Lizard is the presence of enlarged ear scales in the former but no such enlargement in the latter. The dorsum of *X. henshawi* contains large dark spots on a pale background when it undergoes melanophore contraction and assumes its nocturnal light color phase. In its dark phase, normally seen during daylight hours, the dark spots expand to a point where the pale background is reduced to a network of narrow whitish or pale-yellow lines.

The Sandstone Night Lizard color pattern closely resembles the light phase of the Granite Night Lizard, but the dark spots are smaller and more numerous, and apparently no shift in color phase has been observed. The tail color pattern of both species is similar to that of the dorsal body but usually with narrow bars and bands of a dark color rather than spots. The venter of *X. henshawi* is whitish with dark speckling throughout, whereas in *X. gracilis* speckling is present only on the anterior ventral surface. Males of both species have larger femoral pores and preanal scales and a more swollen tail base than females.

RANGE: In California the Granite Night Lizard occurs from the northern slope of the Peninsular Ranges on the south side of San Gorgonio Pass and south through these ranges, including their desert and coastal slopes, into Baja California, Mexico. The Sandstone Night Lizard's range is restricted to sandstone and mudstone outcrops in one small area in Anza-Borrego Desert State Park on the western edge of the Colorado Desert.

NATURAL HISTORY: Essentially nothing is known about the life history of the Sandstone Night Lizard, except that it appears to be more nocturnal than the Granite Night Lizard. The following account is therefore restricted to the latter species.

The Granite Night Lizard inhabits arid and semiarid rocky hillsides, especially granite sites where it hides by day in crevices in rocks. It seems to prefer areas of massive rock rather than scattered small boulders where deep crevices permit retreat from temperature extremes. Such features are most often frequented in the shadier parts of canyons and slopes near water. Where optimal habitat conditions for this species occur, it may be more abundant per unit area than any other lizard in California.

Observations of daily activity reveal that most locomotor movements occur in the afternoon and early evening, when the lizards are still in their crevices. This is the warmest part of the day, and they may be thermoregulating and foraging on crevice-dwelling arthropods. Shortly after sunset they emerge and position themselves upon the exposed rock surfaces. Occasionally some exposure may occur in the daytime near the openings of crevices. Social interactions, including territorial behavior and foraging, probably occur at this time. Like the other California *Xantusia*, this species is also viviparous, and we assume that the Sandstone Night Lizard is also. It mates in spring, and one or two young are born in the fall. Birthlings measure a little over 60 mm (about 1 in.) from snout to vent. Ants, beetles, and spiders seem to be the most important foods, but seeds, stems, and leaves are also occasionally taken.

CONSERVATION NOTE: The CSSC status of the Sandstone Night Lizard is based primarily on its highly restricted range within relatively small sandstone and mudstone sites. Fortunately, these are all located within a state park and will hopefully be carefully monitored and managed.

ISLAND NIGHT LIZARD *Xantusia riversiana*

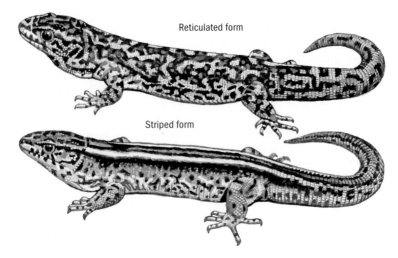

Reticulated form

Striped form

PROTECTIVE STATUS OF X. RIVERSIANA: FT

IDENTIFICATION: This is the largest of the California night lizards, with a snout-to-vent range of 6.3 to 10.6 cm (2.5 to 4.2 in.). In addition to its large size, it differs from other *Xantusia* in having 16 longitudinal rows of scales across the midbelly and in having a double row of supraoculars. The dorsum is gray to brown, spotted or reticulated with sooty gray. Occasionally the dark color may form a dorsolateral stripe on each side, and sometimes there is a brown or rust-colored vertebral stripe edged with black. The ventral surface is often suffused with yellow. Males have slightly larger femoral pores than females.

RANGE: This species is confined to San Clemente and Santa Barbara islands, including the associated islet Sutil, and San Nicolas Island off the coast of southern California.

NATURAL HISTORY: *X. riversiana* is usually found under rocks, pieces of wood, branches, and other surface objects in upland areas as well as on the rocky beaches and in bordering sand dunes of these islands. On well-vegetated islands it frequents chaparral and oak savannah and sandy and rocky dry stream beds. However, San Clemente Island, where the lizards have been taken in considerable numbers, is almost treeless, grassy, and has extensive growth of prickly pear and boxthorn. Here the lizards have been found under rocks in areas where loose rocks, cactus, and sparse brush abound. San Nicolas is even more barren, but this lizard is still found all over this island. Behavioral studies indicate that this species is active in the daytime, belying its common name of "night lizard," but is not often seen exposed.

This viviparous species gives birth to from two to six young in September after a 14-week gestation period. Sexual maturity is probably not reached until the spring of the third or fourth year of life. Food includes both vegetable and animal matter, the latter composed of a variety of small insects and other arthropods, including marine shoreline species. Stomach samples from lizards from San Clemente Island contained about 31 percent plant foods. Like a number of other lizards, it also eats its own shed skin.

Skinks (Family Scincidae)

This is the most widely distributed of all the families of lizards. Skinks occur on all continents except Antarctica, and on many oceanic islands. The body and tail of these lizards are usually rounded, and the head is often little wider than the neck or the same width. The limbs are relatively short, and in some species outside of North America they are reduced or absent, but the shoulder and hip girdles are always present. The skin is firm, usually with hard, smooth scales. There is no dewlap, gular fold, or folds on the sides of the body. Skinks are chiefly ground-dwelling, and many are burrowers. The latter includes slender, cylindrical, snakelike forms that lack limbs completely and instead employ serpentine locomotion. In general, skinks are secretive, diurnal lizards. They are well camouflaged, often with stripes, and blend well with their surroundings. Although many occur in arid regions, most frequent humid tropical and subtropical environments or damp forests and woodlands in the temperate zones.

Perhaps because of their secretive behavior, skinks depend heavily on their sense of chemoreception, as indicated by frequent snakelike tongue flicking. They are able to recognize mates, rivals, and food at a distance by this means and use the tongue–Jacobson organ mechanism in tracking prey and other individuals. In contrast to visually oriented lizards, adult skinks seldom have bright colors, and the sexes usually differ little in coloration. The bright tail coloration of certain young skinks of the genus *Plestiodon* may protect against predation by enticing species such as ground-foraging birds to peck at it instead of the head or body. When grasped, the tail usually breaks off, continues to twitch rapidly, and diverts attention while the lizard attempts to escape. (See figure 29, in the introduction to the Lizards [suborder Sauria] section.) Within the family Scincidae there are oviparous, ovoviviparous, and viviparous species. Among those species whose reproduction habits are known, almost two thirds are oviparous. The whitish, soft-shelled eggs of skinks are often deposited in the ground or rotting logs. In some species, the female stays with the eggs. Dr. Henry Fitch observed that in *Plestiodon* the mother moved the eggs about within the nest, presumably to prevent adherence to the soil and spoilage, shifted their location to avoid desiccation or flooding, and evidently released bladder fluid in the nest when it became dry. Brooding females perhaps also guard the eggs against small predators. This degree of maternal egg care is rare among lizards.

Northern Skinks (Genus *Plestiodon* [*Eumeces*])

This is the largest and most widespread skink genus, occurring along the northern periphery of the world distribution of skinks. It is a cold-adapted group, and ancestral stock presumably dispersed from the Old World across the Bering Land Bridge. This genus is notable in that the female remains in the nest with the eggs throughout their incubation and, after hatching, the female and young remain together for a short time.

Like several other amphibian and reptile taxa, a revision and renaming of the genus *Eumeces* to *Plestiodon* has been proposed (Brandley et al., 2005; Smith, 2005). This is based on the finding that American skinks of the genus *Eumeces* consist of two clades (groups that are each derived from a different recent common ancestor but have the same more ancient ancestor in common) and therefore should be viewed as being paraphyletic. However, as previously stated, we have attempted to use nomenclature that is consistent with that of other recent UC Press natural history guide authors, and therefore use the new genus name *Plestiodon* first in the natural history accounts but also give the former name *Eumeces* in parentheses after it. Fortunately, the common names of both California skink species have not been proposed for change.

WESTERN SKINK *Plestiodon skiltonianus (Eumeces skiltonianus)*

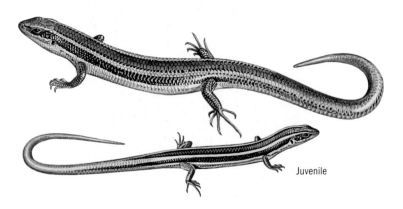

Juvenile

IDENTIFICATION: Its snout-to-vent range is 5.4 to 8.6 cm (2.1 to 3.4 in.), and the tail is one and a half times that range. There are usually seven upper labial scales, and the interparietal scale is short and wedge-shaped with its sides converging abruptly posteriorly. The central area of the back is brown with two longitudinal light stripes that pass from the upper eyelids onto the body and fade out at the base of the tail where they become continuous with the light tail color. The venter is usually pale gray-blue, and sometimes there is a bluish tinge on the throat. Breeding adults develop bright-red or orange-red coloration on the upper and lower labials, chin, and to a lesser extent on the sides of the head. Juvenile coloration is like the adults, but dark and light markings are more contrasting and the tail is brilliant blue. In some individuals partial blue coloration is retained in young adults. Plate 81 presents a bird's eye view of a young Western Skink with its very apparent bright-blue tail and striped body, which blends well with the forest floor litter.

RANGE: The Western Skink occurs throughout the North Coast Range, the Cascade Range, and the Sierra Nevada north of the Yuba River. It also extends southward through the South Coast Range and Coastal Southern California, and part of the Peninsular Ranges. Isolated populations also occur near Mono Lake, in the Owens Valley, and in Sequoia National Park. Santa Catalina Island also supports a population.

Plate 81. Juvenile skink with bright blue tail.

SKINKS (SCINCIDAE)

NATURAL HISTORY: The Western Skink inhabits a wide variety of habitats, including rolling grasslands, broken chaparral, sagebrush, oak woodland, piñon–juniper woodland, open-pine and pine–oak forests, and juniper–sage areas. It tends to occur in colonies where there is an abundance of cover such as rotting logs, flat rocks, leaf litter, clumps of brush, grass or other herbaceous vegetation, soil suitable for digging, and rodent runaways. It is tolerant of low temperatures yet often selects warm south slopes and areas that are moderately humid. It avoids dense forests but may occur in the more open rocky portions of such stands. Riparian habitats are often frequented. This skink is an excellent burrower and constructs tunnels along the side and beneath rocks and logs. When foraging, these lizards move quietly in a slow, jerky, fashion. When moving rapidly they employ sinuous snakelike movements, often with the forelimbs laid back along their sides. During most activities they tend to stay close to bushes, rocks, and other shelter.

Breeding takes place from May into early June. Two-year-old and older breeding males develop intense reddish color on the labials and in the gular area. Eggs are laid in June and July in California, and clutch size ranges from two to six eggs. Females brood their eggs and evidently turn or tumble them. Hatchlings begin to appear in late July. They range from 25 to 29 mm (1 to 1.2 in.) from snout to vent. Some individuals may breed at two years of age but most probably not until three. The normal life span is estimated to be seven to eight years. Western Skinks feed on a wide variety of insects and their larvae that inhabit the ground litter layer. Spiders and sowbugs are also taken, and cannibalism has been reported.

GILBERT'S SKINK *Plestiodon gilberti (Eumeces gilberti)*

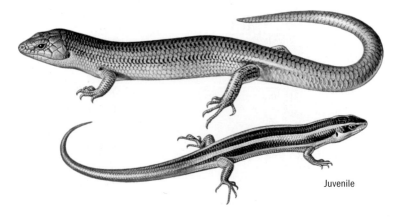

Juvenile

IDENTIFICATION: This skink has a snout-to-vent length of 6.3 to 11.4 cm (2.5 to 4.5 in.) and a tail length of one and a half to two times this range. There are usually eight upper labials rather than seven as in the Western Skink. The young are almost identical in markings to *P. skiltonianus*, except in some regions they have pink or salmon-colored tails, and the dark lateral stripe does not extend as far out on the tail. Fortunately, in most areas of California such as the southern mountain ranges where the two species are known to coexist, young Gilbert's Skinks have pink-salmon tail coloration, as seen in the Plate 83 photo of a juvenile captured in the Tehachapi

Plate 82. Adult Gilbert's Skink.

Plate 83. Juvenile Gilbert's Skink with salmon-pink tail.

Mountains, and therefore can readily be distinguished from the blue-tailed Western Skink young. However, in areas of species overlap in central California, *P. gilberti* young also have blue tails, and here the absence of lateral stripe extension well onto the tail is the only defining color feature. Richmond and Reeder (2002) suggest that some populations curently recognized as Gilbert's Skink may be more closely related to nearby populations of Western Skink than to distant ones of their own species. It is for this reason that we have chosen not to recognize subspecies for either taxon in this reference until further studies resolve this complex taxonomic situation. The adult dorsum color is olive to olive-brown, without markings. Older individuals of both sexes may develop pink or brick-red tails, and adults of some subspecies develop bright-red heads.

RANGE: It occurs mostly west of the deserts and extends from the Yuba River south on the western slope of the Sierra Nevada through the Tehachapi and Transverse ranges to the Peninsular Ranges. It is also found in the Inner Coast Range, from Contra Costa County east of San Francisco Bay southward. There are also scattered populations mostly at higher elevations in the eastern Mojave Desert.

NATURAL HISTORY: The many habitats in which Gilbert's Skink occurs include grasslands, desert scrub, broken chaparral, woodland, open pine forests, spring sites, and streamside growth. In the Sierra Nevada foothills it occurs chiefly in the blue oak zone but also in blue oak–gray pine country and, at higher elevations, in open yellow pine forests. It frequents riparian habitats, salt flats, and grassland of the northern San Joaquin Valley. In the Inner Coast Ranges and the Tehachapi Mountains it is found in rolling grassland with scattered rocks or outcrops and in southwestern California in oak woodland and rocky brushlands. In the desert mountains it occurs in piñon–juniper stands. It is especially attracted to desert spring, riparian, and wash habitats, and seems generally to be more tolerant of drier situations than *P. skiltonianus*. It has even been found near the bases of rocky desert hillsides under fallen Joshua tree branches in association with the Desert Night Lizard (*Xantusia vigilis*).

Like the Western Skink, gravid females are secretive and perhaps largely subterranean. Clutches range from three to nine eggs and are laid in summer. Its food includes insects such as caterpillars, moths, beetles, Jerusalem Crickets, grasshoppers, dipterous larvae, and spiders.

Teiids (Family Teiidae)

This family is confined to the New World and is the counterpart of the Old World family Lacertidae. It ranges from north-central United States through Mexico and Central America and the West Indies south to Chile and central Argentina. Reeder et al. (2002) proposed that certain whiptail lizards of the genus *Cnemidophorus* be placed in the genus *Aspidoscelis*, and once again for consistency with other recent UC Press natural history guides we have followed suit while giving the former Latin name in parentheses. A note to those readers new to the field of herpetology is in order. You may occasionally hear the term "Cnemi" used when referring to whiptail lizards. This is one of a number of herpetological "slang" terms that people have found more convenient than the often longer and more cumbersome Latin proper name. Such terms are also not readily put aside and often serve as a reminder of former names long after an "official" change has been declared.

Whiptails (Genus *Aspidoscelis* [*Cnemidophorus*])

This is the genus of *teiids* that occur in North America and extend as far south as Bolivia. They are slim-bodied lizards with a head about the same width as the neck, a slender snout, and limbs, toes, and claws well developed for digging. The tail is long, rounded, and slender, two or more times the snout-to-vent length, and covered with transverse rows of keeled scales. Tail loss and regeneration occur, but the tail is less fragile than in most iguanid lizards. Scales on the dorsum are granular, and those on the belly are in regular longitudinal and transverse rows. One or more gular folds are present, and the throat expands to form a dewlap. The tongue is long and narrow, deeply forked anteriorly, and strongly heart-shaped posteriorly. Some species are striped, others have a spotted or checkered pattern, and still others are both striped and spotted.

In body form and striping these lizards resemble that of skinks (*Plestiodon*), but the latter have smooth, shiny, cycloid scales on both dorsal and ventral surfaces. These are active, ground-dwelling, diurnal lizards that generally frequent open, rather dry habitats. Whiptails have relatively high thermal preferences and the highest sustained rates of oxidative metabolism currently known among reptiles. In active individuals the metabolic rate is approximately that of a resting mammal of similar body size. The larger species can elevate their body temperature 1 to 2 degrees C (34 to 36 degrees F) above their surroundings through the heat produced by this high metabolism when active. The olfactory epithelium is extensive, and detection of prey and mates by chemoreception with the Jacobson's organ is well developed. Subterranean termites, insect pupae, and other concealed prey are common in the diet. The ability to locate buried invertebrate prey in this manner allows whiptails to occupy a feeding niche not utilized by most other lizard groups.

A foraging whiptail typically moves over the ground in a series of short, jerky spurts, poking its nose into burrows, small crevices, leaf litter, soft earth, and repeatedly flicking its tongue. Even featureless ground surfaces may be inspected. When buried prey is detected, the animal digs with alternate or simultaneous movements of the forefeet, often aided by flipping movements of the snout. These lizards also have acute vision and are capable of catching small insects in flight, leaping their body length into the air to do so. In general, they forage in open, rather sparsely vegetated habitats where they must move considerable distances in search of prey, and hence spatial territoriality appears to be little developed or absent. They do, however, have home ranges.

In temperate areas breeding occurs in spring. All species are oviparous, and often more than one clutch is produced in a season. Whiptails are of special interest, because in addition to bisexual species there are many species that are exclusively female, reproducing by parthenogenesis, in which unfertilized eggs produce only females. Our two California species do not exhibit this feature. In some of the "all female" species, occasional males are found. They may result from backcrossing with a bisexual species.

WESTERN WHIPTAIL	*Aspidoscelis tigris*
	(Cnemidophorus tigris)
CALIFORNIA WHIPTAIL	*A. t. munda*
COASTAL WHIPTAIL	*A. t. stejnegeri*
GREAT BASIN WHIPTAIL	*A. t. tigris*

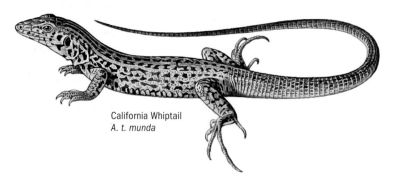

California Whiptail
A. t. munda

IDENTIFICATION: The tail is over twice the snout-to-vent range of 6 to 12.7 cm (2.4 to 5 in.). The body form follows the classic *Aspidoscelis* pattern described in the genus introduction. The dorsal head and shoulders are olive-brown, yellowish brown, or grayish brown, which grades to light tan, dull yellow, or yellow-brown on the back. The tail is brown, gray-brown, or olivaceous, often like the head and shoulders. Certain of the keels of scales of tail are black, giving the tail a dark-flecked appearance. There is a dorsal pattern of numerous closely set transverse bars or spots of black or brown arranged roughly in transverse rows. The ventral surface is light blue, ash white, or blue-gray to yellow and marked with slaty to black spots that are most abundant on the upper abdomen and chest. Both sexes are similar in appearance, but once an individual is in hand, the presence or absence of hemipenis pockets can be determined, and the femoral pores, which are larger in males than females, can be closely observed.

RANGE: This species is found in the Great Valley, Sierra Nevada foothills, the eastern portions of the North and South Coast ranges, the Tehachapi, Transverse, and Peninsular ranges, and the Mojave and Colorado deserts. The California Whiptail occurs in the Great Valley, the Innercoast Range (north of Santa Barbara), and the Sierra foothills. The Great Basin Whiptail occurs in the Mojave and Colorado deserts. The Coastal Whiptail is found in southern San Diego County (excluding Anza Borrego) north to coastal Ventura County.

NATURAL HISTORY: The Western Whiptail ranges from flatlands well up into mountain habitats. The substratum can be of sand, gravel, rocks, or hardpan, but loose soil seems to be preferred. It is most abundant in deserts and semiarid regions, especially in areas of relatively level terrain. It is found in a great variety of plant associations, including creosote bush and burrobush, desert grassland, broken chaparral, gray pine and oak woodland, and open coniferous forest. It generally avoids areas of dense grass and thick shrubby growth. It requires warm and sunny areas for basking, friable soil for burrow construction and foraging, open areas for running, cover of bushes, rocks, or both, and an abundant supply of insects, especially those whose larvae burrow in loose soil. It is uncommon or absent in the cooler fog-belt areas of the California coast.

This is a wary species, seldom allowing close approach. When alarmed it may dash away at great speed, come to a sudden halt, often near a bush beneath which it may go. Frequently such cover is quickly abandoned,

Plate 84. Western Whiptail.

and the lizard darts off again. It may also seek shelter in a rodent burrow. Occasionally it runs bipedally but with the body held nearly horizontally. When running, the tail is usually lifted enough to clear the ground, and the tip is lashed about.

It may dig its own burrow, particularly when few rodent burrows are present. Sand is scraped out by alternate use of the forelimbs with periods of briefly sustained scratching. When sand accumulates at the mouth of the tunnel, the lizard may enter, turn about, and push the sand out with its front feet. It has occasionally been seen to climb into brush or on rock outcrops, possibly in search of food or to escape the heat of the desert floor.

It emerges from hibernation from early March to April or early May depending on the elevation and latitude. In southern areas, a few individuals may be active as early as February. In many parts of the range, activity is greatest from May to June or July. These lizards are active throughout the summer in the Mojave and Colorado deserts, but estivation occurs in July and August in the northern parts of the range. A few juveniles may remain active until mid-August after most adults disappear. Periods of inactivity are spent in self-constructed burrows.

These lizards evidently do not defend territories, and home ranges often overlap extensively. Tail loss frequently seems to be about the same in both sexes, suggesting that males are not aggressive.

Clutch size varies from one to seven eggs, and the number of clutches per season varies from one to perhaps three. Many individuals in southern populations produce a minimum of two clutches a season, whereas northern populations produce only one. Clutch size tends to be larger in the north, which helps compensate for single broods. Hatchlings measure from 32 to 44 cm (1.3 to 1.7 in.) snout to vent. Sexual maturity may be reached the first spring, at an age of eight to 11 months, but many individuals do not mature until their second spring at 20 to 23 months. The life expectancy is three to seven years.

Insect larvae constitute an important food item in the spring. Adult termites, beetles, grasshoppers, and spiders are also an important part of the diet, and small lizards are also taken. One of the senior author's former graduate students, Elinore Benes, found that this species can recognize a great range of colors and even slight differences in a single color, a presumably helpful trait for a lizard that forages mainly on multicolored arthropods.

ORANGE-THROATED WHIPTAIL   *Aspidoscelis hyperythra*
(Cnemidophorus hyperythra)

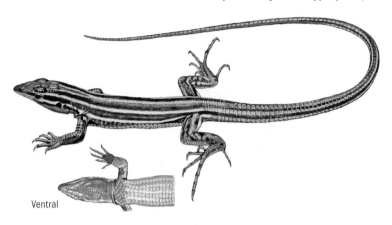
Ventral

IDENTIFICATION: The tail of this species is over twice its snout-to-vent range of 5.1 to 7 cm (2 to 2.75 in.). It occurs only in the southwest corner of California and is the only other whiptail in this state. It differs markedly from the Western Whiptail in that it has five to six light stripes but no spots. The color between stripes is gray, reddish brown, or dark brown to blackish, and the color of the ventral surfaces is uniformly light blue-gray to yellowish white, and often washed with orange. Areas on the underside of the head from the gular fold anteriorly may be intense orange. In males the color of the throat, often the chest, and in the far south of the range the underside of the tail is orange, becoming especially intense in the breeding season. When present, the orange color in females is blotched, faint, or pale, and confined to the extreme sides of chin, throat, and chest.

RANGE: It occurs in Coastal Southern California from near Colton (Reche Canyon), San Bernardino County, Riverside, Riverside County, and Dana Point and Salt Creek, Orange County south to the Cape region of Baja California, Mexico.

NATURAL HISTORY: This whiptail prefers areas of well-drained soil, often where there are firm, somewhat compacted surfaces intermixed with friable, often sandy areas and some rocks. Vegetation is that of the coastal sage scrub or chaparral plant community with intervening open areas of relatively bare soil.

During the season of activity, emergence occurs between 8:00 and 10:00 AM, and peak activity is reached by 11:00 AM. In the hot midday period individuals sometimes climb several feet up in shrubs where they rest in shaded or partly shaded areas. Shallow tunnels are also constructed

Plate 85. Adult Orange-throated Whiptail.

to escape late-afternoon heat. Vibration of the pelvis and tail wagging are employed to compact the sides of the tunnel, and the hind feet are used to remove loose soil. The snout and base of the forefeet are used to push loose soil from the tunnel and to plug the entrance. Home range areas often overlap. Adults appear to move more or less at random throughout their ranges. Threat displays occur in defense of food. In defense posturing the back is humped and the body flattened dorsoventrally; the lizard stands stiff-legged with snout pointed downward at an angle of about 30 degrees. The base of the tail is elevated while the remaining two thirds rests on the ground and twitches laterally. Two males may circle, facing one another, until one withdraws.

Egg-laying occurs throughout June to mid-July. Two years or older females deposit one clutch in June and another in mid-July, but yearlings probably lay only one clutch from late June to mid-July. Hatchlings are around 25 to 29 mm (1 to 1.1 in.) in snout-to-vent length and appear from mid-August through the first week of September. Adult size is reached by the following June. *A. hyperythra* feeds on ground-dwelling insects such as termites, ant lion larvae, the burrowing "sand cockroaches," crickets, grasshoppers, silverfish, springtails, and beetles. Other invertebrates including spiders, mites, and pseudoscorpions are also taken.

Alligator Lizards and Relatives (Family Anguidae)

This family is widely distributed in the Western Hemisphere, Europe, North Africa, Asia, Sumatra, and Borneo. Only the genus *Elgaria* occurs in California.

Alligator Lizards (Genus *Elgaria*)

This genus occurs from the middle and western United States and southwestern Canada southward to Mexico. These are long-bodied lizards with relatively short limbs and a broad, rather triangular head, especially in adult males. The tail is one and a half to two times the body length. The scales tend to be quadrangular and are arranged in longitudinal rows, and the scales on the dorsum are keeled. There is a well-defined lateral fold, consisting of a strip of granular scales between the keeled dorsal scales and smooth ventral scales. They are diurnal, crepuscular, and nocturnal, with nocturnal activity becoming more frequent in warm weather. Alligator lizards are secretive and chiefly ground-dwelling, seldom seen in the open. They move about in grass (sometimes following meadow vole tunnels), leaf and branch litter, and among rocks and logs, searching for spiders, insects, and small vertebrate prey. They are to a large extent thigmothermic, but they also bask. Basking usually occurs in conspicuous locations in sunlit areas, often where there is abundant cover. However, they can be active at low environmental temperatures. When crawling rapidly, they move the body in sinuous snakelike curves but at other times move forward slowly and deliberately, much like a snake, which employs rectilinear movement when stalking prey. Indeed, much of an alligator's foraging behavior is reminiscent of that of snakes, including the frequent protrusion of the moderately forked tongue. These lizards are also good climbers and often use their tail in a prehensile manner to secure a hold on branches. Climbing literally adds a third dimension to alligator lizard foraging and permits the utilization of food items such as small bird eggs and nestlings (figure 31).

When seized by a potential predator, most individuals writhe about besmearing themselves and their captor with urine and feces and the odoriferous contents of their anal scent glands. Alligator lizards usually occur in mesic, well-vegetated habitats. In arid environments, they are nearly always found along stream courses and in the vicinity of springs and seepages. There are both oviparous and ovoviviparous species. The egg-laying species tend to occur in warmer habitats and the live-bearing ones in relatively cool, coastal, or montane habitats. Our three California species exhibit both types of reproduction.

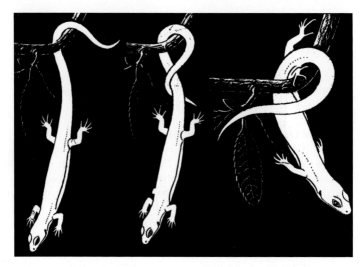

Figure 31. The use of a prehensile tail by an alligator lizard when climbing.

SOUTHERN ALLIGATOR LIZARD   *Elgaria multicarinata*
CALIFORNIA ALLIGATOR LIZARD   *E. m. multicarinata*
OREGON ALLIGATOR LIZARD   *E. m. scincicauda*
SAN DIEGO ALLIGATOR LIZARD   *E. m. webbii*

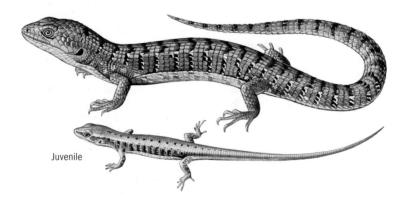

Juvenile

IDENTIFICATION: The unregenerated tail of this species can be over two times its snout-to-vent range of 7.3 to 17.8 cm (2.9 to 7 in.). The dorsal scales are usually in 14 rows, and those on both body and tail are heavily keeled. There is usually only one large interoccipital scale on the top of the head between the eyes. The dorsal color varies from light brown, olive, gray,

Plate 86. Southern Alligator Lizard.

to dull yellow. The body and tail are marked with somewhat irregular transverse bands of dark color. The venter is pale gray to dull pale yellow and usually has longitudinal dark stripes aligned along the center of the longitudinal scale rows. This is the most commonly used feature to distinguish a Southern Alligator Lizard from a Northern Alligator Lizard (*E. coerulea*) when one has an individual in hand. Juveniles resemble adults in form, but their dorsal color occurs as a broad band from head to tail and is uniformly reddish tan, beige, or gray. The head of adult males is somewhat broader proportionately than that of the female.

RANGE: In California it occurs throughout the length of the state, mostly west of the mountain crests in the Cascade and Sierra Nevada ranges and the Mojave and Colorado deserts. To the south it is found near Independence, Inyo County. It is absent from most of the southern Great Valley. It also occurs on Santa Catalina, San Nicolas, San Miguel, Santa Rosa, Santa Cruz, and the Anacapa islands, as well as on Angel Island (in the San Francisco Bay).

NATURAL HISTORY: *E. multicarinata* usually frequents well-wooded areas but also occurs in open grassland with widely scattered shrubs. In coastal areas, it is found in edge situations and openings in Douglas fir and redwood forest, in the oak and chaparral belt of foothills and valleys, and in vegetated dunes and under driftwood along beaches. In the Sierra Nevada it occurs in the blue oak–gray and yellow pine–black oak zones. In the Sacramento Valley it frequents oak woodlands and thickets in riparian habitats. It is fond of cover and often retreats under brush heaps, rocks, logs, and in woodpiles. It is able to live in warmer, drier environments than *E. coerulea*, with which it is sympatric over much of its range.

This species is more crepuscular and nocturnal than *E. coerulea*. It climbs well, and the tail is moderately prehensile. When suspended by the tail from a branch or one's finger, it can draw itself up to its support by tail movements, independent of the limbs (see figure 31). It has been seen to enter streams, irrigation ditches, and conduits, sometimes diving to the bottom and seeking shelter under stones. It may also enter water to catch tadpoles. It swims well using lateral undulations of the body and tail. Occasionally individuals caught in the open play "possum." When encountered away from cover in the field, it sometimes lies quietly, even remaining limp when probed, then suddenly dashes off at high speed. It is often aggressive and may attack animals much larger than it is, especially if cornered. When alarmed it puffs up, hisses, and twitches its tail.

The junior author once observed a juvenile Alameda Striped Racer (*Masticophis lateralis euryxanthus*) attempting to eat a juvenile Southern Alligator Lizard. The lizard immediately grabbed its own tail with its mouth, forming a circle of its entire body. What was originally a slim, easy-to-swallow meal was now a large, unmanageable object, and the young snake soon gave up.

In mating the male holds the head or neck of the female in his jaws and may stroke her sacral region with his hindlimb. In the southern part of the range, the growing season is longer and breeding occurs earlier in the spring. The eggs are probably usually laid in ground squirrel, gopher, and other animal burrows. Clutch size varies from around five to 20 eggs. In southern California females usually deposit two and occasionally three clutches a season, but at high elevations and to the north a single clutch per season is probably the rule. The incubation period is about one and a half to two months. Hatchlings appear from late July to early November and measure approximately 26 to 36 mm (1 to 1.4 in.) from snout to vent. Sexual maturity is reached in the second spring, at an age of about 18 months.

E. multicarinata is a classic feeding generalist. It takes a wide variety of insects, including large species like the Jerusalem Cricket, grasshoppers, crickets, and cockroaches. It also eats scorpions, centipedes, spiders, snails, slugs, bird eggs and nestlings, lizards, small mammals, and carrion. It has been observed to eat the adults and egg cases of the black widow spider. Its ability to be active at relatively low body temperatures probably allows it to catch lizards when they are sluggish in their retreats or just emerging to bask.

NORTHERN ALLIGATOR LIZARD   *Elgaria coerulea*
NORTHWESTERN ALLIGATOR LIZARD   *E. c. principis*
SAN FRANCISCO ALLIGATOR LIZARD   *E. c. coerulea*
SHASTA ALLIGATOR LIZARD   *E. c. shastensis*
SIERRA ALLIGATOR LIZARD   *E. c. palmeri*

IDENTIFICATION: This species resembles *E. multicarinata* but differs as follows. The average adult size is smaller, 7 to 13.6 cm (2.75 to 5.6 in.) from snout to vent, and the tail is shorter, less than two times the snout-to-vent length. The dorsal scale rows are usually 16, and the large interoccipital scale between the eyes is often divided. The dorsum is gray, olive, greenish, brownish, or bluish, with the greatest of these color morphs present in the northern part of its California range. The venter is whitish, cream, or light gray, and dark longitudinal stripes on the venter, when present, are between the scale rows, incorporating the edges of the adjacent belly scales (*E. multicarinata* has dark lines along the middle of the scale rows). In northern California, dark and light color phases coexist with the former usually the more common. In the dark color phase, the dark blotches and variegations greatly reduce the amount of light background color.

RANGE: The species occurs in the Sierra Nevada, Cascade Range, Warner Mountains, Modoc County, North Coast Range, and the South Coast Range north of Big Sur River, Monterey County. It is also found on Angel, Brooks, and Yerba Buena islands in San Francisco Bay, and on Año Nuevo Island north of Monterey Bay.

NATURAL HISTORY: *E. coerulea* inhabits generally cooler and more humid areas than *E. multicarinata*, with which it is associated in many parts of its range. It is usually found within or near coniferous woods under logs and other surface objects in areas where bushes, trees, and open grassy areas afford cover and places for foraging. Open willow–pine forests with abundant rotting logs seem to be especially favored in the southern part of the range. It is often abundant along creeks, and frequents coastal beaches and lake shores where it may be found under driftwood.

Plate 87. Northern Alligator Lizards copulating.

This species is an ovoviviparous live-bearer. It retains the shell-less, yolked eggs instead of laying them. In this manner the female may control embryo development temperature by basking and seeking warm retreats. When born the young are surrounded by a thin, transparent membrane, the amnion, from which they usually soon escape. The mother has been observed to chew through the amnion to free the young and then eat this membrane. The number of young ranges from two to 15, with four to five the average.

PANAMINT ALLIGATOR LIZARD *Elgaria panamintina*

Juvenile

PROTECTIVE STATUS: CSSC

IDENTIFICATION: Its snout-to-vent range is 9.2 to 15.2 cm (3.6 to 6 in.). It resembles *E. multicarinata* in tail length and dorsal scale row count, and it usually has a single large scale between the eyes. It differs in having generally reduced keeling of dorsal scales on the head, limbs, and tail. The dorsum is light yellow or beige with seven to nine regular broad brown crossbands. The top of the head is light olive gray. The venter is gray. There are dark stripes down the center of the ventral scale rows in some individuals, or in others these markings may be scattered. Juveniles have contrasting light and dark crossbanding on body and tail.

RANGE: This is a relict species with a disjunct distribution, found primarily in spring along streams and seepage areas in the desert mountains of Inyo County and southeastern Mono County. It is known from the Panamint, Nelson, Inyo, and White mountains, but probably has a much wider distribution than present records indicate. For instance, there is one record for the Coso Mountains in Inyo County. During the moist pluvial periods of the Pleistocene it may have ranged widely and inhabited much lower elevations.

NATURAL HISTORY: In the Panamint Mountains, conditions in Surprise Canyon, in which the type locality occurs, are representative of its streamside habitat. The canyon is steep-sided with a cold stream fed by springs. Riparian growth consists of willows and a variety of mesic forbs. This lizard has been found in the canyon bottom in or near the riparian growth. Cover is afforded by rock fissures, talus, scattered rocks along the stream course, leaf and twig litter beneath the willows, desert grape, and mats of sage (*Artemisia*). The talus slopes, side canyons, and canyon slopes, especially where seepages occur, provide refuges during the storms that occasionally scour the main canyon bottom.

Little is known about the habits of this lizard. Captives behave like *E. multicarinata* in their manner of feeding, and use of the tail in climbing and sleeping. Both species were observed to sleep with the tail curled forward beside or on top of the head.

Mating has been observed in May. This species presumably lays eggs, possibly up to 12 per clutch. It feeds on insects and other arthropods.

CONSERVATION NOTE: All but two of the approximately 16 known populations of *E. panamintina* occur on private lands where some are at risk from livestock grazing, mining, and off-road vehicle activity. More natural history information is urgently needed so that sound management and habitat enhancement measures along with well-chosen conservation easements can be pursued.

North American Legless Lizards (Family Anniellidae)

The two species that make up this family occur in central and southern California west of the deserts and in Baja California, Mexico. They have the most unique body form of all California lizards. Besides being legless, they have a long, narrow body that blends with an equally slim tail to produce a very snakelike form. Unlike snakes, legless lizards have rudiments of both a pectoral and pelvic girdle. The skin is smooth and glossy and is covered with smooth, cycloid-hexagonal scales. There are no ear openings, and the eyes are small but have moveable lids. This is the best single feature by which a first-time viewer may recognize it as a lizard and not a snake, because the latter have no moveable lids.

CALIFORNIA LEGLESS LIZARD   *Anniella pulchra*

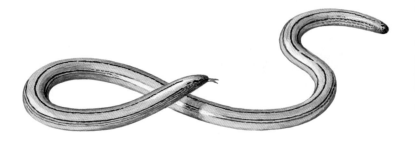

PROTECTIVE STATUS OF *A. PULCHRA*: CSSC

IDENTIFICATION: The snout-to-vent range is 11.1 to 17.8 cm (4.4 to 7 in.), and the tail, which is essentially indistinguishable from the body, is considerably longer than the body length. The dorsal color ranges from silvery, beige, or brownish olive to dark brown or black. Dark individuals (Plate 88) occur in beach dunes of the Monterey Peninsula and the

Plate 88. Dark individual formerly recognized as the subspecies *Anniella pulchra nigra*.

southern coast of Monterey Bay. Dusky populations occur in the Morro Bay area. In light-colored individuals, a thin, dark, median line extends from the head to the tip of the tail. Two to three closely set, fainter longitudinal lines form a band of dark color on the sides. The venter is yellow or light yellow, and the throat is less heavily pigmented. The sexes appear similar, but males have slimmer bodies than females.

RANGE: It has a highly disjunct distribution, because it requires the presence of relatively warm, loamy, or sandy soils. It extends from the sand dune area near Antioch on the south bank of the San Joaquin River, Contra Costa County, south through the South Coast Range, Coastal Southern California, and the Transverse and Peninsular ranges, into northern Baja California, Mexico. There are scattered occurrences on the floor of the San Joaquin Valley, where before agriculture it may have been widespread. It also inhabits sites in the southern Sierra Nevada and the Tehachapi Range.

NATURAL HISTORY: This species occurs in oak woodland, chaparral, riparian woodland, oak–pine forests, and desert scrub. It frequents areas that contain at least some loose fine soil or litter through which it burrows. This includes sand, loam, alluvium, leaf litter, or sand mixed with humus. Other requirements are adequate soil moisture, warmth, and surface cover such as rocks, logs, bushes, or matlike herbaceous growth. The soil is often so friable that, when dry, burrows cannot be maintained and the lizard "swims" through the loose material. Patches of leaf litter are usually present beneath bushes or plant clumps in soils frequented. Much time is spent just beneath the surface, but individuals have been found at depths of up to 60 cm (about 2 ft). In coastal areas they occupy beach sands and dunes and may be found by raking the soil under the canopy of bush lupines, beneath ice plant, other mat-forming species, or under driftwood. In interior areas they may be sought under bushes, cactus clumps, logs, rocks, and other surface objects. The soil on the sunny side of objects seems to be preferred. Alluvial fans and sandy deposits along washes, arroyos, and stream banks are also occupied, but firm soils are avoided.

This lizard is both diurnal and crepuscular, and in warm weather nocturnal, as indicated by their tracks. Movement through the soil and on the surface is by means of lateral undulations of the body. When digging rapidly, lateral movements of the body are employed, and the lizard moves forward by pressing its body loops against the ground. As soon as the head is under the surface it is turned to one side, which begins the sigmoid body curve. When burrowing near the surface in dry sand, the lizard leaves behind a smoothly undulating grooved track about 12 mm (0.5 in.) across as its burrow collapses behind it. In moist sand the tunnels do not collapse. The eyes are closed when the lizards move through dry sand but may be opened when they are in burrows in damp sand.

A. pulchra is ovoviviparous, and the young are born from September to November. Litter size is one to four. At birth the young measure 46 to 67 mm (1.7 to 2.2 in.). Females are at least two years old before they breed.

Its food includes a variety of ground-dwelling insects and their larvae. Feeding takes place on the surface or just below, usually in the leaf litter beneath bushes. In observations of captive specimens, they appeared to perceive movements of beetle larvae crawling on the surface and at times would extend the head vertically out of the sand and lick the air with the tongue while apparently looking around. The olfactory lobes are well developed in these lizards, and chemoreception is presumed to be acute.

CONSERVATION NOTE: A wide variety of human activities that occur often on flat or moderately upland soft soil and sand areas continue to affect this species. These include agriculture, the introduction of exotic pioneer plant species (example: ice plant), urban development, sand mining, golf courses, and the destruction of habitat by off-road vehicles. Areas proposed for any of these land uses must be surveyed by qualified herpetologists to determine if this species is present. If it is, the CDFG must then require meaningful mitigation measures that will ensure the preservation of the population in question.

Venomous Lizards (Family Helodermatidae)

This family also contains only one genus, *Heloderma*, with two species, *H. suspectum* and *H. horridum*. The former is the Gila Monster, which occurs in California, and the other is the Mexican Beaded Lizard, and both produce a neurotoxic venom that flows along grooves in the teeth and is worked into the victim by a chewing action of the jaws. It is apparently used primarily in defense. Helodermatids are terrestrial, chiefly nocturnal lizards of arid, semiarid, and subtropical areas. The teeth tend to be largest and most deeply grooved toward the middle of the lower jaw. The tongue is thick, fleshy, flattened, arrow-shaped, and deeply notched at tip, an indication that the Jacobson's organ provides important sensory input in these species.

GILA MONSTER
BANDED GILA MONSTER

Heloderma suspectum
H. s. cinctum

PROTECTIVE STATUS OF *H. S. CINCTUM*: CSSC

IDENTIFICATION: This is our largest California lizard, with a snout-to-vent range of 22.8 to 35.5 cm (9 to 14 in.). The dorsum has prominent spots, irregular bars, reticulations, and dots of buff, yellow, orange, or salmon and blackish brown. In California the markings on the body may tend to form alternating, irregular light and dark crossbands. The tail is usually marked with rather broad irregular rings of light and dark color. The sexes are similar, but males tend to have a broader head and longer tail than females.

RANGE: In California it occurs in the eastern Mojave Desert in the Providence, Paiute, Clark, Kingston, and perhaps the New York mountains. Old records also include the vicinity of Desert Center in the Chuckwalla Mountains and in the Imperial Dam area, Imperial County.

NATURAL HISTORY: This lizard prefers shrubby, grassy, and succulent desert. Most of its range is in the creosote bush desert, but it avoids the more barren parts. It has been found in the saguaro–ocotillo association, in the vicinity of mesquite clumps and palo verde, and in desert grassland. It appears to be attracted to rocky areas and to the somewhat moister parts of the desert in the vicinity of streams, washes, and irrigated lands. Soils are often sandy or gravelly. The areas where it occurs are subject to summer rainfall. Shelter is sought in dense thickets, under rocks, and in crevices, in self-made burrows or those of other animals, and in woodrat nests.

Tail fat deposits aid these lizards in surviving food shortages and droughts. The color and scales of the Gila Monster resemble beadwork patterns and have inspired designs used in Native American crafts. It is highly cryptic at dusk and at night when it is often active. On predominantly dark backgrounds the black markings blend and the light ones catch the eye, looking like light-colored sticks and rocks. In the daytime the gaudy pattern may serve as a warning. Although often diurnal, it is also crepuscular and nocturnal. In warmer weather diurnal activity is most likely to occur after rains, which seem to stimulate activity, or on overcast days. Hibernation occurs in crevices and burrows usually from October to March.

The usual gait is relatively slow and appears awkward. Although heavy bodied, they are capable of climbing and have been seen in cholla cactus. They are powerful diggers, although the extent to which they construct their own burrows is not known. They are fond of water when in captivity, occasionally lying submerged. In desert towns individuals occasionally have to be removed from swimming pools.

Gila Monsters mate from May to July and perhaps later. Eggs are laid chiefly in the latter part of July or during the first half of August. Four to seven constitute a clutch. The nest hole may be placed in damp soil in a sunny location. There are indications that an individual female lays only every other year. Eggs reared at the San Diego Zoo required around four months for incubation. Hatchlings may measure around 9 to 15 cm (4 to 6 in.) in total length. Captives have often lived eight to 10 years, and two have been reported over 24 years.

The Gila Monster is a true carnivore. Its prey includes juvenile mammals such as the Round-tailed Ground Squirrel (*Spermophilus tereticaudus*), young Audubon's Cottontails (*Sylvilagus audubonii*), and Black-tailed Jackrabbits (*Lepus californicus*), eggs of ground-nesting birds, such as those of the Gamble's Quail (*Lophortyx gambelii*), and eggs of reptiles, including those of Desert Tortoise (*Gopherus agassizii*). It also eats insects and probably occasionally takes carrion.

Gila Monsters are nearsighted and appear to depend heavily on chemoreception for detection of their food, although they become alert and surprisingly quick when confronted with a live rodent and appear to use sight. When they are hunting, the notched, snakelike tongue is frequently protruded. Although they appear sluggish, they are capable of very quick lunges to secure prey or in defense. The function of the venom is unclear, but it is perhaps used primarily in defense, because most prey animals are quickly killed and swallowed, seemingly giving little time for the venom to act. It is a neurotoxin and affects neural transmission at the muscle–nerve synapse. Although formidable in appearance, this lizard is not dangerous unless carelessly handled and certainly should not be killed. There is no record of a person dying from a Gila Monster bite during the past century.

CONSERVATION NOTE: Although the spectrum of human-wrought habitat changes that have affected other special-status reptiles have also come to bear on the Gila Monster as well, its major problem may well be centered on its own unique biology. Its large size, beautiful color patterns, carnivorous feeding habits, long life span, and perhaps above all its status as one of the only two venomous lizards in the world, make it a prime target for illegal collecting and sale on the world reptile black market. The Gila Monster is fully protected throughout its range in the United States and in Mexico. It should also be noted that it is not one of several Species of Special Concern that may be legally collected for the price of a fishing license, and any reader who observes a situation involving illegal collection and possession would be doing this unique animal a favor by calling the CDFG "Poaching Hotline." At the time of this writing the number is 888-DFG-CALTIP.

Snakes (Suborder Serpentes [Ophidia])

Blind Snakes (Family Leptotyphlopidae)	350
Western Blind Snake (*Leptotyphlops humilis*)	351
Boas and Pythons (Family Boidae)	352
Northern Rubber Boa (*Charina bottae*)	353
Southern Rubber Boa (*Charina umbratica*)	353
Rosy Boa (*Lichanura trivirgata*)	354
Colubrids (Family Colubridae)	355
Ring-necked Snake (*Diadophis punctatus*)	356
Common Sharp-tailed Snake (*Contia tenuis*)	358
Forest Sharp-tailed Snake (*Contia longicaudae*)	358
Spotted Leaf-nosed Snake (Western Leaf-nosed Snake) (*Phyllorhynchus decurtatus*)	360
Racers and Whipsnakes (Genera *Coluber* and *Masticophis*)	362
Racer (*Coluber constrictor*)	362
Coachwhip (*Masticophis flagellum*)	364
Baja California Coachwhip (*Masticophis fuliginosus*)	364
Striped Racer (*Masticophis lateralis*)	366
Striped Whipsnake (Desert Striped Whipsnake) (*Masticophis taeniatus*)	369
Western Patch-nosed Snake (*Salvadora hexalepis*)	370
Gopher Snake (*Pituophis catenifer*)	371
Glossy Snake (*Arizona elegans*)	373
Kingsnakes (Genus *Lampropeltis*)	374
Common Kingsnake (*Lampropeltis getula*)	375
California Mountain Kingsnake (*Lampropeltis zonata*)	377
Western Long-nosed Snake (*Rhinocheilus lecontei*)	378
Baja California Rat Snake (*Bogertophis rosaliae*)	379
Garter Snakes (Genus *Thamnophis*)	379
Common Garter Snake (*Thamnophis sirtalis*)	381
Western Terrestrial Garter Snake (*Thamnophis elegans*)	385
Aquatic Garter Snake (*Thamnophis atratus*)	387
Sierra Garter Snake (*Thamnophis couchii*)	388
Giant Garter Snake (*Thamnophis gigas*)	389
Two-striped Garter Snake (*Thamnophis hammondii*)	390
Northwestern Garter Snake (*Thamnophis ordinoides*)	391
Checkered Garter Snake (*Thamnophis marcianus*)	392
Western Shovel-nosed Snake (*Chionactis occipitalis*)	393
Western Ground Snake (*Sonora semiannulata*)	395
California Black-headed Snake (*Tantilla planiceps*)	396
Southwestern Black-headed Snake (*Tantilla hobartsmithi*)	397
California Lyre Snake (*Trimorphodon lyrophanes*)	398

Sonoran Lyre Snake (*Trimorphodon lamba*)	398
Coast Night Snake (*Hypsiglena ochrorhynchus*)	399
Desert Night Snake (*Hypsiglena chlorophaea*)	399
Sea Snakes (Family Hydrophiidae)	400
Yellow-bellied Sea Snake (*Pelamis platurus*)	401
Vipers (Family Viperidae)	401
Pit Vipers (Subfamily Crotalinae)	401
Western Rattlesnake (*Crotalus oreganus*)	410
Western Diamond-backed Rattlesnake (*Crotalus atrox*)	413
Mojave Rattlesnake (*Crotalus scutulatus*)	415
Speckled Rattlesnake (*Crotalus mitchellii*)	417
Panamint Rattlesnake (*Crotalus stephensi*)	417
Red Diamond Rattlesnake (*Crotalus ruber*)	418
Sidewinder (*Crotalus cerastes*)	419

Key to the Families and Genera of Snakes

All snakes lack legs and moveable eyelids (legless lizards have moveable eyelids).

1a Small cycloid scales encircle the entire body; no large ventral scales; eyes reduced to small pigment spots . . . . Western Blind Snake
1b Ventral scales much larger than those on back or side; well-developed eyes 2a

2a Tail with a rattle or single "button" at tip; two pairs of small openings on snout (nostrils and loreal pits) . . . . Rattlesnakes
2b Tail without rattle and snout without paired pits 3a

3a Only small scales on throat on underside of jaw. . . . . Boas
3b Large scales on throat on underside of jaw 4a

4a Rostral scale at tip of snout much enlarged and patchlike or leaflike . . . . Patch-nosed and Leaf-nosed Snakes
4b Rostral scale at tip of snout not greatly enlarged 5a

5a Dorsal scales smooth, not keeled 7a
5b Dorsal scales keeled 6a

6a Usually four prefrontal scales in front of eyes; large brown blotches on a tan-yellow background; eye pupils round, not vertical . . . . Gopher Snake
6b Two prefrontal scales in front of eyes; bold to faint mid-dorsal stripe usually present; single anal scale anterior to vent . . . . . Garter Snakes

7a Adults with olive-green dorsal and pale-yellow ventral color; juveniles with brown blotches on tan background, similar in appearance to Gopher Snake but with two prefrontal scales and noticeably large eyes . . . . Racer
7b Adult coloration not as in 7b 8a

8a Long, slender snakes; large eyes with round pupils; small preocular scale wedged between upper lip scales; lateral stripes or mottled brown-red dorsal coloration. . . . . Whipsnakes and Coachwhips
8b Normal-size eye; no body stripes; lower preocular scale not wedged between upper lip scales 9a

9a Sharp spinelike point at tip of tail; all ventral body scales with regular narrow black crossbands. Sharp-tailed Snakes
9b No spinelike point at tip of tail; ventral scales usually not marked with black crossbands 10a

10a Plain dorsal coloration without pattern; head often darker; narrow orange or white dorsal neck band usually present. . . . .
. . . . Ring-necked and Black-headed Snakes
10b Dorsal color pattern of spots, blotches, or crossbands. 11a

11a Bold dorsal crossbands that extend across ventral body . . . .
. . . . Kingsnakes
11b Dorsal crossbands or with brown and tan mottled pattern but with no bold banding across entire ventral surface. 12a

12a Large red and black dorsal blotches or red mid-dorsal color with or without black blotches. . . . .
. . . . Long-nosed, Shovel-nosed, and Ground Snakes
12b Dorsal color mottled brown to light tan 13a

13a Scales very shiny or glossy; round eye pupil . . . . Glossy Snake
13b Vertical eye pupil; scales not glossy 14a

14a Large, dark saddle-shaped figure on dorsal neck area; eye not noticeably large . . . . Night Snakes
14b Large eye; lyre-shaped figure usually on dorsal neck area . . . .
. . . . Lyre Snakes

Close to 3,000 species of snakes have been described. Despite their lack of limbs, they have become adapted to most major habitats and are distributed nearly worldwide. The majority are terrestrial, but some are arboreal, fossorial, or aquatic. In the absence of limbs there is considerable restriction in the variability of bodily form. Nevertheless there are obvious modifications related to habits. The smallest of California snakes, the Western Blind Snake (*Leptotyphlops humilis*), is cylindrical with a short blunt tail, no ventral plates or scutes, and vestigial eyes. Loose soil burrowers like shovel-nosed snakes (*Chionactis*) have nasal valves, a countersunk lower jaw that prevents "shipping" sand when "sand-swimming," and lateral abdominal ridges that prevent slippage when crawling in sand. Other California species such as patch-nosed and leaf-nosed snakes have the rostral scale modified for digging. Arboreal snakes such as boas and pythons may be slow-moving and heavy-bodied, while others are streamlined and slim-bodied like the vine snakes. Snakes have successfully adapted to life in both freshwater and saltwater. The most aquatic snakes in mainland California are several species of garter snakes, which often forage in water and retreat into it when disturbed. However, within the state's territorial limits are often seen the Yellow-bellied Sea Snake (*Pelamis platurus*), one of the most aquatic of all reptiles, giving birth at sea and thus never actively coming ashore.

Body Adaptations

Skin

As in lizards, the outer epidermis of the skin is periodically shed. Before this the skin becomes lusterless, and the eye appears opaque. The skin begins to loosen about the lips and is turned back as the snake crawls among rocks, branches, or other objects. When shedding is completed, the skin has been turned inside out. The outlines of each scale are evident, and vague indications of the color pattern are sometimes present. In his fieldwork with endangered and threatened snakes, the junior author has used a recovered skin on several occasions to verify the presence of a protected species or subspecies at a habitat site within its known range. Each time a rattlesnake sheds, a horny segment is left behind at the base of the tail, adding to the rattle. The number of segments indicates frequency of shedding but not age, because there may be more then one shedding each year.

Locomotion

Despite their lack of limbs, many snakes can crawl rapidly, some of them considerably faster than a person can walk. Snake locomotion consists of four types: lateral undulatory, rectilinear, concertina, and sidewinding. The first is familiar to everyone as the typical snake wriggling motion. The body is thrown into S-shaped curves, and the snake moves as though it were forced to follow a sinuous stream channel. Backward pressure of the body loops against irregularities on the ground surface propels the snake forward. Substrate irregularities are critical for traction, and a snake on a smooth surface such as glass or a polished floor has difficulty, especially if it attempts to move rapidly. On relatively smooth but soft substrate such as loose soil or sand, traction is achieved by the pressure of the body curves into surface material. This produces small ridges on portions of the curved track left by snakes, and because the body has pushed against these areas to move forward, you can determine the direction in which a snake was traveling.

Rectilinear locomotion entails moving forward with slight or no lateral undulations. The snake seems to flow along an essentially straight course. This locomotion has also been referred to as caterpillar action. The loosely attached skin is advanced in waves as the broad ventral scales or belly scutes obtain purchase on the ground, and the body is shifted forward

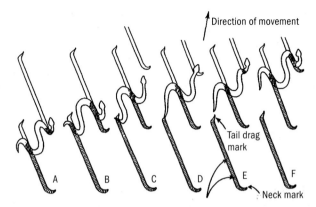

Figure 32. A sequence showing the production of the backwards "J" track of the Sidewinder as it continuously changes the body-ground contact points (dark areas) as it moves.

within the skin. Rectilinear movement is often used when slowly stalking prey. A fully extended snake moving very slowly in a straight line can be very hard to detect, even for the human eye.

Concertina locomotion is often used in climbing steep grades and rough-surfaced tree trunks. The snake extends its body upward, then throws its foreparts into a series of lateral undulations that are pressed laterally against the slope surfaces to secure anchorage. It then draws its posterior body upward and secures it firmly again against the slope surface. By repeating the process of alternately extending and anchoring the forebody and drawing upward the posterior half, the snake hitches its way upward. The locomotion is sometimes used on level ground as well, especially when an initial burst of speed is needed for rapid escape.

Sidewinding locomotion is an adaptation for moving over smooth and often hot surfaces (sand, hardpan) with minimal expenditure of energy. Sidewinding snakes are capable of considerable speed, which may be needed to escape predators in relatively open terrain where retreat sites are not always close by. In sidewinding locomotion the alternate curves of the body are elevated, and thus only those on one side press against the ground so that the snake moves sideways, creating a series of disconnected J-shaped tracks (figure 32). The hook of the J points to the direction of travel. Because the weight of the snake is applied over a restricted length of the body, the pressure per unit area of body that is applied to the substrate is increased, which in turn increases traction and reduces slippage. Note that the track contains the impressions made by the ventral scutes. The reduced and continuously changing body–substrate contact area also reduces thigmothermic heating and overheating on hot desert substrate and allows the Sidewinder (*Crotalus cerastes*) to move across such hot-surfaced areas when other snakes could not.

Tail

Unlike many lizard tails, the tail of a snake is generally a firmly attached structure and cannot be regenerated when severed. Many male snakes use the tail during copulation, curling it about that of the female. Bright coloration of some snake tails serves to distract attention of predators from more vulnerable parts of the body. The red tail coil of the otherwise somber dorsal color of the Ring-necked Snake (*Diadophis punctatus*) apparently serves this purpose. Caudal luring, the waving or vibrating of the tail seen mainly in ambush-foraging snakes, is used to attract prey to within striking distance. This behavior may also have been the precursor to the evolution of the rattle and its apparent function to announce a rattlesnake's presence to predators and other potentially dangerous animals such as large, herd-oriented herbivores that could trample it.

A pair of saclike organs known as scent glands is present in the base of the tail of snakes. The secretion of the scent glands is lipoid in nature and has been variously implicated as a defensive repellent, a sex attractant, an alarm stimulus, a communal retreat indicator, and a pheromone for interactions between blind snakes (*Leptotyphlops* spp.) and ants. It is even possible for humans to learn to recognize some snake genera by their odor.

Sensory Input

Eyesight is well developed in many snakes, but most seem to see objects best at close range. Active diurnal species such as the racers and whipsnakes have large eyes, as do certain nocturnal snakes such as lyre snakes (*Trimorphodon* spp.). The snake eye is moveable beneath a transparent nonmoveable covering. In many snakes the pupil is round, but in California species such as rattlesnakes and lyre snakes it is vertically elliptical. Snakes with a vertical pupil tend to be nocturnal or crepuscular, because a vertical pupil can be more fully closed than a round one and is therefore more effective in preventing dazzle when exposed to light.

The Jacobson's organ is exceptionally well developed in snakes. It occurs as a pair of spherical chambers in the roof of the mouth, opening separately toward the anterior end of the palatine grooves, a pair of furrows that accommodate the forked tip of the tongue. The moist tongue tip transfers molecules from the air or surrounding objects to the palatine region for analysis within the chambers. The Jacobson's organ–tongue transfer system appears to be used in a number of ways, depending upon the species. These include tracking females by males, tracking prey by both sexes, determination of food suitability, location of aggregation sites, and detection of predators. The characteristic response of a rattlesnake to kingsnakes and certain other ophiophagous or snake-eating snakes is abolished by experimentally blocking projection of the rattler's tongue, even though the predator can still be seen.

Snakes are noted for their ability to hiss. Among western North America forms the Gopher Snake (*Pituophis catenifer*) is notable in this regard. Whether such sounds are used in snake communication is unknown. Although snakes were long thought to be deaf, there is now evidence that many are capable of detecting airborne vibrations. Despite the absence of an eardrum and middle-ear cavity, vibrations impinging on the surface of the head are transmitted via the quadrate bone, which suspends the lower jaw and the columella, the middle-ear bone of lizards, amphibians, and birds. As in these groups, one end of the long columella is attached to the oval membrane of the inner ear, and therefore in snakes it transmits vibrations of the skull and jaw. Studies have shown that the quadrate functions to some extent as an eardrum, because breaking its connection causes some loss in responsiveness of the auditory nerve to airborne vibrations. Given this arrangement, it may be that snakes hear sounds made by their jaws when engulfing prey.

The pit organs of pit vipers (family Viperidae) and the boids (boas and pythons, family Boidae) are sensitive to heat radiated from their warm-blooded prey. Pit vipers include the copperhead, moccasin, and the rattlesnakes. The organs are paired and located in the loreal region in these snakes (see in figure 35 in the Vipers introduction) and on the labial scales in boids. Although these pit-bearing snakes are able to find their prey by sight and the Jacobson's organ–tongue mechanism, the pits supplement these other senses and may be especially helpful in aiming the strike when visibility is poor at night or in a rodent burrow. Because the pits are paired and directed forward, they make possible accurate perception of distance, and a blindfolded pit viper with nostrils and tongue blocked can strike a warm object with great precision. Recent studies at the Technical University of Munich, Germany, have demonstrated that a blindfolded snake can consistently strike running rats behind the ears to avoid being bitten. One proposed reason for such accuracy is that the snake's brain is able to process information from the 2000-odd heat receptors on the extremely thin pit membranes so as to filter out background "noise" and create a sharp image comparable to what the eyes see. The thermal sensitivity of the pits is also remarkably high. A heat source only about 1 degree C (2.5 degrees F) above the background temperature can be detected at a distance of 0.3 m (1 ft).

Teeth and Jaw

The majority of snakes have teeth in the upper and lower jaws. In most nonvenomous species there are four longitudinal rows in the upper jaw. These lie in a row on each maxillary bone and a row on each side of the midline of the roof of the mouth. The teeth are typically slender, tapered, recurved structures that are broadly anchored, not in sockets. Replacement teeth in a graded series lie on the inside of the jawbones, but on the palate they are located on the outer side of the teeth. In venomous snakes certain of the teeth are modified for the injection of venom. Venom-conducting teeth may be grooved or hollow. Fangs, like other teeth, are also replaced, and in most venomous snakes the replacement series can be seen in the open mouth. A pair of sockets at the front of each maxillary bone allows for a new fang to become functional while the old is being shed. Sometimes two fangs may be in operation at the same time, and one or both may conduct venom.

The absence of limbs and cutting or slicing teeth has led to a snake's astonishing ability to swallow objects several times their own diameter. This is made possible by several unique anatomical features. Unlike most other vertebrate jaws, each side of a snake's lower jaw is not attached directly to the skull but instead to the quadrate bones, which in turn articulate with the skull, and it is this double-hinged connection that is primarily responsible for a snake's enormous gape. In addition, the two halves of the lower jaw are not fused as in mammals but instead are connected by ligaments, which allow each half to move independently. These halves also have a middle joint that permits each to bulge outward to accommodate wide-bodied prey. The upper jaw is also quite mobile, and even the attachment of the palate and snout to the braincase is moveable.

As the jaw elements are slowly worked around a food item, the recurved teeth ensure firm anchorage of the prey in the jaws and allow only a one-way posterior movement of the prey body. During swallowing, the glottis, which opens near the tip of the lower jaw in snakes, may be protruded from time to time to permit breathing. This is a very important adaptation, because the swallowing of a large prey item can take many minutes, during which a snake with a body temperature within its preferred thermal range must breathe. The trachea is also reinforced to help prevent its collapse under the pressure of feeding on large objects. The actual physical passage of a large prey down the esophagus is possible because the integument is greatly expansible by virtue of flexible skin between the scales. The absence of a sternum in snakes is also important here, because if present it could combine with the ribs to form a non-expansible space through which a large prey mass could not pass. Once a large prey body has finally reached the stomach, restriction of the anterior portion of the long, single snake lung occurs, because it is situated adjacent to the stomach. This is compensated for by the extensive posterior air sac portion of the elongate lung, which stores air and through muscular action periodically moves it anteriorly to the vascular portion.

Prey

Snakes feed on animals, living and dead, and on eggs, the diet depending upon the species. There are no herbivorous snakes. Studies of recently born garter snakes (*Thamnophis* spp.) feeding for the first time indicate that the recognition of specific prey types is in part innate and that chemoreception by the Jacobson's organ is the primary sense used to identify a preferred food item.

Several methods of subduing prey are employed. Garter snakes simply seize the prey in their jaws and begin swallowing it, either anterior or posterior end first. Racers pinion the prey with their coils while continuously squeezing hard with their strong jaws, and then begin a head-first swallowing process once the prey is subdued. The constrictors (boas, gopher snakes, kingsnakes, and others) seize their prey in their jaws, throw several coils about it, and squeeze it into submission or to death through lung collapse and/or extreme pressure on the heart. Rear-fanged venomous snakes like the Night Snakes (*Hypsiglena* spp.) chew their venom into the flesh of their prey while it is held in the mouth by the recurved teeth. Rattlesnakes, with their hypodermic injection method of administering poison, strike the prey to introduce venom but not to hold the prey, and during the dying process the prey may move away from the initial contact point. It is then found by the use of olfaction, the Jacobson's organs, the heat-sensing pits, or all of these. Swallowing usually begins at the prey's head after an extensive inspection of the item by the forked tongue.

Sex and Reproduction

In snakes there are little or no external differences in sex. Color and pattern differences are rare and when present are usually slight. Males of many species are somewhat smaller than females and have a longer tail. The tail base tends to be broad in mature males because of the presence of the inverted hemipenes. The posterior margin of the vent can be probed with a narrow, rounded object such as a bobby pin to detect the presence of these organs. However, you should be instructed in this technique by an experienced handler before attempting it yourself.

Given that vision is probably not used for sexual discrimination because of the near absence of sexual dimorphism in snakes, olfaction appears to be the chief sensory cue for sex recognition. The odor and behavior of a receptive female indicates her sex. In many snakes the male crawls along the back of the female, advancing in short jerks and often following the curves of her body while he attempts to twine his hindquarters around hers. In copulation, which in some snakes may last several hours, the tail of the male may be twined around that of the female. The hemipenes are provided with spines that ensure firm anchorage, and sudden withdrawal might cause injury to the female. The cloaca of some female snakes is thick-walled, evidently to cope with the hemipenial spines. Males of some species such as the garter snakes (*Thamnophis* spp.) secrete a substance, apparently from the kidneys, which forms a copulatory plug in the oviductal part of the female's cloaca for a few days, preventing loss of seminal fluid and also negating successful fertilization by rival males that may later breed with her.

The majority of snakes lay eggs. These are usually white or cream-colored, and most have a flexible membranous shell impregnated with

Plate 89. Partially hatched Gopher Snake egg clutch.

lime. They usually swell following laying because of absorption of moisture from their surroundings. Often embryonic growth is already under way at the time of deposition. In California, snakes that give birth to their young are ovoviviparous and include the boids, garter snakes, rattlesnakes, and sea snakes. The young are born in the amniotic sac, which may rupture upon extrusion or is broken by the young snake shortly after birth. In oviparous snakes the young typically have an egg tooth situated at the lower edge of the rostral scale for use in slicing open the leathery shell. In the young of ovoviviparous species the tooth is reduced or absent.

As in other reptiles, most newly hatched or born young are completely precocial and able to feed and function without parental aid. However, the Black-tailed Rattlesnake (*Crotalus molossus*) female defends her brood until after their first molt, and this behavior may be present in other species as well. Juvenile mortality appears to be generally high in all snake species. Carnivorous lizards, large anurans, small mammals, avian predators, and many adult snake species exert a formidable predatory force on these worm-sized young. During several decades of snake trapping in the greater Bay Area by the junior author, the capture of a young-of-the-year snake has always been a rare event.

Fear of Snakes

One cannot conclude an introduction to this group of reptiles without a few comments on ophidiophobia, the fear of snakes. Snakes have the dubious distinction of being the most universally feared group of animals.

Plate 90. The junior author introduces his son Ross to a kingsnake.

Only a few other taxa such as spiders and to a lesser extent bats and large sharks even begin to affect many people in the way snakes do. However, the cause of this phenomenon is not clear. The junior author is convinced that one factor that is not involved is heredity. His father had a severe case of ophidiophobia, but to his credit he allowed his non-ophidiophobic 12-year-old son to keep a small colony of the Common Garter Snake (*Thamnophis sirtalis*) in a large box pen in a far corner of their Wisconsin backyard. A couple of decades later this author had an opportunity to further test the inheritance theory on his own three-year-old son Ross, but no initial fear of snakes was ever noted.

However, it is at this early formative age that a lifelong fear of these reptiles is probably instilled. Several years ago researchers at the Department of Psychology at the University of Miami published the results of a study in which a variety of possible sources of ophidiophobia based on hundreds of interviews were explored. Low on the list of causes were negative reactions by a child's peers to snakes, whereas an expression of fear by the mother ranked second only to the most common source of snake fear—the Bible. The authors note that this survey was conducted in the area of the United States often referred to as the "Bible Belt," and thus the ranking of ophidiophobia sources may be different in other areas.

Perhaps one further reason why snakes have been so negatively regarded for so long is that a moderate number of species (about 400) are venomous. This number is second among vertebrates to fishes, which include about 700 poisonous forms. Subduing prey with venom is a very effective feeding method, especially in snakes wherein paws and claws cannot assist in prey capture. Those who have witnessed a raptor capturing and then tearing apart a living rodent or rabbit may also agree that this form of "death by injection" is a far more humane form of predation. The problem arises when venomous snakes are stepped on or grabbed by humans, at which time the venomous bite is used for defense. This places some snakes in that small category of animals that have the potential of killing people, but this rather rare ability of a few forms is usually applied to all species. Perhaps the most common question asked of both authors when exhibiting a snake to a group of people is: "Is it poisonous!?" It is interesting to note that inanimate objects such as cars and guns, which kill thousands of people each year, do not instill the fear that snakes do but instead are admired, if not loved, by most Americans.

Blind Snakes (Family Leptotyphlopidae)

These are extremely small snakes and are often called "worm snakes" because of their superficial resemblance to earthworms. Indeed, upon first encountering our California species, the Western Blind Snake (*Leptotyphlops humilis*) in the field, you must remind yourself that this is not a worm but a higher vertebrate with basically the same internal anatomy (except for one less lung) as a human. The body of blind snakes is very slender with no neck constriction, and the smooth, shiny cycloid scales appear moist, which further adds to the resemblance to worms. Unlike most other snake groups, there are no enlarged ventral scales or scutes. Most species have a greatly enlarged rostral scale, which in some covers most of the front of the head. This and the underslung jaw are adaptations for a burrowing existence. Blind snakes are the most primitive of the snake groups, and several of species possess remnants of a pelvic girdle, a leftover from their legged ancestors.

Most species remain in burrows in loose substrate or in crevices during the day and emerge at night or on cloudy days when the relative humidity is high. Their large surface-to-mass ratio makes them highly susceptible to rapid desiccation, and thus they are nearly always found where there is adequate soil moisture. These are feeding specialists that forage mostly on ants, termites, and their larvae and pupae. Many are capable of following pheromone scent trails left by these insects to their nest sites. The very small body diameter of smaller species permits easy entrance into such nests, and their tough, tightly set scalation protects against ant and termite bites. Some Asian species may be viviparous, but all Western Hemisphere species are oviparous. One African species, the Flowerpot Snake (*Ramphotyphlops braminus*), is parthenogenic, the only such example of this mode of asexual reproduction known in snakes. Its unusual common name derives from its accidental worldwide distribution in potting soil, and only one of this all-female species is needed to start a new population in a new land.

WESTERN BLIND SNAKE　　　　　　*Leptotyphlops humilis*

IDENTIFICATION: Adult total length ranges from 18 to 41 cm (7 to 16 in.), and the body is usually no more that a few millimeters (about 0.1 inch) in diameter. The head is blunt with a large rostral scale, and there is no neck constriction. The eyes appear as minute spots and are vestigial, and there is a tiny spine at the tip of the tail. Dorsal coloration ranges from a purple-brown to black in southern coastal populations to pink with a silvery sheen in desert-dwelling individuals. The venter may be a light cream, pink, purple, or gray.

RANGE: In California it currently occurs from Coastal Southern California (except the greater Los Angeles area) eastward through the Tehachapi, Transverse, and Peninsular ranges and on through the Mojave and Colorado deserts.

NATURAL HISTORY: The Western Blind Snake is found in a variety of habitats, ranging from deserts to brush-covered mountain slopes. The common denominators at most areas are loose and relatively moist soil for burrowing or rock crevice retreats, and a permanent or seasonal water source nearby. It remains sequestered during the day and emerges at night when it may be found moving over exposed areas. The specimen (see Plate 91) was

Plate 91. Western Blind Snake hatchling crawling across a penny.

seen crawling out of a small crack in a high rock wall while the junior author's natural history class was having a cookout at the Desert Study Center at Zzyzx Springs in the Mojave Desert. In contrast to desert sites such as this, it may also be found along the southern California coast in beach sand above the high-tide mark.

A clutch of two to six eggs is laid in July and August. Detailed information on this species' reproduction is currently unavailable, but females of a closely related species, the Texas Blind Snake (*Leptotyphlops dulcis*) have been observed coiled around their eggs during the incubation period.

The Western Blind Snake is a feeding specialist that eats ants, termites, and the larvae and pupae of both groups. It appears to be able to use its Jacobson's organ to follow the pheromone trails of ants to their nests. Once a nest is located, it is able to pass through the ant tunnel system to the nest areas, since its body diameter is about the same as that of a large ant. This ability to go into a prey species' burrow system to forage is reminiscent of the feeding strategy of slender salamanders (*Batrachoseps* spp.), which forage in earthworm burrows. However, extremely small size does have its disadvantages, because the number of potential predator species, especially ground-feeding birds, is greatly increased. This potential is partially reduced by *L. humilis*'s nocturnal and burrowing habits, but night-active predators such as toads and even small owls still pose a threat. When molested, its defensive behavior includes rapid writhing, smearing fecal material over its body, and the assumption of a tonic immobility state (playing dead).

Boas and Pythons (Family Boidae)

Family Boidae contains the world's largest snakes, although the majority of species are of average snake size. It is a primitive, relatively unspecialized family that still possesses remnants of a pelvic girdle and hindlimbs in the form of small spurs on each side of the vent. Unlike most other snakes, they also have two well-developed lungs instead of one. Pythons occur mainly in Asia, Africa, and Australia and are oviparous. Boas have a worldwide distribution and are live-bearing (ovoviviparous). One Central and South American species, the Boa Constrictor (*Boa constrictor*), is especially well known because of its use in movies where a large serpent-supporting actor is required. Its generally placid behavior is ideally suited for co-starring with even the most sensitive stars and starlets. That some large boid species are indeed capable of killing a human most likely explains the human fascination with them, and there are a couple of fairly well-documented accounts of the Reticulate Python (*Python reticulatus*) of Asia swallowing a small person. This attribute along with large size and a usually calm disposition has made boids the most popular group of snakes in the multimillion-dollar serpent trade in the United States and Europe.

Boas and pythons have infrared-sensitive pits on their labial scales that are used to detect warm-bodied mammalian and avian prey. Once located and grabbed by the jaws, body coils are wrapped around the prey, and it is killed by exerting extreme pressure on the lungs and heart. Only three members of family Boidae are native to North America, and all three occur in California.

NORTHERN RUBBER BOA	*Charina bottae*
SOUTHERN RUBBER BOA	*Charina umbratica*

PROTECTIVE STATUS OF *C. UMBRATICA*: ST

IDENTIFICATION: These are stout-bodied snakes with a total length range of 35 to 84 cm (14 to 33 in.). Southern Rubber Boas occupy the lower portion of this size range. The body has a rubbery feel and appearance, and the tail is blunt and resembles the head. The top of the real head is covered with large symmetrical plates, and the pupil has a vertical shape. The dorsal color ranges from light brown or tan to olive green, and the venter may be various shades of yellow, orange, or cream with little or no mottling. Anal spurs are present in males but small or absent in females.

RANGE: *C. bottae* occurs throughout the North Coast Range, Cascade Range, and most of the Sierra Nevada. *C. umbratica* is found in the northern part of the South Coast Range and at selected sites in Kern, San Bernardino, and Riverside counties. These include Mount Pinos, Mount Abel, and the Tehachapi, San Bernardino, and San Jacinto mountains, and are indicated by the ovals and circles on the lower third of the combined range map.

NATURAL HISTORY: These are chiefly woodland and forest species that often bury themselves in loose soil or seek retreats beneath rotting logs, rocks, and fallen tree bark. A rocky stream with banks of sand or loam in a coniferous forest with meadows and numerous rotting logs is prime habitat for this snake. It is adapted to cool weather and is often active above ground when other snake species are in retreat. The first specimens of *C. bottae* ever observed in the field by the junior author were crawling between snow patches at dusk in late May at Kennedy Meadows in the Sierra Nevada.

Like all other boas, these are ovoviviparous and give birth to from two to eight young between August and November. This is a relatively low number of young compared with other live-bearing groups such as the garter snakes and rattlesnakes, but it is compensated for by an amazingly long life span of 50 years or more in the wild. They feed on a variety of small vertebrates, including other snakes, and they have been observed elevating and striking with their blunt, headlike tails at mice whose young they were in the process of consuming. When discovered under a ground-cover object, they often hide their heads under their bodies while elevating the headlike tail in an apparent attempt to entice a predator to peck or bite this less vital area.

ROSY BOA *Lichanura trivirgata*

IDENTIFICATION: This is a relatively large snake with a total length range of 43 to 112 cm (17 to 43 in.). Like most other boid species it is heavy-bodied with smooth, shiny scales. The head is only slightly wider than the neck, and the eye pupil is vertical. Unlike the rubber boas (*Charina* spp.), it has no large head plates. The most common dorsal color pattern consists of three longitudinal stripes, a feature described by its Latin species name. The color of the stripes varies from black to brown, orange, rust, or shades of red, depending on locality. The base dorsal color also varies greatly from shades of blue and gray to tan, yellow, or cream. In arid habitats the edges of the stripes are well defined, but in coastal California populations they are very irregular and may connect one stripe to another. Young snakes have a lighter coloration, but the striped pattern is more distinct. As with most boids, males are recognized by their well-developed anal spurs.

RANGE: *L. trivirgata* is found throughout the Mojave Desert from the Death Valley region southeast to the California–Arizona border and beyond into western Arizona, south to Joshua Tree National Park, and on through the lower slopes of the Transverse and Peninsular ranges. It is also present in Coastal Southern California from Orange County south to and beyond the United States–Mexican border. It appears to be absent from the Colorado Desert from the Coachella Valley southward except for an area in the extreme southeastern corner of the state.

NATURAL HISTORY: The Rosy Boa is usually associated with shrublands in rocky uplands and desert areas, often near permanent or seasonal water sites. However, it is a true xeric species and does not require a water source nearby. It is mainly nocturnal and crepuscular, and like most other boids is a good climber. It gives birth to from three to 14 young in October and November. It is a feeding generalist and will capture most vertebrates within its consumption size range that it encounters. Its infrared-sensing pores are apparently very useful in locating warm-bodied prey such as nestling Desert Woodrats (*Neotoma lepida*). After being grasped, prey is killed by constriction in true boid fashion.

CONSERVATION NOTE: Currently, only the small Southern Rubber Boa (*C. umbratica*) that occurs in the San Bernardino and San Jacinto mountains has been granted protection under the State Endangered Species Act. The Rosy Boa is quite popular in the pet trade, although currently most of the pet demand for this species appears to be met through captive propagation by individual breeders under a Native Reptile Propagation Permit, issued by CDFG. Both the Rosy Boa and Northern Rubber Boa (*C. bottae*) continue to be listed among the many California reptiles of which the holder of a freshwater fishing license can take two each from the wild. Given the annual multimillion-dollar trade in boid species alone, the consequences of this CDFG regulation, especially with respect to the eventual disposition of some if not most of such collected native boid species, must be seriously considered.

Colubrids (Family Colubridae)

This is an extremely large family, containing two thirds of all snake species in California. Members of this assemblage exhibit tremendous variation in size, color, scalation, body form, habitat preference, and feeding behavior. Indeed, it is difficult to present even one unifying characteristic for the group. It does not appear to represent a natural taxonomic group, and we may expect to see it divided into smaller family units as was recently done with the former large lizard family Iguanidae. However, within this family are a number of large genera such as that of the garter snakes (*Thamnophis*

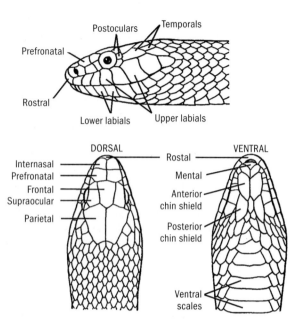

Figure 33. The head anatomy and scale nomenclature of a colubrid snake.

spp.), which exhibit great consistency throughout their species range and for which we present a somewhat extensive set of introductory comments. There are several other species that are simply unique by themselves and for which we therefore give no generic introduction.

The head scalation of an average colubrid snake shown in figure 33 will often be referred to in the following species descriptions. Snake scale types, other than the broad ventral scutes, are the same as those shown for lizards in figure 28 (see the "Key to the Families and Genera of Lizards").

RING-NECKED SNAKE *Diadophis punctatus*

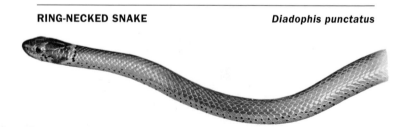

IDENTIFICATION: This is a relatively small, slender snake with most individuals at the lower end of its length range of 20.3 to 63.5 cm (8 to 25 in.), although snakes from the California part of the range are much smaller.

It has a conspicuous orange or orange-red neck band. The dorsal color ranges from olive to brown, blue-gray, or nearly black. The venter varies from orange to red near the tail and is flecked with small black spots.

RANGE: It occurs throughout much of California except for the Great Valley, the eastern slopes of the Cascade Range and the Sierra Nevada, and most of the three desert regions except for spring areas in the eastern Mojave. It is also found on Santa Catalina Island.

NATURAL HISTORY: The Ring-necked Snake is usually associated with moist habitats where much of its preferred food such as small salamanders, slugs, and earthworms occur. In California it is quite seasonal in its occurrence aboveground and becomes scarce once the hot, dry summer season sets in. In arid areas it is only found near springs or water courses. In all habitats where it occurs it relies heavily on various ground-surface cover objects such as large pieces of fallen tree bark, discarded boards, sheet plastic, and so forth, for protection as well as foraging, because it is under such cover that many of its prey species also conduct their aboveground activities. Carefully lifting and replacing such objects during the wet season is usually the best way to see this snake. When disturbed, it often tightly coils its tail while inverting it so that the intense red color is presented to a potential

Plate 92. Ring-necked Snake displaying ventral tail color.

predator, while at the same time the head is usually placed on the ground within or beneath the body coils. This behavior appears to be an attempt at diverting a strike or bite to the less vulnerable tail region. Because this snake, when disturbed, releases from its anal scent glands a strong-smelling substance that may also be distasteful, perhaps the small red neck band signals its identity to an experienced predator.

One or two clutches ranging from two to 10 eggs are laid in June and July, often in communal nest sites. Perhaps its most preferred prey are various species of slender salamanders whose aboveground activity also coincides with the rainy season. At this time other favored foods such as small frogs, slugs, earthworms, and tadpoles are also available. The saliva of the Ring-necked Snake appears to contain a toxin that helps to subdue large prey. It has no grooved or hollow fangs, but the enlarged rear teeth may aid in abrading the skin of prey so that the toxin can enter its capillary bed. Although this species may technically be classified as a "venomous snake," it poses no harm to humans.

COMMON SHARP-TAILED SNAKE   *Contia tenuis*
FOREST SHARP-TAILED SNAKE   *Contia longicaudae*

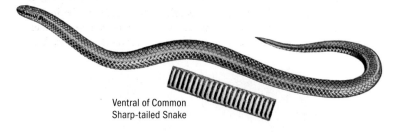

Ventral of Common Sharp-tailed Snake

TAXONOMIC NOTE: Recently (2010), populations of sharp-tailed snakes occurring along the outer Coast Ranges and Klamath Mountains of northern California, as well as the western slopes of the Santa Cruz Mountains in central California, have been described as a new species, *C. longicaudae* (Forest Sharp-tailed Snake). This new species is distinguished from the more widespread *C. tenuis* on the basis of a noticeably longer tail (and higher caudal scale counts, 43–48 in *C. longicaudae*, 24–43 in *C. tenuis*), distinctive ventral markings, greater adult size on average, and mitochondrial and nuclear DNA differences. Our range map reflects the composite distribution of both species.

IDENTIFICATION: Collectively, the two species of *Contia* are among California's smallest snakes, with a length range of 30 to 46 cm (12 to 18 in.), with *C. longicaudae* attaining larger average and maximum size compared to *C. tenuis*. The dorsal color is a uniform red-brown or gray, while the venter has alternating crossbars of black and light gray or cream (prominent in *C. tenuis*, covering one half to one third of each ventral scale; reduced in *C. longicaudae*, covering only one fourth to one third of each ventral scale). This color pattern closely resembles that of noxious millipedes with which these snakes often coexist, thus perhaps providing some protection from predators. The namesake feature is a small sharp spine at the tip of the tail.

RANGE: Collectively, the two species of sharp-tailed snakes occur throughout the North Coast Range and Cascade Range, the northern part of the South Coast Range, and much of the Sierra Nevada. They are absent from the Great Valley, areas east of the Sierra Nevada and Cascade Range, and all of the southern one third of California.

NATURAL HISTORY: These secretive species occur in a wide variety of mesic vegetation types but almost always near streams, springs, or seepages. They are usually seen out from under cover only after periods of rain and then usually at night. They may be active at ambient temperatures as low as 10 degrees C (50 degrees F), which, along with wet ground-surface conditions, approximates the microclimate at which their preferred prey species such as slugs are also abroad. Sharp-tailed snakes have noticeably long, curved, needlelike teeth for their diminutive size, and these probably allow the snakes to hold on to such mucus-laden prey (figure 34). With the beginning of the California dry, hot season, these

Plate 93. Common Sharp-tailed Snake.

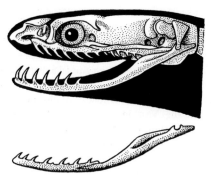

Figure 34. Common Sharp-tailed Snake head and jaw skeletal anatomy and the lower jaw bone of the Ring-necked Snake, both showing the recurved teeth that prevent moist prey from escaping when grasped.

snakes seek underground retreats where they remain until the return of wet weather. One or two clutches of from two to 10 eggs are laid in June and July. Larger clutches have been found, indicating communal nesting. The function of the namesake feature of sharp-tailed snakes, the tail spine, is not clear. These snakes have been observed to probe in soft substrate with the tail spine and also on one's hand when they are being handled. One possible reason for this behavior is that it helps anchor the snake when it is pulling on slug or worm prey.

SPOTTED LEAF-NOSED SNAKE   *Phyllorhynchus decurtatus*
(WESTERN LEAF-NOSED SNAKE)

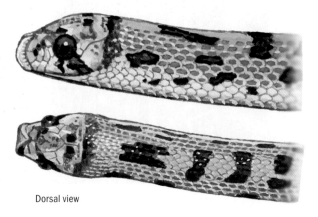

Dorsal view

IDENTIFICATION: This a moderate-sized 30- to 50-cm (12- to 20-in.) pale, blotched snake with a blunt snout covered by a much enlarged rostral scale with free edges. The only other California snake with an enlarged and backward-folded rostral scale is the Western Patch-nosed Snake (*Salvadora hexalepis*). However, once the Spotted Leaf-nosed Snake's very large rostral scale with its free edge is viewed at close range, its leaflike appearance is hard to forget. The background dorsal color may be pink, tan, yellowish, or pale gray with 17 or more mid-dorsal brown blotches between the head and the region above the vent. These blotches also extend to the tail. The venter is white and unmarked, and the pupils are vertical.

RANGE: In California it extends from Inyo County south through the Mojave and Colorado deserts.

NATURAL HISTORY: This snake is usually found in desert areas with sandy or gravelly substrate where creosote bush is the dominant shrub species. Its greatly enlarged rostral scale protects the snout area when it burrows in abrasive ground material. It is nocturnal and quite secretive. The individual pictured was captured by one of the junior author's past herpetology classes as they walked abreast across a creosote bush flat with flashlights just after sundown. It feeds primarily on the Western Banded Gecko (*Coleonyx variegatus*) and its eggs. The eggs of other lizards such as the Zebra-tailed Lizard (*Callisaurus draconoides*) and insects are also taken. A small clutch of two to six eggs are laid in early to midsummer.

Plate 94. Spotted Leaf-nosed Snake.

Racers and Whipsnakes (Genera *Coluber* and *Masticophis*)

If the large family Colubridae is eventually subdivided, these two genera will most likely constitute one of the new groupings. These are sleek, slender snakes with broad heads and large eyes. They are fast, diurnal sight hunters that often forage with the neck and head held erect and occasionally moving side to side, presumably to aid in making accurate distance judgments when searching for prey. They are very aggressive snakes and subdue their prey by applying sustained pressure to the body with their strong jaws. When hand capturing any species within this group, you will most likely be bitten, and although a little blood (yours) may be let, these species are not venomous.

RACER   *Coluber constrictor*
WESTERN YELLOW-BELLIED RACER   *C. c. mormon*

Juvenile
Adult

IDENTIFICATION: In California this snake grows to a maximum adult size of about 90 cm (36 in.), although east of the Rockies it may attain up to twice this length. One of two subspecies, the Western Yellow-bellied Racer (*C. c. mormon*), occurs in California and several other western states. It has the classic racer conformation. The dorsal color is brown, olive, or sometimes a bluish green, and the venter ranges from light yellow to white. The coloration of the young differs markedly, with brown saddles on the midline and smaller blotches on the side, set against a light-brown background color.

RANGE: It occurs throughout most of California except in the Mojave and Colorado deserts and closely associated areas. Distribution is also spotty in the southern Great Valley and parts of San Diego, Orange, and Riverside counties in Coastal Southern California.

NATURAL HISTORY: The Racer prefers open grassland, coastal scrub, chaparral, and woodland. It is not found in arid habitats. In many areas, especially those in the North and South Coast ranges containing chaparral or coastal scrub, it is often

Plate 95. Western Yellow-bellied Racer in defensive coil.

the most abundant snake species. In a three-month trap-mark-recapture survey on a 2-hectare (5-acre) area of coastal scrub and grassland at Año Nuevo State Reserve in southern San Mateo County, the junior author captured 114 individual Western Yellow-bellied Racers. The majority of these snakes most likely did not reside permanently on the survey plot but were instead passing through the area when captured. However, such results reflect how abundant it can be when habitat conditions are apparently optimal. Despite this abundance, people rarely observe this snake in the field. Its rapid and often rectilinear movement coupled with its basic cryptic coloration and its apparent ability to see you before you see it all account for its elusiveness.

In addition to being at home in a wide variety of open habitats, its role as a feeding generalist is a major reason for its abundance. Its prey preferences range from frogs, small snakes, lizards, and hatchling turtles to mice, birds and bird eggs, and a variety of large insects, a staple food for the hatchling Racer. In true feeding generalist fashion it can switch from one major prey source to another as seasonal abundance of these various foods changes. Our California subspecies lays from three to 11 eggs in midsummer, sometimes in communal nests. Hatchlings are often numerous in late summer and at first appearance may be confused with young Gopher Snakes (*Pituophis catenifer*), which have a similar coloration and pattern. However, the large eyes and smooth dorsal scales of the young Racer readily distinguish it.

COACHWHIP
SAN JOAQUIN COACHWHIP
RED RACER
BAJA CALIFORNIA COACHWHIP

Masticophis flagellum
M. f. ruddocki
M. f. piceus
Masticophis fuliginosus

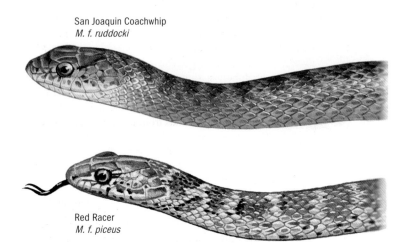

San Joaquin Coachwhip
M. f. ruddocki

Red Racer
M. f. piceus

PROTECTIVE STATUS OF M. F. RUDDOCKI: CSSC

TAXONOMIC NOTE: Grismer (2002) recognized the Baja California Coachwhip, a former subspecies of the Coachwhip, as a separate species, based primarily on the absence of any intergrades at points of contact in their respective ranges in southern San Diego County and northern Baja California, Mexico. Because of the presumed similarity in their life histories, we discuss both in one natural history account.

IDENTIFICATION: The Coachwhip is the second longest snake in California, with an adult length range of 91 to 167 cm (36 to 66 in.), exceeded only by the Gopher Snake (*Pituophis catenifer*) in some parts of its range. Given such length and its slender racer-type build, the common name of Coachwhip is appropriate, even though this essential item for drivers in the horse and carriage days may be unfamiliar to modern readers. Two subspecies color variations exist in this state. The Red Racer (*M. f. piceus*) has wide, dark brown, black, or pink crossbands on a tan background in the dorsal neck region. The San Joaquin Coachwhip (*M. f. ruddocki*) has a reddish, light-yellow, or olive-brown body color with only faint, dark crossbands on the dorsal neck area. The coloration of the young generally resembles that of the adult with dark crossbanding on a brown background.

The Baja California Coachwhip has a smaller size range than the Coachwhip (62 to 132 cm or 24 to 52 in.) and has two color phases. The light phase ranges from light gray to pink or yellow, and the dark phase is gray-brown or even black. There are white spots on the body and neck, and the venter is cream-colored with dark spots that occur in pairs toward the head.

RANGE: One of the two California subspecies of Coachwhip, the Red Racer (*M. f. piceus*), occurs throughout the southern deserts and mountain ranges plus parts of Coastal Southern California. The other California subspecies, the San Joaquin Coachwhip (*M. f. ruddocki*), is found only in the southern half of the Great Valley and the eastern slopes of the South Coast Range. An isolated population occurs on the Sutter Buttes, Sutter County. The range of the Baja California Coachwhip extends from just across the United States–Mexico border into southern San Diego County in Coastal Southern California. From there it extends south through the entire length of Baja California.

NATURAL HISTORY: Desert and open-brushland habitats are prime for these species. The Coachwhip also occurs in open woodland but avoids dense vegetation where one of its main attributes, speed, is not possible. These are primarily diurnal snakes with a relatively high thermal tolerance, a feature that allows them to hunt when heliothermic lizard species are abroad. With their good speed they can usually "run down" such fast prey species. The photo of a Red Racer eating a Desert Iguana (*Dipsosaurus dorsalis*) was taken near midday on the Mojave Desert where air temperature near the ground was well above 40 degrees C (104 degrees F). When desert day-

Plate 96. Coachwhip swallowing an adult Desert Iguana.

time temperatures get very high, these snakes retreat into rodent burrows or climb into brush where both shade and cooler air are present. Like the Racer (*Coluber constrictor*), their highly aggressive hunting behavior lapses into defensive behavior when captured, and human captors usually receive a sharp bite or two that in most cases lacerates the skin. Clutch size ranges from four to 20 eggs, which are laid in June or July. Both species are feeding generalists. Their prey includes lizards, snakes (including rattlesnakes), small mammals (including bats), birds and their eggs, frogs, hatchling turtles and tortoises, and large insects. They have also been known to eat carrion.

CONSERVATION NOTE: Like several other vertebrate populations that once thrived in the scrub and grassland habitats of the pristine Great Valley, San Joaquin Coachwhip numbers have been dramatically reduced by the advance of large-scale agriculture throughout most of its range. Even isolated populations in deep side canyons on the western side of the valley have been all but negated by overgrazing and other human land misuse practices. Over the past several decades the authors have personally witnessed what appears to be the elimination of this snake from Corral Hollow, a desert relic habitat on the Alameda–San Joaquin County border where a state park devoted exclusively to off-road vehicles and dirt bikes has destroyed much of its prime habitat.

STRIPED RACER	***Masticophis lateralis***
CALIFORNIA STRIPED RACER	*M. l. lateralis*
ALAMEDA STRIPED RACER	*M. l. euryxanthus*

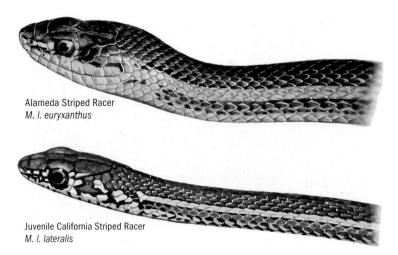

Alameda Striped Racer
M. l. euryxanthus

Juvenile California Striped Racer
M. l. lateralis

PROTECTIVE STATUS OF *M. L. EURYXANTHUS*: FT, ST

IDENTIFICATION: The Striped Racer has a length range of 76 to 152 cm (30 to 60 in.). Throughout most of its range, the dorsum of this snake is dark brown or black above with a thin pale-yellow or white stripe extending from behind the head along each side of the body. The ventral body is cream to pale yellow, and the underside of the tail is usually a coral pink. A Federal and State Threatened subspecies, the Alameda Striped Racer has comparatively wide orange lateral stripes that span a full scale row plus half a scale row on each side. This orange stripe extends anteriorly above the angle of the jaw. The anterior venter area is also orange.

RANGE: This snake is appropriately named, because it occurs only in California and in Baja California, Mexico. In California it occupies the eastern slopes of the North Coast Range, the southern portion of the Cascade Range, the western Sierra Nevada, and the Peninsular Ranges and Coastal Southern California. The threatened subspecies, *M. l. euryxanthus*, occurs only in portions of Alameda and Contra Costa counties.

NATURAL HISTORY: One of the two subspecies of Striped Racer, the California Striped Racer has also been called the Chaparral Whipsnake, because its primary habitat is open-canopy chaparral or coastal scrub, often with scattered grassy patches, rocky gullies, or streamcourses. It is chiefly a foothill species but may be found in mixed deciduous and pine forests in mountain areas. Like the other racer species, this is a diurnal species that maintains a high body temperature by periodically basking in the sunny patches within its open-canopy brush habitat.

Plate 97. Periscope-like neck and head of the Alameda Striped Racer.

Plate 98. Alameda Striped Racer swallowing a Western Fence Lizard.

The authors and former graduate student Karen Swaim used radio-telemetry to track the other subspecies, the threatened Alameda Striped Racer, in an upland coastal-scrub area in Tilden Regional Park, Contra Costa County, and found an average home range area in that habitat to be 4.7 hectares (11.6 acres). Often when these snakes patrol their home range area the head and neck is held high off the ground in periscope fashion, which permits a better view of the field ahead for this sight hunter. In a prey preference study, McGinnis and Swaim presented hatchlings of *M. l. euryxanthus* with a variety of small vertebrates and large invertebrates, and found that only lizards consistently elicited a rapid strike response. Both hatchling Alameda Striped Racers and hatchling Western Fence Lizards (*Sceloporus occidentalis*) usually hatch in the mid- to late-summer period, which is fortunate timing for these young lizard-feeding specialists. Adult Alameda Striped Racers also have a marked preference for lizard prey, but additional foods such as frogs, snakes, small mammals, birds, and large insects have also been recorded for the California Striped Racer. Both subspecies lay a clutch of six to 11 eggs in May or June.

CONSERVATION NOTE: Through trapping surveys at 22 upland sites within the range of the Alameda Striped Racer, the junior author and former graduate students found that this threatened species usually occurs on south-, southeast-, and southwest-facing slopes that support an open coastal scrub or chaparral plant community and also contain rock crevice or rodent burrow retreat sites plus at least one abundant lizard species. In most areas of the state these habitat requirements would not by themselves be the cause of a species or subspecies threatened status. However, in this case, such habitat conditions occur in the limited range of this subspecies in the greater San Francisco Bay Area where our own species also prefers this habitat complex, mainly for the views it affords. Fortunately, environmental impact reports for proposed development in such areas now concentrate heavily on determining Alameda Striped Racer presence or absence. When it is found to be present, the relatively large home range of this snake must be a prime consideration in setting aside appropriate habitat for its continued existence.

STRIPED WHIPSNAKE (DESERT STRIPED WHIPSNAKE) *Masticophis taeniatus*

IDENTIFICATION: This is another slim, long racer with a length range of 76 to 183 cm (30 to 72 in.). It is a close arid-land relative of the Striped Racer (*M. lateralis*). It is the least strikingly colored of the California racers. The dorsal color is black, dark brown, or gray, with a whitish lateral stripe on each side, bisected by a dashed or continuous black line. The lower sides are usually a pale gray with additional dashed or longitudinal stripes. The venter is white or light shades of yellow, grading to coral pink toward the tail.

RANGE: In California this species occurs in the Great Basin Desert both north and south of Lake Tahoe and extends south to the northern segment of the Mojave Desert. In Inyo County it can be found up to 3,077 m (10,100 ft) in the White Mountains.

NATURAL HISTORY: This is a racer of the arid sagebrush flats, grassland, piñon–juniper woodland, and open pine–oak forests of the western edge of the Great Basin and northern Mojave deserts. Deep crevices in rock outcrops and rodent burrow retreat sites are important habitat features. It is an excellent climber and may glide through the brush canopy when foraging or being pursued. When moving on the ground it often keeps the head and neck elevated. Like other racers, it is a diurnal predator on lizards, snakes, rodents, nestling birds, frogs, and large insects. A clutch of three to 12 eggs is laid in June or July.

Plate 99. Juvenile Whipsnake.

WESTERN PATCH-NOSED SNAKE — *Salvadora hexalepis*
COAST PATCH-NOSED SNAKE — *S. h. virgultea*
DESERT PATCH-NOSED SNAKE — *S. h. hexalepis*
MOJAVE PATCH-NOSED SNAKE — *S. h. mojavensis*

PROTECTIVE STATUS OF S. H. VIRGULTEA: CSSC

IDENTIFICATION: This species has a very racerlike build but with a smaller body-length range of 51 to 117 cm (20 to 40 in.). Its big eyes and the large patchlike rostral scale, which folds back on the snout, are its two most striking anatomical features. The venter is plain white but sometimes washed with dull orange, especially toward the tail. This basic color pattern varies slightly in the three subspecies found in California. Because of its greatly enlarged rostral scale, it may be initially mistaken for the Spotted Leaf-nosed Snake (*Phyllorhynchus decurtatus*). The most apparent difference is that this species has spotted markings as opposed to the stripes of the Western Patch-nosed Snake. In addition, the latter's internasal scales are only partially separated by the large rostral scale, whereas this scale completely separates the internasal scales in the Spotted Leaf-nosed Snake.

RANGE: Except for its presence in the Honey Lake Basin of eastern Lassen County, this species is confined to the southern one third of the state in the Mojave and Colorado deserts, the southern tip of the South Coast Range, and most areas in the southern mountain ranges and Southern Coastal California. The range of the Coast Patched-nosed Snake (*S. h. virgultea*), the subspecies that has been designated as a CSSC, is confined to the southern part of the South Coast Range and Coastal Southern California, two adjacent regions that have experienced extensive land use changes.

NATURAL HISTORY: The Western Patch-nosed Snake inhabits a variety of arid and semiarid habitats ranging from chaparral and piñon–juniper woodland to annual grassland, sagebrush and creosote bush plains, and dry mesquite stands. Its physical habitats vary from rocky lower mountain slopes to low desert plains. It is a diurnal forager but also may be abroad early on warm nights. Its racerlike build and speed allow it to chase down lizard prey and also quickly escape into brush when pursued. This snake does not burrow in the substrate as its enlarged rostral scale would initially indicate. Instead, this structure apparently aid it in digging out buried snake and lizard eggs, which appear to be its preferred food. Given the relatively large populations of lizard and snake species in many xeric habitats, all of which deposit one or more clutch of eggs in substrate nests each year, this rather specialized food preference may be quite lucrative. Other foods consist of small mammals, adult lizards, and nestling birds and bird eggs. A clutch of four to 12 eggs is laid between May and August, depending on the elevation at a given habitat.

GOPHER SNAKE	*Pituophis catenifer*
GREAT BASIN GOPHER SNAKE	*P. c. deserticola*
PACIFIC GOPHER SNAKE	*P. c. catenifer*
SAN DIEGO GOPHER SNAKE	*P. c. annectens*
SANTA CRUZ ISLAND GOPHER SNAKE	*P. c. pumilus*
SONORAN GOPHER SNAKE	*P. c. affinis*

PROTECTIVE STATUS OF *P. C. PUMILUS*: CSSC

IDENTIFICATION: This is the longest snake in California, with a subspecies length range of 76 to 279 cm (30 to 110 in.). It is also one of the most widely distributed snakes in the western and central United States. Throughout this range the dorsal coloration pattern of large black, brown, or reddish brown blotches is quite consistent. In most localities a dark stripe across the front of the head anterior to the eye is also very apparent. The dorsal background color ranges from yellow to cream in most areas but may be a light tan in some desert populations. Some specimens in the central and west-central part of the state have a longitudinally striped-unblotched or striped-blotched pattern. The venter in all individuals grades from white to yellow and is often spotted with black. The mid-dorsal scales are keeled, a feature that is helpful in identifying a recovered shed skin.

RANGE: Except for an area on the eastern slopes of the Sierra Nevada between Sierra and Fresno counties, the Gopher Snake occurs throughout California, including Santa Cruz and Santa Rosa Islands. Its range beyond this state extends east to western Wisconsin and Illinois, north into parts of southern Canada, and south into the northern states of Mexico.

NATURAL HISTORY: The Gopher Snake is a classic example of a habitat generalist. It occurs in desert, prairie, brushland, woodland, open conifer forest, and farmland habitats. Its ability to readily adapt to the latter situation that has replaced most of the pristine grassland and brushland throughout its range has been a key feature for its persistence in the face of human-wrought environmental changes. In California it occurs throughout all 12 geographic regions, where it occupies the feeding niche that includes large rodents and rabbits. It is primarily diurnal but tends to be crepuscular and even nocturnal during hot weather. In farmland this snake is often seen dead on the road surface, especially during the first spring warm spell when individuals may be forced to move between crop fields as they are tilled and planted. Tillage and harvesting equipment also account for many Gopher Snake deaths in agricultural land. The junior author is fortunate to have a small resident group of this snake on his farm but also has had the sad experience of killing one while disking a weedy orchard row. Unlike a racer, which usually makes a speedy exit upon the approach of farm equipment, the Gopher

Plate 100. Gopher Snake in annual grassland habitat.

Snake instead tends to "hunker down" in the grass–weed cover and remain in harm's way, apparently relying on its cryptic coloration for protection.

As the resident herpetologists at their respective universities, both authors have received calls asking for urgent advice on how to cope with a "rattlesnake" in the yard or garden. Such alarms have usually turned out to be false, because the snake in question was really a Gopher Snake, which, when disturbed, often flattens its head, hisses, and vibrates its tail. When this defensive behavior takes place where dried leaves or other similar ground cover is present, the tail striking these objects can produce a sound very similar to that made by a real rattlesnake. The general similarity in coloration between the Gopher Snake and most rattlesnake species also helps to account for such misidentifications and can sometimes lead to the unwarranted killing of the former. When this happens, the land owner has lost a most effective rodent control specialist. When the senior author was a youngster on his family's farm in California, such a rattling sound was heard in an alfalfa patch. A neighbor was called, and all watching were impressed with his bravery as he drew close to the sound with a raised hoe. He always advanced with his left leg foremost. Finally, he struck with the hoe and brought up a Gopher Snake! He then revealed that he had a wooden left leg.

In addition to a variety of rodents, including ground squirrels, this snake takes small rabbits and young hares, moles, birds and their eggs, and lizards and large insects. It kills larger prey by constriction. Its chief predators in many areas are large hawks and eagles. One or two clutches of from two to 24 eggs are laid between June and August.

CONSERVATION NOTE: Decades of habitat destruction by introduced ungulates have severely impacted the Santa Cruz Island Gopher Snake (*P. c. pumilus*) on Santa Cruz, Santa Rosa, and possibly San Miguel islands off the central California coast. The current efforts to remove domestic and feral stock from these islands will permit the recovery of the native scrub–grassland island flora and that of this subspecies as well.

GLOSSY SNAKE	***Arizona elegans***
CALIFORNIA GLOSSY SNAKE	*A. e. occidentalis*
DESERT GLOSSY SNAKE	*A. e. eburnata*
MOJAVE GLOSSY SNAKE	*A. e. candida*

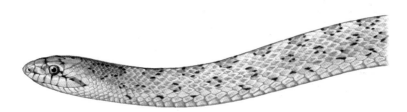

IDENTIFICATION: This is a moderately slender snake with very shiny smooth scales that impart the glossy appearance for which it is named. Its adult length ranges from 66 to 178 cm (26 to 70 in.). In the southern desert areas its coloration is a pastel version of that of a Gopher Snake (*Pituophis catenifer*), with light shades of brown, gray, yellow-gray, pink, or cream on the dorsum, and light tan, brown, or gray blotches edged in black. In the less arid portions of its range the base colors are darker and more closely resemble that of the Gopher Snake. The venter is pale buff or white. The bleached appearance of this species in arid regions apparently affords good protection when moving over light desert substrates at night. The lower jaw is moderately inset, an adaptation for burrowing.

RANGE: It occurs from Mount Diablo in Contra Costa County south through the western edge of the San Joaquin Valley, the South Coast Range, and Coastal Southern California, the Mojave and Colorado deserts, and the southern mountain ranges. It is absent in coastal areas from Los Angeles northward.

NATURAL HISTORY: The Glossy Snake's many habitats include scrubby to relatively barren deserts, sagebrush flats, annual grassland, chaparral slopes, and open woodland. It is an excellent burrower and prefers areas with a sandy or loamy substrate. It is crepuscular and nocturnal, remaining mostly underground during the day. Given this behavior, it is somewhat surprising to note that lizards, a primarily diurnal group, are one of its preferred foods. The junior author recalls extracting four whiptail lizards (*Aspidoscelis* spp.) from the stomach of a large adult Glossy Snake. It may be very successful at capturing such prey during the dusk and dawn periods when the neuromuscular coordination of most lizards is suboptimal. It also feeds on rodents, snakes, and occasionally birds. Like other large, heavier bodied colubrids, it usually kills prey by constriction. Five to 12 eggs are laid in June or July.

Kingsnakes (Genus *Lampropeltis*)

This is another closely knit colubrid subgroup that some day may be assigned its own family designation. These are medium-size snakes with smooth scales and, in California, some have striking body bands and blotches. They are found in a variety of woodland, scrub, and desert habitats, and like the Gopher Snake (*Pituophis catenifer*), have also adapted moderately well to croplands and orchards. Most species are noted for their calm, placid behavior when handled by humans, and thus have become pet store favorites. However, in their natural habitat they are a formidable predator on small vertebrates, including rattlesnakes and coral snakes.

COMMON KINGSNAKE
CALIFORNIA KINGSNAKE

Lampropeltis getula
L. g. californiae

IDENTIFICATION: This a relatively large, medium-bodied snake with an adult length range of 76 to 216 cm (30 to 85 in.). Over most of its range it has alternating bands of black, brown, or dark brown and pale yellow or white, with the latter widening on the venter. A striped phase that has a pale-yellow or white mid-dorsal stripe occurs in scattered locations in southern California and may result from a single clutch that also contained young with the normal banded pattern. In coastal Los Angeles County a dark phase of the banded pattern also occurs, and in the northern part of the San Joaquin Valley there are black-bellied individuals in which the pale crossbands broaden to form a lateral stripe. All scales are smooth and rather glossy.

RANGE: The Common Kingsnake occurs in all 12 California geographic subdivisions. It is absent only in the northern segment of the Great Basin Desert, the extreme northeast corner of the state, at higher elevations in the Sierra Nevada from Sierra County to Fresno County, and in the greater Colorado River area.

NATURAL HISTORY: This is the only other species besides the Gopher Snake (*Pituophis catenifer*) that occurs in essentially all habitats and regions of California. Its classic disruptive coloration pattern is apparently effective in a wide variety of habitat substrates. As with the stripes of the zebra, the bold dark and light stripes break up the standard snake body form. This is probably most effective in dim morning or evening light and at night, periods when this snake is most active. During the day it tends to remain inactive under ground-cover objects, and the turning and then replacing of such items is one of the best ways to find it.

Its role as a feeding generalist also accounts for its widespread distribution. Prey ranges from small mammals, frogs, birds and their eggs to lizards, reptile eggs, hatchling turtles, and other snakes. The inclusion of rattlesnakes in its dietary preference is usually of particular interest to

Plate 101. Disruptive coloration of a kingsnake.

many people. The Common Kingsnake is relatively immune to rattlesnake and even coral snake venom, but a number of other snake species possess such immunity as well. It apparently does not have a marked preference for rattlesnakes, although this prey may be easier to subdue by constriction due to its more lightly muscled body wall compared to that of snakes that kill prey by constricting. Rattlesnakes are apparently a long-standing prey of kingsnakes, because the former have developed an innate response to the odor of kingsnakes and other snake-eating colubrids. When presented with a kingsnake or even a snake stick that has been used to handle one, a rattlesnake stops rattling, presses its head and neck to the ground, and elevates the central part of its body, occasionally striking the ground hard with it. Because a kingsnake typically attempts to seize a rattlesnake by its neck, this lunging movement of the elevated body mass may function to thwart or break a neck hold.

Although the Common Kingsnake usually exhibits the calm, placid behavior for which the kingsnake group is noted, some individuals will occasionally vibrate their tail, strike, and hiss in Gopher Snake fashion when disturbed. This may be an important second line of defense that is occasionally triggered when this snake's disruptive coloration fails to deter a predator. A clutch of usually six to 12 eggs is laid between May and August, depending on the latitude and elevation where populations occur.

CALIFORNIA MOUNTAIN KINGSNAKE
COAST MOUNTAIN KINGSNAKE
SAN BERNARDINO MOUNTAIN KINGSNAKE
SAN DIEGO MOUNTAIN KINGSNAKE
SIERRA MOUNTAIN KINGSNAKE
ST. HELENA MOUNTAIN KINGSNAKE

Lampropeltis zonata
L. z. multifasciata
L. z. parvirubra
L. z. pulchra
L. z. multicincta
L. z. zonata

Coast Mountain Kingsnake

PROTECTIVE STATUS OF *L. Z. PARVIRUBRA* AND *L. Z. PULCHRA*: CSSC

IDENTIFICATION: This is one of the most strikingly colored snakes in North America. Its shiny, smooth scales are covered with black, white, and red crossbands, with the red bordered on each side by the black. This is in contrast to coral snakes, where red bands are bordered by yellow or whitish ones. Coral snakes do not occur in California. This is a medium-size snake, with an adult length range of 51 to 123 cm (20 to 48 in.). Unlike a somewhat similarly marked species, the Western Long-nosed Snake (*Rhinocheilus lecontei*), the snout of *L. zonata* is usually black, and the white crossbands do not broaden conspicuously on the lower scale rows.

RANGE: True to its name, the California Mountain Kingsnake occurs in nearly all mountain ranges of the state, including the inner coast segments of the North and South Coast ranges. It is absent from the major nonmountainous regions, especially the Great Valley and Great Basin, Mojave, and Colorado deserts.

NATURAL HISTORY: This is an upland kingsnake that inhabits most foothill and mountain habitats from coastal sage scrub and chaparral to conifer and oak–pine forest. It favors open areas where there are rotting logs, rock cover, or both. This snake is mainly diurnal but despite its bold markings may be overlooked, because the broken color pattern often obscures its outline on substrates littered with fallen leaves, pine needles, or dry-streambed and talus rubble. It also climbs into brush, possibly to search for one of its preferred foods: bird eggs and nestlings. Other prey includes snakes, lizards, and small mammals. A clutch of from three to nine eggs is laid in June or July.

CONSERVATION NOTE: The two subspecies that have been designated as CSSC, the San Bernardino and San Diego Mountain Kingsnakes, have a comparatively limited range within their namesake areas as compared with other California subspecies and intergrades. These areas are also some of the most heavily populated in the state, which probably results in a higher percentage of reptile collectors who desire this spectacularly colored California snake The current (2011) CDFG freshwater fishing regulations prohibit the collecting of the latter subspecies but not the former.

WESTERN LONG-NOSED SNAKE   *Rhinocheilus lecontei*

IDENTIFICATION: This is a colorful, moderately slim snake with a length range of 51 to 152 cm (20 to 60 in.). It does have a slightly longer snout than most other snakes, which probably is a result of a countersunk jaw, but this subtle feature is not a good defining characteristic. Far more apparent are its black saddlelike markings with white flecks on the sides. These are bordered by thin white strips, and a broad red, pink, yellow, or cream band separates each white-lined saddle. These bands have dark lateral flecks. The venter is white to yellow with scattered dark blotches along the sides. In more arid parts of its range the red and orange coloration is often absent, resulting in a cream-to-white banding on a black background, a pattern similar to that of the Common Kingsnake (*Lampropeltis getula*). All scales are smooth.

RANGE: It occurs from Mendocino and Lake counties south through the Great Valley and adjacent foothills to Southern Coastal California, the southern mountain ranges, and the Mojave and Colorado deserts. It has also been reported from the Sutter Buttes in the Sacramento Valley, and there is an old record for Mount Sanhedrin, Mendocino County.

NATURAL HISTORY: This is a snake of the deserts, annual grassland, and shrubland. It is crepuscular and nocturnal, often engaging in thigmothermic warming on dark road surfaces after dark. It is a good burrower and uses its projecting snout to push into soft substrates. Among its favorite prey are whiptail lizards (*Aspedoscelis* spp.), which it may catch as they begin to retreat for the night. Additional prey includes other lizard species and their eggs, which are found by burrowing, small snakes and mammals, birds, and large nocturnal insects. One and sometimes two clutches of from four to 11 eggs each are laid between June and August.

Plate 102. Scale mosaic of the Western Long-nosed Snake.

BAJA CALIFORNIA RAT SNAKE *Bogertophis rosaliae*

PROTECTIVE STATUS: CSSC

IDENTIFICATION: This is a large and relatively slender snake, with an adult length range of 85 to 152 cm (34 to 60 in.). It is one of the few snakes that occur in California that have one uniform dorsal color with no markings. That color ranges from olive to yellowish or reddish brown, depending on locality. The venter grades from light yellow to pale rust or tan posteriorly.

RANGE: It occurs near Mountain Spring at the junction of Imperial and San Diego counties, and on south to the tip of Baja California Sur, Mexico.

NATURAL HISTORY: This Baja California "endemic" species has the northern tip of its range in the fan palm and creosote bush associations of southwestern Imperial County. It is primarily nocturnal and an excellent climber. True to its name, it appears to prey heavily on rat-type rodents, which in fan palm groves may include both woodrat species (*Neotoma*) and introduced European rats (*Rattus*), all of which are also good climbers. Clutches of from two to 10 eggs have been reported.

Garter Snakes (Genus *Thamnophis*)

In discussing the large family Colubridae, we mention several genera that may possibly emerge as new families if in the future this taxon is divided. If this does happen, the garter snakes are perhaps the most deserving of such recognition, because they are such a large and relatively uniform group. These are slender snakes with keeled scales and often a striped pattern, which usually consists of a pale stripe down the middle of the back and one low on each side of the body. These stripes approximate the color of dried grass stems and afford good concealment in the grassland habitats that many species prefer. The color between the stripes may be uniform or may consist of a blotched or checkered pattern. This patterning resembles a clothing accessory that was a popular necessity for the well-dressed person throughout the first part of the twentieth century—the garter. Although essentially obsolete today, the name persists in this well known group of snakes.

These are the most aquatic of all mainland California snakes, although the more terrestrial species may wander considerable distances from water. Other species center their lives around aquatic habitats, where they obtain their preferred fish and/or amphibian prey. They swim by lateral undulations of the body, with the head elevated somewhat above the surface. Garter snakes also swim well under water, and several species employ a quick water retreat as a major defense. Another defensive behavior that is universal in garter snakes is the release of a foul-smelling fluid from cloacal scent glands along with watery excrement when grabbed by a predator's mouth or a human hand.

Garter snakes are the most abundant and most widely distributed group of serpents in North America. Except for true desert habitats void of any water site, they exist in essentially all other major habitat situations. This includes much of the boreal forests of Canada, where the range of one species, the Common Garter Snake (*Thamnophis sirtalis*), extends into the southern area of the Northwest Territories. One key for survival in northern latitudes is the habit of several species to seek out deep, cavelike winter retreats well below the frost line. Such sites are rather scarce, which has led to communal denning behavior where thousands of snakes come from some distance to spend the winter in a thermally safe site and then disperse, usually after breeding, in spring. Their live-bearing (ovoviviparous) reproductive mode is another key to the success of garter snakes in a wide range of habitats, because specific microhabitat situations are not needed for egg incubation.

Garter snakes are perhaps the most challenging snake group for the taxonomists. The 1972 first edition of this book lists five species within California. This second edition has eight, and as studies of the molecular makeup of some of the current subspecies are conducted, we may expect to see more added to this list. One problem is that there is often great variation in coloration and color patterns, at both species and subspecies levels. It is therefore best to rely on geographic location in addition to color description when attempting to determine what species you are observing in the field.

COMMON GARTER SNAKE
SAN FRANCISCO GARTER SNAKE
CALIFORNIA RED-SIDED GARTER SNAKE
VALLEY GARTER SNAKE

Thamnophis sirtalis
T. s. tetrataenia
T. s. infernalis
T. s. fitchi

San Francisco Garter Snake
T.s. tetrataenia

Valley Garter Snake
T.s.fitchi

California Red-sided Garter Snake
T.s. infernalis

PROTECTIVE STATUS OF *T. S. TETRATAENIA*: FE, SE
PROTECTIVE STATUS OF *T. S. INFERNALIS*: CSSC in southern portion of range
IDENTIFICATION: This is a relatively large garter snake, with a total-length range of 46 to 140 cm (18 to 55 in.). The coloration between and even within subspecies is highly variable, but in most cases the dorsal and lateral stripes are well defined. There are usually red spots or blotches, and a double row of alternating black spots on the sides between the stripes. In some individuals the dark blotches may fuse to form vertical dark bars or horizontal dark stripes as in the San Francisco Garter Snake subspecies. Dorsal head color varies from black, brown, gray, or olive to a vivid red-brown. There are usually seven upper labials, and the rear chin shields are usually longer than the front pair.

RANGE: This is one of the most widely distributed snakes in North America, and it extends farther north than any of our other reptiles. It occurs throughout northern California and south through the coastal ranges to San Diego County and the south end of the Great Valley and the Sierra Nevada. It is absent from the desert regions. The endangered San Francisco Garter Snake occurs only in San Mateo County and the northern edge of Santa Cruz County.

NATURAL HISTORY: With the exception of deserts, this species occupies nearly every other conceivable habitat type in North America, and because of this and its high reproductive output it is perhaps the most numerous snake species on this continent. It is usually associated with various types of wetlands, where some subspecies feed almost exclusively on amphibians and fish. It also has adapted well to farms and city lots where some sort of wetland habitat remains. The junior author has especially fond memories of this species, because it was his first reptile pet. While attending a spring Sunday family gathering in Wisconsin in the late 1940s, an era when children were expected to sit quietly in the parlor and "be seen but not heard," he and a cousin managed to sneak out the back door of their uncle's house in Milwaukee and go exploring the large vacant lots that were still present at that time. In one they found a large pair of "long johns" that some amply proportioned person had discarded after it saw extensive service during the long, cold winter just past. His cousin had seen snakes at this site before, so they quickly lifted the soggy, massive garment. Beneath was a wiggling mass of several dozen *T. s. sirtalis*, of which each boy managed to grab a handful.

Plate 103. A frog's view of a hunting San Francisco Garter Snake.

This introduction to the Common Garter Snake proved to be prophetic, and for the past two decades another subspecies, *T. s. tetrataenia*, has been the main subject of that author's natural history research.

The Common Garter Snake exhibits most of the features described in the introduction to genus *Thamnophis*. In addition, it appears to be capable of activity at lower body temperatures than other garter snake species. This ability can be especially important at northern latitudes where in the spring it may have to forage in and out of very cold water. Our California subspecies also have a feeding advantage over the one or more other garter snake species with which they share a habitat area. This is the ability to eat an abundant prey group, the Pacific newt species (*Taricha* spp.), without suffering lethal poisoning. A complete prey list for this species is quite extensive and includes fish, anurans and their larvae, salamanders and the larvae of aquatic-breeding species, birds and their eggs, small mammals and reptiles, worms, slugs, and leeches. However, a given subspecies may specialize in feeding on only a few of these items. This snake does not employ constriction to subdue prey but instead simply grabs it and begins the swallowing process, which occasionally results in the hind legs entering the throat first. Some subspecies are very prolific for a livebearing snake, having up to 85 young, but a range of 12 to 18 is far more common.

CONSERVATION NOTE: The current ecological status of the endangered San Francisco Garter Snake presents a prime example of how multiple features of natural history and unnatural human-wrought conditions can combine to assign a species or subspecies this protective status. Perhaps the

Plate 104. San Francisco Garter Snake with bullfrog metamorph.

most important factor here is that this snake is a feeding specialist, preferring amphibians above all other available prey. When the junior author and former graduate student Sheila Larsen presented approximately 40 newborn *T. s. tetrataenia* with both a wide selection of appropriate-size potential prey species plus cotton swab tips containing these various prey species' scent, only amphibians or swabs containing amphibian odor received a strike or bite. Further observations with adult snakes showed that this restricted prey preference can be modified later on to include items such as small fish and earthworms, but the overriding preference for amphibians, especially frogs, appears to remain throughout life.

Given these findings, this subspecies' success depends heavily on the availability of pond and marsh habitats where its two principal prey species, the Pacific Chorus Frog (*Pseudacris regilla*) and the California Red-legged Frog (*Rana draytonii*), reproduce and thrive. Here is where humans enter this equation, because small wetlands are usually the first habitat to be destroyed to make way for agricultural or urban development, and this has happened extensively throughout this endangered species' San Mateo County range. A trap–release survey for this snake conducted by the junior author for the CDFG in the mid-1980s revealed its presence at 32 pond or marsh sites throughout the county. A few more sites have been discovered since then, but an equal number may have been lost.

One further detrimental factor also stems from a combination of natural history and human behavior. This is one of the most beautiful snakes in North America, and when such a feature is combined with its great scarcity and endangered status, the door is wide open for illegal collecting for the reptile trade in both this country and even more so in Europe. Current estimates for the price of one prime adult specimen in Europe are $250 and upward.

Despite this rather dismal current picture, there are a few reasons for optimism concerning San Francisco Garter Snake survival. One is the basic nature of the Common Garter Snake as a "city lot snake." A relatively small acreage that supports a permanent wetland habitat where the Pacific Chorus Frog thrives will usually support a small *T. s. tetrataenia* population as well. Indeed, the largest existing population of this snake persists in a very large "vacant lot" habitat surrounded by extensive urban development. During the development of agricultural acreages in coastal San Mateo County, low-lying wetlands were often drained to make way for crops. However, farmers soon found that water storage was a key to their success, and most farms now have irrigation ponds, many of which have acquired small populations of frogs and endangered garter snakes.

Finally, although both the USFWS and CDFG have been diligent in monitoring land use changes within or adjacent to the remaining sites where this endangered snake occurs, more emphasis should be placed on biologically meaningful restoration and expansion of the endangered species' habitat once such projects are completed.

WESTERN TERRESTRIAL GARTER SNAKE   *Thamnophis elegans*
COAST GARTER SNAKE   *T. e. terrestris*
MOUNTAIN GARTER SNAKE   *T. e. elegans*
WANDERING GARTER SNAKE   *T. e. vagrans*

Coast Garter Snake
T. e. terrestris

Mountain Garter Snake
T. e. elegans

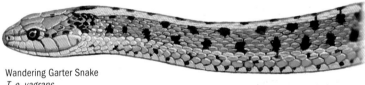

Wandering Garter Snake
T. e. vagrans

IDENTIFICATION: This is a medium-size garter snake with a length range of 46 to 109 cm (18 to 43 in.). Most individuals have a well-defined yellow or white mid-dorsal stripe plus a pale stripe on each side. The background color between the stripes varies between subspecies from black with white flecks to brown, gray, olive, or red-brown. Small, scattered dark blotches are often present between the stripes in the *T. e. vagrans* subspecies, and in the coastal subspecies, *T. e. terrestris*, large dark blotches often form a checkered pattern interspersed with bright orange or red. This subspecies occurs with the San Francisco Garter Snake (*T. sirtalis tetrataenia*) in San Mateo County, and highly colored individuals are sometimes mistaken for the endangered snake. However, the dorsal head color of *T. e. terrestris* is brown, not red as in *T. s. tetrataenia*, and it usually has eight upper labial scales as opposed to seven in *T. sirtalis*. Both pairs of its chin shields are of even length. The venter color ranges from brown or gray to a dull blue, but never the sky-blue color found in prime specimens of the San Francisco Garter Snake.

RANGE: *T. elegans* occurs throughout the North Coast and South Coast ranges, the Cascade Range, the Sierra Nevada, the northern segment of the Great Valley, and the southern segment of the Great Basin Desert in California. Isolated populations are also present in the San Bernardino Mountains of southern California and along the Amargosa River in the eastern Mojave Desert in California.

NATURAL HISTORY: In California this garter snake is even more of a habitat generalist than *T. sirtalis*, because in some areas its food preference may be slanted heavily toward upland prey such as small rodents and lizards, which often occur well away from wetland sites. In numerous trapping surveys in upland coastal scrub habitats in the greater San Francisco Bay Area, the junior author has found this species to be one of the two most common snakes, along with the Racer (*Coluber constrictor*). Given this terrestrial orientation, the upland-oriented Coast Garter Snake (*T. e. terrestris*) and Mountain Garter Snake (*T. e. elegans*) usually retreat to dense plant growth or under ground-cover objects when disturbed rather than going into water as is often the case with the more aquatic-oriented Wandering Garter Snake (*T. e. vagrans*).

T. elegans is a classic feeding generalist that takes advantage of the variability in prey species availability that often exists in California's highly variable climatic conditions. The junior author recovered food items from Coast Garter Snakes by stomach palpation throughout several spring-to-summer trapping seasons and found that it often consumes large numbers of slugs and earthworms when these are active on the surface of wet soil in early spring. As the habitat dries in late spring, stomachs often contained

Plate 105. Brightly colored Coast Garter Snake.

numerous larvae of the Pacific Chorus Frog (*Pseudacris regilla*), which are usually caught in the drying basins of shallow pools and seasonal ponds. Then, as spring gave way to summer, it was more common to recover small rodents from the stomachs of captured specimens than the former prey items. A moderately complete prey list for this species includes earthworms, slugs, snails, leaches, small fish, salamanders, anurans, tadpoles, lizards, small snakes, rodents and shrews, birds and their eggs and nestlings, and occasionally carrion.

The Western Terrestrial Garter Snake gives birth to from four to 27 young from July through September. Recently born snakes are often found under ground-cover objects in moist-substrate areas where small earthworm and slug prey are available in the late-summer months.

AQUATIC GARTER SNAKE   *Thamnophis atratus*
DIABLO RANGE GARTER SNAKE   *T. a. zaxanthus*
OREGON AQUATIC GARTER SNAKE   *T. a. hydrophilus*
SANTA CRUZ AQUATIC GARTER SNAKE   *T. a. atratus*

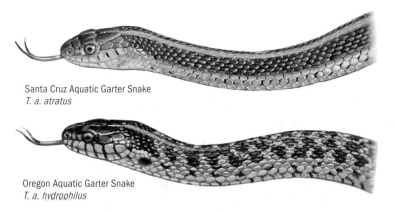

Santa Cruz Aquatic Garter Snake
T. a. atratus

Oregon Aquatic Garter Snake
T. a. hydrophilus

IDENTIFICATION: The adult size range of the Aquatic Garter Snake is 46 to 102 cm (18 to 40 in.). The coloration varies greatly between the three subspecies. The Santa Cruz Aquatic Garter Snake (*T. a. atratus*) usually has a yellow or yellow-orange dorsal stripe running the length of the body that is flanked by black or dark brown; light lateral stripes may or may not be present, and the throat ranges from yellow to white. The Oregon Aquatic Garter Snake (*T. a. hydrophilus*) has a less distinct dorsal stripe, and there are two alternate rows of dark blotches on the gray or olive-gray sides. The venter of both subspecies is light-colored and unmarked, and there are eight upper labial scales.

Plate 106. Santa Cruz Aquatic Garter Snake.

RANGE: The Santa Cruz Garter Snake occurs in the Greater San Francisco Bay Area south to Santa Barbara County. The Oregon Aquatic Garter Snake is found in Mendocino County northward. Intergrades occur in Sonoma County.

NATURAL HISTORY: The Aquatic Garter Snake can be found in a variety of plant communities ranging from grassland–woodland edges to brushland and forests, but it is nearly always associated with ponds, small lakes, streams, and rock-laden creeks. This is because it is an aquatic-vertebrate feeding specialist, very adept at catching small fishes and amphibian larvae. In very shallow water it often does this by encircling several of such prey with its body coils and then striking rapidly as they try to escape from this living holding pen. It also eats anurans, at which it lunges in the water rather than using the stalk-and-strike technique of subspecies such as the San Francisco Garter Snake (*T. s. tetrataenia*). This snake sometimes engages in extensive basking on open shoreline areas where it may be easily viewed. This is possibly a means of regaining an adequate body temperature after a round of foraging in cool water. It has a small litter size (three to 12 young) compared with other garter snakes but may have up to five litters, especially in the San Francisco Bay area, between August and mid-October.

SIERRA GARTER SNAKE *Thamnophis couchii*

IDENTIFICATION: This is one of the largest of our California garter snakes, with a length range of 46 to 124 cm (18 to 49 in.). It also differs from most in having little or no dorsal striping. The dorsal color ranges from olive brown to dark brown or black with two rows of alternating dark blotches on each side. The throat in most specimens is a pale yellow, and the light-colored venter is blotched with gold or salmon. There are eight upper labial scales.

RANGE: True to its name, this garter snake is found mainly in the Sierra Nevada. However, it ranges into drainages of the Pit and Sacramento rivers, in the south into the Tehachapi Mountains, and east of the Sierra Nevada into the Owens Valley. A population of garter snakes at California City in the western Mojave Desert may also be of this species.

NATURAL HISTORY: This is a garter snake of the mountain and foothill wetland habitats. It occurs within a number of plant communities from sagebrush and chaparral to mountain forests, but a common denominator at nearly all sites is the presence of a stream, creek, meadow pond, or lake. It is usually seen along the shores of such sites and retreats into the water when disturbed. This is an aquatic-vertebrate feeding specialist of the Sierran uplands, with small fishes, frogs, toads, and amphibian larvae composing its main food. It will also take adult salamanders when their activity period overlaps its own. From five to 38 young are born between July and September.

GIANT GARTER SNAKE   *Thamnophis gigas*

PROTECTIVE STATUS: FT, ST

IDENTIFICATION: This is indeed a giant among garter snakes, with a length range of 94 to 165 cm (37 to 65 in.). In addition to this great length, adults have a rather massive head. There is usually a yellowish dorsal stripe and a lighter side stripe on each side of the body. The color between the stripes is olive or brown with alternating rows of well-separated small dark blotches. The venter ranges from light gray to light brown. Like most other garter snake species, it has eight upper labial scales.

RANGE: This species was once widespread throughout most of the Great Valley. It now occurs in select wetland areas from Glenn and Butte counties south to northern Merced County. A few may still persist in the San Joaquin Valley west of Fresno.

NATURAL HISTORY: The Giant Garter Snake is our most aquatic mainland California snake. The common Native American names for it usually translate to "water snake." It is always found along the shores of marshes, sloughs, mud-bottom canals, and ditches in rice-growing areas, and occasionally in slow-flowing streams. The tule and cattail stands that are common at such sites are used for retreat, and the dead mats that they form provide good basking sites. It is an aquatic-vertebrate feeding specialist that eats small fishes, anurans and their larvae, and occasionally birds and nestlings that occur in the shoreline vegetative cover.

CONSERVATION NOTE: This was once a common snake throughout the great wetland complex that once covered much of the Great Valley. However, the upstream damming of the Kern and San Joaquin rivers, various water diversion and flood control projects, and the drainage of massive, shallow lakes such as Tulare Lake for agriculture have resulted in a loss of about 98 percent of its former range. It has adapted moderately well to artificial irrigation ditches and canals as long as they support an emergent vegetation stand. However, when these are lined with concrete, this snake no longer remains. One seemingly unlikely supporter of the Giant Garter Snake throughout this long history of habitat loss has been the waterfowl hunter and huntress. Permanent backwater sloughs and channels at the several federal and state waterfowl refuges within its current range plus other wetlands supported in part by Ducks Unlimited have proven to be an important though originally unintended refuge for this threatened species.

TWO-STRIPED GARTER SNAKE *Thamnophis hammondii*

PROTECTIVE STATUS: CSSC

IDENTIFICATION: This is a relatively small garter snake with a length range of 61 to 102 cm (24 to 40 in.). It has no dorsal stripe, and its lateral stripes are also often faint or absent. However, there are usually dark spots above the lateral stripes that tend to make them more obvious. The dorsal color is olive, brown, or brown-gray. Venter color ranges from yellow to red-orange or salmon. Melanistic individuals occur in coastal areas of the southern portion of the South Coast Range and on Catalina Island. Along the Piru River in northeastern Ventura County dark-greenish and dull-reddish morphs occur.

RANGE: It is found from northern Monterey County south through the South Coast Range and Southern Coastal California and on into Baja California, Mexico. A population exists on Santa Catalina Island.

NATURAL HISTORY: This is another aquatic-feeding specialist, inhabiting permanent and intermittent drainages of the seasonally arid regions of southwestern California. It prefers watercourses with good riparian stands. Unlike most garter snake species, it is often active at night as well as during the day. It is a feeding specialist on aquatic vertebrates, including anurans and their larvae, larval newts, small fish and fish eggs, and earthworms. Four to 36 young are born during summer.

CONSERVATION NOTE: Urban and suburban development in much of its range in the southern end of the South Coast Range and Coastal Southern California has reduced the historic range of this snake by an estimated 40 percent. Carefully constructed and reviewed environmental impact reports for all proposed land use changes involving riparian areas are a crucial first step in curtailing further habitat loss.

NORTHWESTERN GARTER SNAKE   *Thamnophis ordinoides*

IDENTIFICATION: This is one of the smallest garter snakes in California, with a length range of 33 to 96 cm (13 to 38 in.). The coloration of this species is variable. Individuals usually have a well-defined dorsal stripe of yellow, orange, or red, but the stripe may be faint or absent in some populations. The dorsal color ranges from black or brown to a greenish or bluish hue.

The venter may be yellow, olive, or slate, and often has red blotches. This species has seven upper labial scales.

RANGE: This is a garter snake of the Pacific Northwest, whose range extends south to the Mad River in the northwest corner of the North Coast Range of California.

NATURAL HISTORY: The Northwestern Garter Snake inhabits forest meadows and clearings where there are clumps of low-growing vegetation. It occasionally occurs near water but is chiefly a terrestrial species. Its preferred foods include earthworms, slugs, land snails, amphibians, and other such prey that are most abundant aboveground when the substrate is wet after rain or during heavy fog. It is during such times that this snake is most active. Three to 20 young are born between June and August.

CHECKERED GARTER SNAKE *Thamnophis marcianus*

IDENTIFICATION: This species has a wide adult-length range of 32 to 107 cm (12.75 to 42.25 in.) and a checkered pattern of large square blotches on a brown-yellow, brown, or olive background. There is a cream-to-yellow dorsal stripe and a white-to-yellow crescent between the dark side blotches and the corner of the mouth. There are also paired dark blotches at the back of the head. The venter is various shades of off-white. It has eight upper labial scales.

RANGE: This garter snake occurs in the extreme southeast corner of the state along the Colorado River and associated agricultural land. There are also records of this species in the southern Coachella Valley.

NATURAL HISTORY: The Checkered Garter Snake occurs in grasslands, farmlands, and streamcourses in desert lowlands. In undeveloped desert areas it centers its activities around streams, rivers and natural springs. Where agriculture has moved in, it has adapted to the irrigation canals, ditches, and artificial ponds that result. This accounts for its presence in the southeastern corner of the Imperial Valley of California, and given this association with irrigation

Plate 107. Juvenile Black-necked Garter Snake (*T. cyrtopsis*), which resembles the Checkered Garter Snake.

canals, it may continue to expand through the Imperial Valley. It feeds on a wide spectrum of small vertebrates such as anurans and their larvae, lizards, small snakes, and a number of small rodent species that are also attracted to these watercourse-edge habitats. Here it also finds earthworms and slugs when they are active above ground. It gives birth to from three to 35 young during the late-spring through early-fall periods.

WESTERN SHOVEL-NOSED SNAKE	*Chionactis occipitalis*
COLORADO DESERT SHOVEL-NOSED SNAKE	*C. o. annulata*
MOJAVE SHOVEL-NOSED SNAKE	*C. o. occipitalis*
NEVADA SHOVEL-NOSED SNAKE	*C. o. talpina*

IDENTIFICATION: This is a relatively small snake with a length range of only 25.4 to 43 cm (10 to 17 in.). The dorsal color is yellow or cream, contrastingly marked with black or brown broad saddles or crossbands that may encircle the body. Orange or red saddles are sometimes present between the dark bands. The lower jaw is deeply countersunk, and this along with well-developed nasal valves, a concave belly, and smooth scales adapts it well to sand burrowing.

RANGE: It occurs in the Mojave and Colorado deserts.

NATURAL HISTORY: This is a true desert snake and may be found wherever pockets or extensive areas of sand or loose soil exist. It is highly adapted for burrowing in such substrates and centers most of its life activities around this ability. It burrows rapidly by using lateral movements of its shovel-like head. At such times the head is tipped downward at a 45-degree angle to create a pocket in the sand below the head that makes possible the throat movements necessary for breathing. It actually wriggles through sand rather than tunneling through it, a type of movement called "sand swimming." During the desert day it remains underground where temperatures are considerably lower than on the surface. However, at night it engages in extensive surface roaming, especially after a rain. Its presence in an area can usually be detected by the small, smoothly undulating tracks on bare sand between desert shrubs. When foraging it will occasionally climb into bushes where its banded pattern matches the dark nodes and light-colored internodes of the branches. Its burrowing ability allows it to exploit food sources such as insect larvae and the buried chrysalid pupae of moths that are unavailable to most other snakes. It also eats spiders, scorpions, and centipedes. In spring or summer it lays a small clutch that usually contains from two to five eggs but can contain as many as nine.

Plate 108. Western Shovel-nosed Snake.

WESTERN GROUND SNAKE   *Sonora semiannulata*

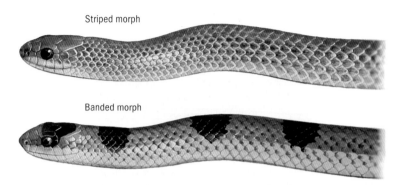

Striped morph

Banded morph

IDENTIFICATION: This is another small desert snake with a length range of 20.3 to 46 cm (8 to 18 in.). The dorsal color varies from brown, olive, or gray to orange or shades of red. There is a dorsal saddle pattern that also varies greatly. In some individuals dark crossbands encircle the body or may form saddles on the back. In others there may be only a single neckband or all banding may be absent. These variations are not a result of an adaptation to a particular habitat but instead may all occur in the same population. Some populations along the Colorado River have a broad beige, red, or mid-dorsal stripe and greenish gray or bluish gray sides. The venter is white or yellow and may contain the ventral portion of dark crossbands. It has a smooth and glossy scalation like shovel-nosed snakes (*Chionactis* spp.), but the jaw is not countersunk.

RANGE: This snake has a spotty distribution in the Great Basin, Mojave, and Colorado deserts.

NATURAL HISTORY: The desert habitat of the Western Ground Snake is similar to that of the Western Shovel-nosed Snake (*Chionactis occipitalis*), but it is less tolerant of extreme aridity. It frequents river bottoms, borders of cultivated fields, desert flats, sand hummocks, and rocky hillsides where there are pockets of loose soil. It is nocturnal and highly secretive but does not engage in "sand swimming." Instead it forages on surface-oriented invertebrate prey such as grasshoppers, crickets, scorpions, and centipedes but will also seek out buried insect larvae. A clutch of three to six eggs is laid between June and August.

Plate 109. Western Ground Snake.

CALIFORNIA BLACK-HEADED SNAKE *Tantilla planiceps*

IDENTIFICATION: This small snake has a length range of 12.7 to 39 cm (5 to 15.5 in.). It has smooth scales and a flattened head, the top of which is black or dark brown. The dark cap extends downward behind the corner of the mouth, and behind the cap is a narrow white or cream collar that may or may not be bordered on its posterior edge by dark dots. The venter is white with a broad orange or reddish stripe down the center. Because of its similarity to the other black-headed snake in California, the Southwestern Black-headed Snake (*T. hobartsmithi*), hemipene configuration may be helpful in identifying captive male specimens. In this species they are nearly cylindrical, without small spines near the enlarged basal spine.

RANGE: It occurs in portions of the South Coast Range and most of Southern Coastal California but may no longer persist in the greater Los Angeles area. In southern California it ranges to the desert side of the mountains as at Whitewater Canyon, Riverside County.

NATURAL HISTORY: This is a secretive, ground-dwelling snake that apparently spends much time underground in crevices and rodent burrows, because surface observations of it are rare, except on warm nights. It occurs in annual grassland, chaparral, oak and oak–pine woodland, desert edge and thorn-scrub habitats. It feeds almost exclusively on insects, especially beetle larvae, and centipedes. Its orientation toward subsurface retreats complements this food preference, because these types of invertebrates also spend much of their time below ground. This and other black-headed snake species are venomous and have slightly enlarged and grooved teeth at the back of each upper jaw that may aid in "chewing" the poison into prey. However, because of its small size and secretive nature, it is not harmful to humans. This is one of the least fecund snakes in California, with a single clutch of one to three eggs laid in May or June.

SOUTHWESTERN BLACK-HEADED SNAKE   *Tantilla hobartsmithi*

IDENTIFICATION: This species closely resembles the California Black-headed Snake (*T. planiceps*). It length range is essentially identical at 13.2 to 38 cm (5.5 to 15 in.), and it also has a dark brown or black head cap. However, its cap does not extend below the corner of the mouth. It has a narrow white or cream collar, but there are usually no dark spots along its posterior border. The dorsum is a plain brown or beige, and the white venter has a broad coral red or rufous stripe down the center. The hemipenes are club-shaped with an enlarged globular tip and two medium to large spines at the base.

RANGE: Its distribution is very spotty, with populations occurring in the southern San Joaquin Valley portion of the Great Valley, including the Tulare Lake area, the southern Sierran foothills, the Great Basin Desert, the desert side of the Tehachapi Range, and the northern and western portions of the Mojave Desert, including Jawbone and Hogback canyons.

NATURAL HISTORY: Habitats where it has been found include annual grassland, brushland, sagebrush–greasewood, mesquite, yucca, and creosote bush, open chaparral, thorn-scrub, piñon–juniper woodland, and open coniferous forests. It is usually found under dead plant and rock ground cover, where it presumably forages for invertebrate prey including beetle larvae, caterpillars, centipedes, and millipedes. As in other black-headed snake species, venom is often used to subdue prey. A clutch of one to three eggs is laid between June and August.

CALIFORNIA LYRE SNAKE	*Trimorphodon lyrophanes*
SONORAN LYRE SNAKE	*Trimorphodon lamba*

IDENTIFICATION: These snakes have a broad, flat, triangular head with a lyre-shaped marking on top. Their length range is 46 to 121 cm (18 to 47.75 in.). They have large eyes with vertical pupils. Short, grooved fangs toward the rear of the upper jaw assist in chewing venom into prey. The dorsal color varies from light brown to pale gray with brown hexagonal-shaped blotches that are split by pale crossbars. The venter is cream or pale yellow and often has scattered brown dots.

RANGE: They occur in Southern California from the vicinity of Santa Barbara, the Tehachapi Range, and the Amargosa River area southward. They are absent from most areas in the central portion of the Mojave Desert and the centrally developed areas of Los Angeles and San Diego.

NATURAL HISTORY: These are rock-dwelling snakes of mesas and lower mountain slopes. They often frequent massive rock areas where they retreat in deep crevices during the day. However, they are occasionally found in rockless sites. They often prey on rock-dwelling lizards that they capture in crevice retreats. Here they can subdue them by chewing in their venom and then waiting for the lizard to die while remaining hidden from their own predators. Other prey include small snakes, birds, small rodents, and bats, which are caught in crevice roosts and also immobilized by their venom. Prey may be restrained by constriction while the venom is being introduced. The bite of these snakes may produce mild swelling in humans, but no serious effects or deaths have been reported. Clutches of seven, 12, and 20 eggs have been reported to date.

COAST NIGHT SNAKE
DESERT NIGHT SNAKE

Hypsiglena ochrorhynchus
Hypsiglena chlorophaea

Coast Night Snake

TAXONOMIC NOTE: A recent genetic study of this species proposes that the original species, the Night Snake (*Hypsiglena torquata*), be divided into six new species, with two of these, the Coast Night Snake (*H. ochrorhynchus*) and the Desert Night Snake (*H. chlorophaea*), designated for Calfornia. As with most newly declared "genetic species," the question of whether or not these new species are reproductively isolated populations that cannot produce viable offspring with each other remains unanswered. However, given the current trend to accept most new species designations without such information, we include the two newly designated California night snake species and their ranges here but discuss their natural history as one, because that remains essentially the same, regardless of taxonomic changes in this very uniform population of snakes.

IDENTIFICATION: This is a relatively small, pale-gray, light-brown, or beige snake with a length range of 30 to 60 cm (12 to 26 in.). It has dark-brown blotches on the back, smaller ones on the sides, and usually a pair of large dark-brown fused blotches on the neck. There is a dark-brown stripe above the mouth set off by pale upper labial scales. The head is flat, scales are smooth, the pupils are vertical, and grooved rear fangs are used to chew a neurotoxic venom into prey.

RANGE: The Coast Night Snake occurs from the northern part of the Great Valley and southern part of the North Coast Range south through the South Coast Range of Southern California, where it is widespread throughout the southern third of the state. It extends throughout the foothills of the Sierra Nevada but is absent in the central portion of the Great Valley. There is also an old record of its presence on Santa Cruz Island. The newly designated range for the Desert Night Snake includes the Mojave and Colorado deserts, plus the segment of the Great Basin Desert in the southeast corner of the state.

NATURAL HISTORY: This is a habitat generalist, existing in grassland, chaparral, sagebrush flats, deserts, woodland, and moist mountain meadows. It is crepuscular and nocturnal, and is often seen on blacktop roads where it lingers in the early evening to obtain warmth from the surface. When it bites its prey, venom is secreted with its saliva, and the mixture passes down grooves on the enlarged rear teeth. It feeds on lizards and their eggs, small snakes, and a variety of amphibians. A clutch of two to nine eggs is laid during spring or summer. Although venomous, it poses no real threat to adult humans because of its small size and generally placid nature. However, children could be affected if one of their small fingers were bitten and chewed.

Sea Snakes (Family Hydrophiidae [Elapidae])

The authors are fortunate that one member of this unusual group of marine snakes, the Yellow-bellied Sea Snake (*Pelamis platurus*), occasionally enters the offshore waters of San Diego and Orange counties, thus allowing them to make it a legitimate inclusion in this book. These are the most marine of all living reptiles, because unlike sea turtles, many species give birth to their young and thus remain at sea their entire life. Most are moderate-size snakes with lengths of around 1 to 1.5 m (3 to 4 ft), but some grow as large as 2.75 m (about 9 ft). The tail is vertically flattened and used as a sculling organ to gracefully propel these snakes at both the surface and during dives. A body flattened from side to side in many species also enhances this locomotion. In a manner similar to cetaceans, valvular nostrils are positioned on the dorsal anterior portion of the head to facilitate surface breathing. The sea snake lung is also longer than that of land snakes, and large, posterior air storage sacs contain muscular tissue that can pump air forward where oxygen can then be utilized in the vascular anterior lung segment. This along with the ability to take up oxygen through the skin in some species permits submergence for up to two hours when resting.

Sea Snakes live in an "aquatic desert," because no freshwater exists in the sea, and body water can pass out of their bodies through the skin. Sea water may also pass down their throats when prey are swallowed. Like terrestrial desert reptiles, they have salt-secreting glands, which in this family are located at the base of the tongue. This gland removes and concentrates excess sodium from the blood and then passes it out into the water when the tongue is extruded. The body water that follows its concentration gradient out through the skin to the sea is replaced by the preformed and metabolic water gleaned from their prey. Like their close relatives the coral snakes (Elapidae), sea snakes employ a potent neurotoxic poison to subdue their prey, which consists of small fishes, squid, cuttlefish, and crustaceans. Hollow, immobile fangs located in the front of the mouth inject the venom while the prey is being held. Approximately 160 species occur throughout all oceans that have tropical and subtropical segments, except the Atlantic.

YELLOW-BELLIED SEA SNAKE *Pelamis platurus*

IDENTIFICATION: This species has a total-length range of 51 to 114 cm (20 to 45 in.). It is dark brown or black above and bright yellow or cream-colored below. This presents a classic aquatic countershading pattern, with the dark dorsal and light ventral colors joining in a sharp line with no fusing or blending. In the tail region the dark color extends ventrally in a series of scallops with additional spots and blotches. This apparently functions as a warning coloration by which would-be predators may recognize and avoid this venomous potential prey. The tail and to some extent the body are vertically flattened for sculling. Valved nostrils are positioned high on the snout, but the eyes are situated on the sides of the head instead of high up on the head, an adaptation that enables these predominantly subsurface feeders to better see fish and other marine prey. Short fangs are present in the front upper jaw. There are no belly scutes for crawling as in terrestrial snakes, and this species is essentially helpless if stranded on a beach.

RANGE: This is one of the most widely distributed snakes in the world. It occurs in or occasionally enters tropical and subtropical segments of all oceans except the Atlantic. It has been reported off the southern California coast as far north as San Clemente in Orange County.

NATURAL HISTORY: The Yellow-bellied Sea Snake is completely aquatic and apparently never comes onshore where it, like most fishes, is essentially helpless. This species of sea snake specializes in feeding along ocean slicks, calm areas of water where debris such as loose algae, driftwood, flotsam, plastic items, and so on form a floating mat. Such areas afford good cover to small surface-dwelling fishes, which constitutes this snake's principal prey. These are usually grabbed with a sideways strike and then held while the neurotoxic venom is injected. Despite the potency of this venom, the snake is usually reluctant to strike when handled by humans, and when it does, venom is seldom injected. It gives birth to from one to eight young, perhaps throughout the year. Given the ability to bear their young alive, the Yellow-bellied Sea Snake and its relatives join cetaceans, manatees, and dugongs as the only completely marine higher vertebrates.

Vipers (Family Viperidae)

Pit Vipers (Subfamily Crotalinae)

The North American pit viper group, which also occurs in Central and South America, and Southeast Asia are heavy-bodied snakes with broad

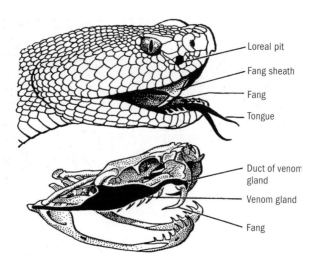

Figure 35. Upper: Head anatomy of a rattlesnake, showing the heat-sensing loreal pit and the scent-gathering forked tongue. Lower: Rattlesnake head skeletal anatomy showing the moveable fang and location of the venom gland and duct.

triangular heads, keeled scales, elliptical pupils, and large, hollow fangs. All have one unique anatomical feature in common: the loreal pit. This is a depression on the snout area posterior to the nostril that contains a membrane that supports nerve endings sensitive to infrared radiation. The pits, one on each side of the head, are directed forward in a binocular fashion so that depth perception can be obtained when detecting heat radiated from warm objects. These and the pits in the labial scales of family Boidae are the only infrared-sensing structure in vertebrates, although some insects also have this ability. These pits along with the Jacobson's organ, the loosely hinged jaw, a trachea that opens at the tip of the lower jaw, and forward-moving hollow fangs with attached poison gland and duct, constitute one of the most unique set of adaptations of the head for feeding in the vertebrate world (figure 35).

TAIL RATTLES: Except for the Copperhead (*Agkistrodon contortrix*) and the Cottonmouth (*Agkistrodon piscivorus*) of the eastern United States, all other North and Central American pit vipers are further distinguished by the presence of a rattlelike structure at the tip of the tail, and hence the common name "rattlesnake" or often just "rattler." This is a series of highly modified interlocking semispherical shed tail scales that are shaken by specialized tail-shaker muscles to produce a rattling (slow vibration) or buzzing (fast vibration) sound when the snake is disturbed by a potential danger source. Tail vibration is not unique to rattlesnakes and also occurs in other species such as the Gopher Snake (*Pituophis catenifer*), but no sound is produced unless the vibrating tail touches dried leaves or

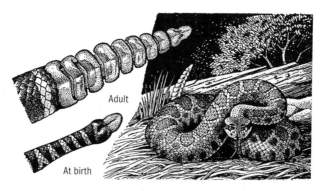

Figure 36. Comparative rattle size of an adult and newborn rattlesnake and the prestrike posture when the rattle sound is often heard.

similar objects. The addition of the rattle structure ensures sound production with every tail shake. A rattle segment is added at the tail base each time a snake sheds its skin, and a young snake may add three or four segments each year. However, a mature one may shed only once a year, and so the idea that one can tell the age of a rattler by the number of its rattles is incorrect (figure 36).

The actual rattling process probably had its evolutionary origin in caudal luring in which many ambush-foraging snakes elevate and vibrate their tails, which often have bold markings, in an apparent attempt to lure potential prey within effective striking distance. This has been observed in young rattlesnakes whose future rattle has only developed to the "button" stage and thus does not produce a sound. Adult rattlesnakes only rattle in defensive contexts, and this function of the rattle to alert potential predators to the snake's presence and identity as a venomous species is the most widely accepted reason for its evolution. Its use as an auditory warning is especially intriguing, because snakes are a nonvocal group and at best can produce only short bursts of hissing or rasping sounds. However, the rattling sound does not rely on the sporadic expulsion of lung air but instead is continuous over a relatively long period of time, which allows a potential predator to better locate and then avoid it. Supporting the theory of rattle evolution as a defensive behavior, it has been observed that in several rattlesnake populations that occur only on islands where there are neither predators that could eat them nor large ungulates that could trample them, the rattle structure is highly reduced or absent.

CAPTURING PREY: Because of the unique structural modifications at both ends of this snake group, some confusion has arisen as to how they are linked in the overall adaptation and survival of this subfamily. The best way to understand a pit viper's highly specialized head is perhaps to visualize a rattlesnake in its feeding mode. These are primarily nocturnal rodent feeders and usually use an ambush strategy to capture prey. Input from its Jacobson's organ–tongue mechanism along with its sense of smell allows a rattler to detect often-used small-mammal pathways and select an ambush site adjacent to one. A rattlesnake's vertical eye pupils are indicative of reptiles with good night vision, and so we may assume that when lying in ambush it may be able to visually detect the distant approach of a rodent or small rabbit. Diurnal snakes such as racers (*Coluber*) and whipsnakes (*Masticophis* spp.) also use vision to locate moving prey, and their great speed then enables them to pursue it if the first strikes fails to grasp it. However, a relatively slow predator that hunts from ambush usually has only one opportunity for a successful strike, and if that misses the mark, the prey escapes. It is here that a rattlesnake's focused infrared-detection ability may enable it to more accurately judge its distance from a warm prey animal at night and successfully inject its venom during its one strike opportunity. Those who have used "night vision" infrared binoculars or cameras have a good appreciation of the extent to which this is an advantage. Recent studies indicate that 2,000 or so receptors in the heat-sensitive membrane of a loreal pit may produce very precise images.

Envenomated prey are usually not killed instantly but may move some distance before dying from the usually hemotoxic poison, which negates the ability of both blood and muscle tissue to obtain and utilize oxygen. At this time a rattlesnake's olfaction and Jacobson's organ inputs again come into play during the poststrike tracking process. Studies have shown that a pit viper can distinguish the scent of a bitten prey from nonaffected ones. This ability is crucial in many areas where numerous fresh rodent and rabbit trails may pass through the ambush site.

DEFENSIVE STRATEGIES: As useful as a pit viper's infrared detection system appears to be in its ambush-feeding behavior, this system may not have evolved solely for mammalian prey detection. These snakes rely on good cryptic coloration coupled with a static position within vegetative or rock cover to avoid detection by potential predators. In contrast, European vipers, which lack loreal pits, have striped body patterns and usually flee rapidly at the approach of potential danger. Given these differences, it is possible that a pit viper's infrared picture of the general size and shape of a potential predator may determine whether or not to initiate its rattle vibration warning system, because to begin rattling at any slight disturbance would be counterproductive for a predator that relies on concealment and silence to ambush prey. However, if an approaching animal is considerably larger than the rattlesnake's preferred prey and might indeed be a snake-eating predator, signaling its presence could be a distinct advantage for a rattlesnake. Such "advertisement behavior" in many animals is usually coupled with some sort of "punishment" for those individuals that do not heed its warning. One familiar example is the brazen black-and-white coloration of the Striped Skunk (*Mephitis mephitis*). Once a potential predator has been sprayed, it usually avoids any close contact with a small mammal displaying that color pattern and may even impart such a precaution to its young. In the case of rattlesnakes, the punishment is of course the pain associated with a venomous bite, and continuous buzzing and rattling could deter a predator from pursuit as the snake seeks a safe retreat site.

The "don't tread on me" rattler warning system may have arisen in ancestral rattlesnake stock in the rocky upland habitats of the Mexican Plateau, where there was little dried vegetation against which a vibrating snake tail could produce a sustained sound. Here mammalian predators such as the White-nosed Coati (*Nasua narica*) and American Black Bear (*Ursus americanus*) could learn to avoid a pain-inflicting snake that advertised its presence with tail vibrations that do not rely on dried vegetation for sound production. Later in the Pleistocene when the northward radiation of rattlesnakes from ancestral populations in the Mexican uplands occurred, the open plains of North America were inhabited by large herds of ungulates that posed the potential of death by trampling to these relatively large, slow-moving snakes. However, these pioneer rattlesnakes had already acquired a tail rattle mechanism and were preadapted to cope with this new threat to their existence. A rattlesnake bite can inflict pain but not usually death upon large mammals, and this scenario requires that such ungulates hear and associate the rattling of the snake with either the infliction of pain or the detection of fear of another herd member that has been bitten. It also relies on the retention of such an event in the ungulate memory bank, and indeed many ungulate species seem to have the ability to remember a disturbing experience throughout life, a fact to which most horse owners will readily testify. In this dual-selection process a rattlesnake advertises its position as a grazing herd approaches by vibrating its tail, which, because of the rattle structure, always makes a noise whether or not dried vegetation surrounds it. The grazing ungulate herd responds to the rattling by moving away from it as the memory of some past fright response to this sound is recalled. The successful outcome to such a scenario is that not only is the snake not trampled, but one or more ungulates escape being bitten.

In support of the theory that the rattle evolved as a warning system to ward off predators and other potentially dangerous animals is the previously mentioned occurrence of "rattleless" rattlesnakes on the large islands, such as those in the northern Sea of Cortez (which separates Baja California from the Mexican mainland), where neither ungulates nor mammalian predators originally occurred. Here, in the absence of such forms, these pit vipers either lost their rattles in recent evolution through a lack of continuing selection for them or were separated from the evolving rattlesnake stock on the Mexican mainland in the early Pleistocene before the selection for rattles was under way. However, scenarios describing past events that directed natural selection for the rattlesnake's rattle are still only theories, and the exact evolutionary pathways may never be fully revealed.

RATTLESNAKES AND HUMANS: Either of the preceding two theories of warning-sound production by rattlesnakes also applies to potential human predators, except that we communicate knowledge of this event without having to first witness an acquaintance being bitten or sustaining a bite ourselves. Unfortunately the sound-producing ability of rattlesnakes coupled with widespread ophidiophobia has often resulted in a complete reversal of this otherwise efficient protection system, because one of the most common reactions of people in many parts of the country to sound of a rattlesnake is to find it and kill it by whatever means are available. There is no denying that a large amount of hemorrhagic venom injected into a human, especially a child, can result in death, especially if the fangs pierce a major blood vessel and medical treatment is not readily available. However, such occurrences must be put into perspective. Statistics compiled by the U.S. Center for Disease Control serve this purpose well.

Since 1960 there have been only a few documented fatal pit viper bites per year, and most of these have been by the Western Diamond-backed Rattlesnake (*Crotalus atrox*), a species that is hunted and often carelessly handled in many parts of the west, especially during environmentally destructive events known as "rattlesnake roundups." However, recent statistics on annual deaths from other sources greatly overshadow this number. In just the category of other venomous animals, bee and wasp stings kill about 50 people each year, and fire ant bites average about 100 annual deaths. The weather also takes it toll. In 1995 lightening killed 89 people, and heat-related deaths in 2006 numbered more than 500.

Over his many years of taking natural history students for extended field trips to the state university's Desert Study Center in the Mojave Desert, the junior author was often warned by colleagues and acquaintances that he was running quite a risk by leading his charges into "rattler-infested" terrain. His counterargument, that natural history students are among the safest persons in the field because they are actively looking for snakes, nearly always seemed to fall on deaf ears. Ironically, it was rarely noted that other activities in which young people also engage have an element of risk. Tragically, during one past high school football season, 15 student players died and another 29 suffered catastrophic injuries that left them paralyzed or seriously disabled as a result of playing America's most popular school team sport. In 2007, the last year for which medical records are currently available, there were 29,846 accidental poisoning deaths, none of which were the result of rattlesnake "poison." And of course Americans' love affair with their cars often masks any realization that the only real danger involved in a trip to areas where rattlesnakes may occur is the ride to and from such sites. In 2003, 42,643 people died in the United States as a result of vehicle-related accidents, and of these 4,749 were pedestrians!

Despite the extremely low annual mortality from venomous snake bites in the United States, the California Poison Control System reports approximately 250 cases of rattlesnake bites each year in the Golden State and about 8,000 bites from all types of venomous snakes countrywide. Some of these bites result in painful hospital stays and certainly must put a dent into what would have otherwise been an enjoyable outing. The following few additional statistics and suggestions may hopefully keep your field outings "bite free." First, a word of warning to all male readers is in order. Of the 78 people bitten by rattlesnakes in California in 2004, 83 percent were male, and 81 percent of all bites were on the hands and forearms. Perhaps we fellows should think less of promoting our macho image when in the field and more about safety before plunging our forelimbs into cover areas where a rattlesnake may be sequestered or, worse yet, handling a live rattlesnake barehanded. When in a retreat mode in brush, log, or rock cover, a rattlesnake may not be stimulated to rattle, so you cannot always rely on receiving a warning signal. Indeed, a young rattler with only one or two buttons on its rattle may produce very little sound, as does an adult that has recently lost a large rattle complement and is beginning to grow a new one.

An also distressing possibility is that often the rattle warning may not be given at all. The junior author vividly recalls a night in the Chihuahuan Desert when, while walking with a lantern behind a colleague, he saw a coiled Mojave Rattlesnake (*Crotalis scutulatus*) just before his friend's foot came down upon it. This normally aggressive species neither rattled nor struck, and because there was a cushion of loose dried plant mat on the ground, it appeared none the worse for wear because of the encounter. However, that cannot be said for my colleague, whose nerves were pretty well shattered for the rest of the evening.

Even when a rattlesnake does bite a human, the results can be quite variable. Several recent studies suggest that pit vipers are capable of venom metering and may actively regulate the dosage of the injection. However, an opposing view holds that venom expenditure may instead be determined by other factors such as the surface features of the target. It is also possible that strikes at a large object are often meant to deter and not to kill. This may be the reason that approximately 20 percent of all bites reported annually are "dry bites." Many other reported "rattler bites" are in reality those inflicted by large, nonpoisonous similar-appearing species like the Gopher Snake (*Pituophis catenifer*). If indeed it was a pit viper that produced a bite, there should be one or most likely two distinct puncture wounds, because no other mainland California snakes have these large, forward-positioned, piercing fangs, and only if there is a burning pain and eventual swelling around the bite is it a "wet bite." However, both types deserve immediate treatment.

RATTLESNAKE BITES AND TREATMENT: Certainly the best way we can avoid a rattlesnake bite when in the field is through the use of two features with which most healthy humans are endowed: excellent diurnal vision and common sense. However, when partially masked by vegetation, a rattler's excellent cryptic coloration can make it very hard to detect, especially under less than perfect light conditions. Two pictures of the Sidewinder (*Crotalus cerastes*) in its species account (see Plates 114 and 115) illustrate this point. When coiled in an open sandy area, it was very easy to detect. However, when this same snake retreated to edge of a nearby brush stand, it became all but invisible. Here is where the "common sense" rule of never stepping or reaching into vegetation when in rattlesnake country without carefully visually searching your intended path, probing ahead with a long stick, or both, is always the best way to prevent a rattlesnake bite.

Stepping or jumping over a fallen tree trunk without inspecting the other side first is a common shortcoming of the novice foothill or coast range hiker. And of course a rattlesnake in the hand is the most dangerous rattler of all. The basic rule here is simply to not handle a rattlesnake except with a well-designed, long mechanical snake stick with which the user has practiced on nonvenomous species or flexible tubing of the appropriate diameter first. The high percentage of males being bitten indicates that many of these are young men attempting to demonstrate their manly powers and abilities. Also, remember that a freshly road-killed rattler may still have the potential to strike reflexively as a result of the anaerobic nature of reptile muscle.

As for what to do if, after all of the preceding precautions, you are or a companion is bitten by a rattlesnake, there is a long history of snake bite cures and remedies, nearly all of which ultimately worsen the patient's condition. For instance, up through the mid-1920s, alcohol was an accepted medical cure. It was thought to neutralize the venom, and the general belief was that a bitten man could drink unlimited amounts without ill effect. If snake bite statistics were kept during that era, there would have probably been a great disproportion of reports and treatments on Saturday nights! In the mid-1930s the medical profession finally got on the right track when a Brazilian physician developed the technique of producing antivenin, and today it remains the only effective cure when properly administered. Antivenin therapy is now a standard service of every California hospital, where a patient's state of health can be monitored and the appropriate dosage of bovine-derived antivenin administered. The replacement of horse-derived antivenin with bovine-based serum in the United States in 2000 represents a major advance in snakebite treatment, in that the former occasionally caused an allergic reaction that was often more severe that the trauma produced by the venom. After antivenin has been administered, physicians and the nursing staff should also watch for and be ready to treat hypotension and shock, which often set in two to three days after a snakebite has occurred.

However, there may be considerable travel time between the field site where the bite occurred and the hospital emergency room, and during this interval it is most important to keep the bitten person as calm and inactive as possible so as not to hasten the spread of venom through the body. Your calmness will also help greatly here. After all have settled down in the transport vehicle, any constricting clothing and jewelry, including rings, should be removed to prevent prolonged concentration of venom in the bitten area. Once under way, it's time to utilize the cell phone and, if available, a global positioning navigation unit to locate the nearest hospital and confirm the shortest route to it. Of equal importance is to call and alert the emergency room that a snakebite victim will be arriving and to make sure that this facility has enough antivenin on hand to fully treat the bitten person. First-aid measures en route such as cutting across the bite site, attempting to suck the venom out, or both, are no longer recommended, although they are often perpetuated by western movie reruns.

One former first-aid measure was to apply a tourniquet to the bitten limb between the bite and the body. Most bite venom enters the lymphatic system, which passes material to the venous system shortly before it reaches the heart. A "lymphatic tourniquet," which restricts only lymph flow, can reduce the effects of the venom during the period between bite and antivenin application if it is applied properly. However, that is rarely the case. A study by the Stanford University School of Medicine found that 20 lay volunteers who were given instruction in snakebite lymphatic tourniquet application were able to apply it correctly only 5 percent of the time during 500 trials. Inevitably either arterial or venous flow or both were restricted along with lymph flow. In an actual rattlesnake bite situation this would concentrate the venom in one area where severe tissue breakdown may then result. A new lymphatic tourniquet using a wide bandage that avoids blood vessel constriction has been tested in Australia on people bitten by snakes with neurotoxic venom that must be kept from reaching critical organs such as the heart. However, the hemotoxic venom of North American pit vipers can cause extensive tissue damage when confined to a small body area, and therefore tourniquets of any kind should not be used.

Bovine-derived antivenin serum, currently marketed under the trade name CroFab by BTG plc, is the only sure cure for envenomation, and it should be given in a fully equipped medical facility. Over the past decade fewer than 1 percent of bite victims treated in this manner have died, but 25 percent of those bitten who did not receive antivenin and proper medical care did die.

CONSERVATION NOTE: Given the preceding summary of what potential harm rattlesnakes can cause humans, a few words must be said about the harm many people continue to cause this group of reptiles. Rattlesnakes do have the potential for killing humans, but unlike other animals that occasionally kill people, such as a large shark, they themselves can be easily killed in return. In past decades this has gone far beyond the killing of a "nuisance rattler" that may refuse to leave a backyard area where children play. In many states "rattlesnake roundups" have become an annual event for the entire family. The sources of these snakes are primarily communal dens and burrows, into which gasoline is poured. Vaporization of the gasoline eventually results in the emergence of dazed rattlesnakes and all other creatures that may inhabit a burrow–den retreat. These "roundups" usually culminate in a "rattlesnake fry," the sale of skins, and a number of bites, the only protest that these snakes can register. Recently local conservation groups in states such as Kansas and Texas, where these events appear to be the most popular, are beginning to speak out for the recipient of this carnage whose only voice is its rattle.

So far California has had relatively few such events, possibly because its most abundant rattler, the Northern Pacific Rattlesnake (*C. o. oreganus*), occurs mainly in warmer regions in this state where prolonged low winter temperatures do not require deep communal denning for survival. This makes the "roundup" portion of such events far more work than most macho males who participate are apparently willing to endure. Despite this, California still maintains an open season on rattlesnakes with no "bag limit," regardless of species or protective status. Unlike the 58 species of reptiles that the CDFG allows to be taken from the wild by anyone who buys a freshwater fishing license, the 2011 *California Freshwater Sport Fishing Regulations* state that "No sport fishing license is required for the **sport take** of **any** rattlesnake." It fails to define the term "sport take," apparently leaving that definition open to each citizen of or visitor to California. Also overlooked is that one California species, the Red Diamond Rattlesnake (*Crotalus ruber*) is designated CSSC.

Fortunately, each year more and more Californians have come to realize that there is really no "sport" in killing a major rodent predator that is a key component of the ecosystem in which it resides. Unfortunately, the people who propose and pass on our sport fishing regulations are apparently not part of this group. These current CDFG regulations go on to state that "Rattlesnakes may be taken by any method," which apparently could include the pouring of gasoline down burrows. It should be noted that it is actually not the CDFG that used this language; instead, it is part of Title XIV, California Code of Regulations, and therefore it is our state representatives who vote on these regulations and allow this language to persist. Given such unfortunate statements, we could indeed see more rattlesnake roundups in the Golden State in the future—that is, as long as the "supply" lasts!

WESTERN RATTLESNAKE	*Crotalus oreganus*
GREAT BASIN RATTLESNAKE	*C. o. lutosus*
NORTHERN PACIFIC RATTLESNAKE	*C. o. oreganus*
SOUTHERN PACIFIC RATTLESNAKE	*C. o. helleri*

Northern Pacific Rattlesnake

TAXONOMIC NOTE: Ashton and de Queiroz (2002) concluded that based on mitochondrial DNA the former wide-ranging Western Rattlesnake (*Crotalus viridis*) should be recognized as two distinct species: *Crotalus viridis*, now called the Prairie Rattlesnake, and *Crotalus oreganus*, which now becomes the Western Rattlesnake. Douglas et al. (2002) offered further molecular analysis in support of this division and also recognized the California subspecies of the Western Rattlesnake as separate species: *Crotalus oreganus, C. helleri,* and *C. lutosus.* This taxonomy and nomenclature have been adopted by two other authors of recent UC Press natural history guides. However, we have chosen to maintain a conservative view and retain these taxa as subspecies of *C. oreganus.*

IDENTIFICATION: In California, the Great Basin Rattlesnake averages 61 to 101.6 cm (24 to 40 in.). Adults are lighter than the other two subspecies; coloration varies from buff or tan to pale yellowish or brown. The dorsal blotches often are light-centered, outlined by dark brown.

The Northern Pacific Rattlesnake is one of our larger rattlesnakes, with a total-length range of 38 to 165 cm (15 to 65 in.). It has brown or black dorsal blotches, usually with light-colored edges that give way posteriorly to crossbands. The general coloration is variable, but in most areas it usually harmonizes with the predominant soil color. Adults have well-defined dark tail rings, while the young have a bright yellow tail, a possible adaptation for caudal luring.

The Southern Pacific Rattlesnake has a smaller size range, 55.9 to 137.2 cm (22 to 54 in.), and is a relatively dark rattlesnake, with some individuals nearly black at higher elevations. The last tail ring on adults is poorly defined, and the young also have a yellow tail. Both the Southern and Northern Pacific Rattlesnakes are the only crotalids that usually have more than two internasal scales or granules touching the rostral scale.

RANGE: The Great Basin Rattlesnake occurs east of the Cascades and Sierra Nevada and throughout the Great Basin Desert. The Northern Pacific Rattlesnake occurs throughout the North Coast Range, all but the southern end of the South Coast Range, Cascade Range, the western slope of the Sierra Nevada, and the Great Valley. The Southern Pacific Rattlesnake occurs from the southern tip of the South Coast Range southward through Coastal Southern California, parts of the Transverse and Peninsular ranges, and western segments of the Mojave and Colorado deserts. It also inhabits Santa Cruz and Santa Catalina islands. Beyond California its range extends south to central Baja California, Mexico.

NATURAL HISTORY: Among the rattlesnakes, these closely related subspecies are the consummate habitat generalists. Within their geographic range, which encompasses most of California, they occur in such contrasting habitats as coastal sand dunes, prairies, desert-edge sites, montane forests,

Plate 110. Northern Pacific Rattlesnake.

and timberline areas. They are often found in rock outcrop areas, rocky slopes, rocky streamcourses, and rock ledge areas. They den in deep crevices provided by such sites and also in rodent burrows. They are also two of the most prolific rattlesnake species, with up to 25 young born from August to October. However, an average litter range is four to 12. Like most habitat generalists, they are also feeding generalists, with prey ranging from rodents and small rabbits to birds and their eggs, lizards, snakes, and amphibians. Newborn snakes also take large insects.

Surprisingly few prey items per year are needed to maintain an adult, at least in captivity. Professors of herpetology occasionally have pet reptiles thrust upon them by students, or more often their parents. Over two decades ago the junior author received a large, well-constructed terrarium from a student with a pleading note from his mother to "please give Jake a good home." Jake is a Southern Pacific Rattlesnake that was captured on Santa Catalina Island and eventually overstayed his welcome in the captor's home, and in a weak moment the gift was accepted. However, Jake turned out to be a superb performer, illustrating the venom-based prey capture method of white mice to countless herpetology and natural history classes. At an age of at least 25 years, Jake recently retired from teaching along with the junior author, but he still gets his "mouse a month," which is all that he will usually accept.

CONSERVATION NOTE: During the past several decades, human encounters with the Western Rattlesnake have not all been as amicable as that with the junior author and his longtime boarder. In areas such as the western portions of Kansas and Texas, *C. viridis* engages in communal denning and for decades has been a prime target of "rattlesnake roundups." It is especially vulnerable in these regions because of its high numbers in winter dens. Recovery from such wasteful collecting is also slow, primarily because of the relatively late reproductive maturity of females (four to five years of age). In 1972 the noted rattlesnake biologist Laurence Klauber reported on the drastic decline and in several cases complete extermination of *C. viridis* at some sites in Kansas as a result of annual "roundups" over many years.

Fortunately, this destructive activity has never really caught on to any great extent in California, even though the state's freshwater sport fishing regulations sanction the "sport hunting" of all rattlesnake species within its jurisdiction. However, in many areas of the Golden State there exists an ongoing threat to this and other species that, though far less obvious than "roundups," can be equally devastating. This is the ever-expanding suburban development in this state, which continues to invade prime Northern Pacific and Southern Pacific Rattlesnake habitats, especially in coastal scrub, chaparral, desert scrub, and foothill woodland plant communities. Unlike endangered or threatened species, these snakes have no protective status, and thus their presence on a proposed development site has no deterrent effect except perhaps for a warning to construction workers to be careful. Soon after people move into their new homes, stories often appear in the media about rattlesnakes invading a family's backyards, when in reality it's the family who have invaded the snake's backyard!

One solution to this problem is to evoke the wisdom of the old adage, "Good fences make good neighbors." Snake exclusion fences around the border of a developed area in known Western Rattlesnake habitat can be quite effective and can be constructed at relatively low cost. Compared with most other snake groups, rattlesnakes are poor vertical climbers, and a 1.0-m (39-in.) smooth-faced fence with the bottom one fourth buried below grade and a 15-cm (6-in.) top overhang on the side facing the natural area can keep a rattler from directly entering one's property. Annual weed control on the snake's side of the fence should be conducted to prevent "snake ladders" from breaching the barrier. Developers often produce a line of houses bordering a natural area, and in such cases a continuous barrier fence can be constructed between the human and crotalid communities. The comparatively small additional development cost for such an amenity might be easily offset by the added attraction to home buyers of a property that offers a "rattlesnake security system."

WESTERN DIAMOND-BACKED RATTLESNAKE *Crotalus atrox*

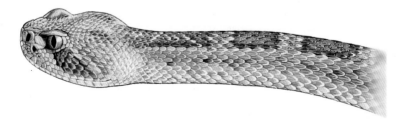

IDENTIFICATION: With an adult length range of 76 to 152 cm (30 to 60 in.), this is the largest rattlesnake species in the west and second only to the Eastern Diamond-backed Rattlesnake (*C. adamanteus*), for which a maximum length of 243.8 cm (96 in.) has been reported. This is often a relatively light-colored snake, with dorsal color ranging from yellow and pink hues to gray and brown. The "diamonds" on the dorsum are not always diamond-shaped but instead may be oval or hexagonal dark-brown or black blotches with light edges. There are also fainter and smaller blotches on the sides. The tail has very pronounced black-and-white or gray rings that are of approximately equal width, creating an effect that is often referred to as a "coon tail" because of its resemblance to that of a Raccoon (*Procyon lotor*).

RANGE: In California this snake is found only in the eastern portion of the Colorado Desert and the most southeastern tip of the Mojave Desert. However, it has an extensive range east of the Colorado River through southern Arizona and New Mexico, most of Texas, and Mexico.

NATURAL HISTORY: This largest of the six California rattlesnake species is also one of the most dangerous because of its bold defense and often its refusal to retreat when encountered. Because of its long-standing role as a "supporting actor" in countless western movies and novels, it is perhaps the best-known rattlesnake in the United States. Indeed, the term "diamond-backed" has become synonymous with rattlesnakes in general and is the only snake name to be adopted by a major professional sport team. Unfortunately, nearly all

Plate 111. Defensive posture of Western Diamond-backed Rattlesnake.

Plate 112. Western Diamond-backed Rattlesnake.

screenwriters and novelists, including such noted ones as James Michener, have given in to the temptation to demonize this snake, with the result that somewhere in the book or movie it is ultimately killed.

Desert scrub and riparian brush–woodland constitute its habitats in California, but to the east it can also be found in grassland, shrubland, woodland, and open pine forests. Mammals are its primary prey, and *C. atrox*'s large adult size may permit reduced competition with other rattlesnake species by allowing it to take larger prey such as ground squirrels, prairie dogs, rabbits, and hares. It is also diurnal as well as nocturnal, a feature that is very important for ambush-type predators of ground squirrel species that are also diurnal.

The Western Diamond-backed Rattlesnake is one of the most prolific crotalid species, with litter sizes ranging from four to 25 young. The large numbers of young for brief periods in some areas have given the proponents of "rattlesnake roundups" the false idea that the supply of this snake is limitless, and like similar assumptions concerning formerly abundant animals like bison over a century ago, that the supply will last forever. That belief is rapidly proving false; the only real question now is whether entire populations in some western states can ultimately survive this annual, destructive onslaught.

MOJAVE RATTLESNAKE *Crotalus scutulatus*

IDENTIFICATION: This is a medium-size rattlesnake, with an adult-length range of 61 to 129 cm (24 to 51 in.). The dorsal background color varies from gray to shades of brown or yellow, and olive-green. The latter coloration is prevalent in California. The diamond or hexagon patterns on the back are dark gray to brown and edged in a light hue of the background color. A light stripe, somewhat more distinct than in other California rattlesnakes, extends from behind the eye to a point behind the corner of the mouth. The tail has relatively bold contrasting light and dark stripes, but unlike the "coon tails" of *C. oreganus* and *C. atrox*, the dark rings are narrower than the light ones. There are usually only two large internasal scales between the supraocular scales and larger scales on the top of the snout than in other rattlesnakes.

RANGE: True to its name, *C. scutulatus* occurs only on the Mojave Desert in California, but it extends southeast from that point through southwestern Arizona and on across much of Mexico.

NATURAL HISTORY: This is a rattlesnake of barren desert areas. It also occurs in annual grassland, scrubland, and open juniper woodland. Areas of scattered scrubby brush and mesquite are preferred habitats, but it tends to avoid areas of dense vegetation and broken rocky terrain. Like most other rattlesnake species, its prey ranges from desert rodents, rabbits, and small hares to lizards, snakes, birds, and bird eggs. It gives birth to anywhere from two to 17 young between July and September.

This species is of special interest to many, because some populations have both a hemotoxic and neurotoxic venom. Unlike hemotoxin, which destroys the oxygen-retaining capacity of blood and muscle, causing the "suffocation" and eventual breakdown of tissue where it is concentrated, the neurotoxic element affects bioelectrical conduction at the nerve cell and nerve cell–muscle cell synapses. It is therefore similar to the venom of coral snakes and thus far more dangerous to humans. In the rare incident when someone is bitten by a rattlesnake on the Mojave Desert, it is important to get the best possible description of the snake so that the proper antivenin can be administered. Fortunately, the range of the closest look-alike species, the Western Rattlesnake, may only slightly overlap with this species where some presumed hybrids have been found. That of the Speckled Rattlesnake (*C. mitchellii*) does, but the latter does not have the bold diamond markings of the Mojave Rattlesnake, and the range of the Western Diamond-backed Rattlesnake (*C. atrox*) barely overlaps.

As to why some rattlesnake species should have neurotoxic populations within its species complex, one can only speculate. One guess is that for an ambush hunter in desert habitats where the majority of rodent prey species are burrow dwellers, the use of a neurotoxin, which has a higher probability of immobilizing prey before they reach deep burrow retreats than a hemotoxin, may have been a significant selective advantage.

| SPECKLED RATTLESNAKE | *Crotalus mitchellii* |
| PANAMINT RATTLESNAKE | *Crotalus stephensi* |

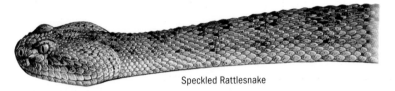

Speckled Rattlesnake

IDENTIFICATION: Its length range is 58 to 132 cm (23 to 52 in.), and the dorsal color varies greatly, harmonizing with that of a given habitat's substrate. This range includes hues of cream, yellow, pink, tan, gray, brown, or in some lava flow areas, black. The dark dorsal markings are often vague in comparison with the well-defined hexagon or diamond patterning in other California crotalid species. In some areas specimens have abundant fine speckling, and this along with a rough scalation creates a picture of decomposed granite. There are dark rings on the tail but no bright white bands as in *C. atrox*. In the southern half of the Mojave Desert, the Colorado Desert, and Peninsular Ranges, the prenasal scales are usually separated from the rostral by small scales (subspecies *C. m. pyrrhus*). The Panamint Rattlesnake lacks these small scales, and the supraoculars are often pitted or furrowed or with irregular outer edges. The dark dorsal hexagon or diamond patterning that is often vague in the Speckled Rattlesnake is more regular and better defined by light colored edges.

TAXONOMIC NOTE: Douglas et al. (2007) presented mitochondrial and nuclear DNA evidence supporting the recognition of the Panamint Rattlesnake as a distinct species, *Crotalus stephensi*. Meik (2008) presented morphological evidence, also concluding separate species status.

RANGE: The Speckled Rattlesnake occurs in the southern portion of the Mojave Desert, the Colorado Desert, the Peninsular Ranges, and Southern Coastal California. The Panamint Rattlesnake is found in the northern segment of the Mojave Desert.

NATURAL HISTORY: These are primarily snakes of the rocky uplands, but they may also occur in loose soil and sand areas populated by creosote brush, sagebrush, thorn-scrub, and desert succulents. In areas of higher rainfall they occur in chaparral and piñon–juniper woodland. Their general behavior is much like that of the Western Diamond-backed Rattlesnake (*C. atrox*), in that they tend to hold their ground

Plate 113. Speckled Rattlesnake in shade retreat.

when approached, especially when in large-rock areas. Their prey includes the usual spectrum of small desert mammal species plus lizards and birds. Compared with other California species, the Speckled Rattlesnake and the Panamint Rattlesnake produce a relatively small litter of from two to 12 young in July or August.

RED DIAMOND RATTLESNAKE *Crotalus ruber*

PROTECTIVE STATUS OF *C. RUBER*: CSSC

IDENTIFICATION: This is a larger reddish version of the Western Diamond-backed Rattlesnake (*C. atrox*), with a length range of 76 to 165 cm (30 to 65 in.). Dorsal hues range from red-brown to pink or tan, and it has a boldly marked black-and-white "coon tail" like *C. atrox*. The red-brown dorsal diamonds have light edges and are therefore quite apparent. The young are dark gray at birth but acquire a reddish brown color as they mature.

RANGE: *C. ruber* occurs only in Coastal Southern California, the Peninsular Ranges, and the western segment of the Colorado Desert within California. It is also found throughout Baja California, Mexico.

NATURAL HISTORY: This rattlesnake is found in plant communities common to the southwestern corner of California. These include chaparral, coastal sage, desert scrub, woodland, and sometimes grassland and agricultural areas. On the desert slopes bordering the Colorado Desert it may often range well out on the desert floor. Although its appearance is much like that of *C. atrox*, its behavior, at least with respect to humans, is quite different. This is a relatively docile rattlesnake, far less aggressive than the Western Diamond-backed. It is quite secretive and will usually retreat rather than maintain its ground when encountered. As with the Western Diamond-backed Rattlesnake, its large size apparently allows adults to feed on relatively large prey such as ground squirrels and rabbits. It also takes lizards, birds, and occasionally carrion. From three to 20 young are born from July through September.

CONSERVATION NOTE: The Red Diamond Rattlesnake has the smallest California range of all of the state's rattlesnake species, and over half of that range includes the heavily populated areas of San Diego and Riverside counties and all of Orange County. It has been all but eliminated from the densely populated coastal segments of this range, and as land development continues to push eastward, more habitat will surely be lost. This situation more than justifies its listing by the CDFG as a Species of Special Concern. However, this is apparently only "paper protection," because this same state agency's current (2011) freshwater fishing regulations state that no license is required for the "sport take" of **any** rattlesnake.

SIDEWINDER *Crotalus cerastes*

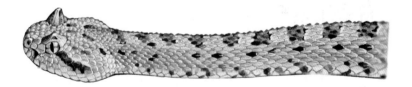

IDENTIFICATION: This is our smallest California rattlesnake, with an adult-length range of 43 to 84 cm (17 to 33 in.). When seen in the field, its unusual sidewise form of locomotion is one of the best identifying features. The dorsal color may be various pale hues of cream, pink, tan, or gray, interrupted by blotches of yellow-brown or tan along the dorsal midline. Two distinct head features are a dark, diagonal stripe below the eye and hornlike supraocular scales. These are flexible and may fold down over the ocular scale when the snake enters burrows.

RANGE: This is a true desert species, occurring almost exclusively in the Mojave and Colorado deserts in California.

NATURAL HISTORY: The Sidewinder is a rather appropriate snake with which to end the serpent section of our book, because its name is familiar to most people. A defensive ground-to-air missile bears its name, along with numerous desert-edge cafes and bars, and what would a "B" western movie be without one actor drawing his six-shooter and calling another one a "dirty sidewinder?" The real Sidewinder is a desert habitat specialist, well adapted to sand–brush flats, sand hummocks, and barren dunes. Its sidewinding locomotion permits fast movement in open terrain, even on soft substrate. By continuously changing the small area of ventral body surface that touches the hot midday desert substrate, it can forage for diurnal prey such as lizards without quickly overheating (see figure 32). However, it is mainly nocturnal and spends much of the day in rodent burrows or coiled in a shallow circular pit that it constructs in shade at the base of a desert shrub. During this process the soft substrate shifts over the snake as it burrows, and in this partially buried state it is often essentially invisible to the untrained eye. This trait, referred to as "cratering," is yet another reason not to step or reach within the shade canopy of desert shrubs, even when no snake is visible. In soft-sand areas, it is often possible to follow the J-shaped track left by a Sidewinder the evening before. The hook of the J points to the direction of travel and may lead the tracker to its shallow pit retreat or a kangaroo rat burrow where it is escaping the heat of the day.

Plate 114. Highly visible Sidewinder on open ground.

Plate 115. Barely visible Sidewinder in desert brush.

Plate 116. Sidewinder leaving its reverse "J"-shaped track.

Adult Sidewinders feed on the smaller array of desert rodents such as kangaroo rats and pocket mice, and lizards, birds, and their nestlings round out its prey list. Young Sidewinders often engage in caudal luring by slowly waving their tail with its buttonlike rattler to entice prey such as lizards close enough for a strike. Two to 18 young are born from July to September.

Tortoises and Turtles (Order Testudines)

Land Tortoises (Family Testudinidae)	426
Desert Tortoise (*Gopherus agassizii*)	426
Pond Turtles (Family Emydidae)	430
Western Pond Turtle (*Actinemys marmorata*)	430
Pond Slider Turtle (*Trachemys scripta*)	435
Western Painted Turtle (*Chrysemys picta*)	437
Softshell Turtles (Family Trionychidae)	439
Spiny Softshell Turtle (*Apalone spinifera*)	440
Snapping Turtles (Family Chelydridae)	442
Common Snapping Turtle (*Chelydra serpentina*)	443
Sea Turtles (Family Cheloniidae)	448
Green Sea Turtle (*Chelonia mydas*)	448
Black Sea Turtle (*Chelonia agassizii*)	448
Loggerhead Sea Turtle (*Caretta caretta*)	451
Olive Ridley Sea Turtle (*Lepidochelys olivacea*)	452
Hawksbill Sea Turtle (*Eretmochelys imbricata*)	453
Leatherback Sea Turtle (Family Dermochelyidae)	454
Leatherback Sea Turtle (*Dermochelys coriacea*)	454

Key to the Species of Tortoises and Turtles

Compared to many lizard and snake species, turtle coloration is rather drab, and therefore physical features such as the shape and position of the epidermal shields of the carapace or their absence along with limb anatomy are usually used to differentiate families, genera, and species.

1a Forelimbs paddlelike, elongate, and flattened; 0-2 claws on each limb. 7a
1b Forelimbs not paddlelike; three or more claws on each foot 2a

2a Carapace flattened and without epidermal shields; three claws on each foot . . . . Spiny Softshell Turtle
2b Carapace covered with epidermal shields 3a

3a Hind legs stumplike; hind feet not webbed. . . . . Desert Tortoise
3b Hind legs not stumplike; hind feet with some webbing. 4a

4a Carapace with raised ridges on midline shields; highly reduced plastron. . . . . Common Snapping Turtle
4b Carapace with relatively smooth midline shields 5a

5a Head and neck without yellow or red markings. . . . . Pacific Pond Turtle
5b Head and neck with yellow and/or red markings. 6a

6a	Head and neck with yellow longitudinal stripes; reddish oval behind eye; concentric yellowish lines on epidermal shields..	Pond Slider Turtle
6b	Head and neck with yellow longitudinal stripes but no reddish oval behind eye; no yellow striping on carapace; reddish carapace edge and plastron...................................	Painted Turtle
7a	Claws absent; back covered with dark, leathery skin..	Pacific Leatherback Turtle
7b	One or two claws present on each limb.......................	8a
8a	One pair of prefrontal (between eyes) scales; usually only one claw on each limb.....................................	Green Sea Turtle
8b	Two pairs of prefrontal (between eyes) scales; two claws on each limb...	9a
9a	Overlapping, shinglelike carapace shields; hawklike mandibles..	Hawksbill Sea Turtle
9b	Nonoverlapping carapace shields; no hawklike mandibles	10a
10a	Carapace usually red to orange brown with ridged midline shields..	Loggerhead Sea Turtle
10b	Olive-brown carapace; long rectangular lateral (costal) carapace shields...................................	Olive Ridley Sea Turtle

About 175 million years ago in the Jurassic Period, this group of reptiles acquired through the natural selection process a structure that has been so successful that their basic body form has changed little since that time. That structure is a shell composed of two parts: an upper rounded carapace and a flatter ventral plastron, connected on each side by a bridge. It accrued to this group a universal mode of protection that has effectively deterred predators from the age of reptiles through the advance of the mammals and birds. There are currently about 335 living species of turtles arranged in 12 families. They occur in both freshwater and salt water, and some are completely terrestrial.

The shell is composed of a series of sutured bony plates overlaid by horny epidermal shields in a definite pattern. Its form is quite variable depending on the species group. Some turtles have the bony shell reduced in thickness or in number and size of the plates, whereas others lack the horny epidermal shields altogether and instead have a shell covering of leathery skin. In many species the head, neck, limbs, and tail can be retracted beneath the shell edge, and in the box turtles a plastron hinge permits complete enclosure of all appendages within a sealed shell. Growth rings are present in the shields of some species, but these rings seldom can be relied upon to determine age, except during early growth stages.

Most turtles have somewhat flattened limbs, with five toes on both the front and hind feet. Usually all the digits bear claws except the inner one of the hind foot. In marine turtles the limbs are modified as flippers, and claws are reduced in number or absent. In the gopher tortoises the hindlimbs are elephantoid and the toes are short.

Male turtles usually have longer tails than females and they may be partially prehensile. A single copulatory organ lies in the anterior wall of the cloaca and when erected protrudes through the vent. One way to sex turtles is to pull the tail straight back. In males the vent will be situated beneath or beyond the edge of the carapace. In females the vent usually lies within the carapace margin. In most species males also have a concave plastron, while females have a flat or slightly concave one. These variations accommodate the dorsal mounting position of the female by the male.

All turtles are oviparous and lay their eggs on land in substrate nests where a moderate amount of soil moisture is present. Thus, sea turtles, which we often think of as completely marine, begin their life on land, and the females must come back to rookery beaches a fair number of times during their lifetime. Although the shell affords adult turtles effective protection, it is quite soft in hatchlings and yearlings, and thus vulnerable to the jaws and beaks of predators. This condition, plus their small size, results in high mortality rates for the young in most species, which is offset by large clutch size and/or great longevity.

Turtles have very good vision. Tests have shown that some species have a considerable power of pattern discrimination and are able to detect differences between lines varying by only one millimeter in thickness. The eyes have both lids and a nictitating membrane, and when closed, the lower lid moves up to meet the upper one. The sense of smell appears to be moderately well developed, and the Jacobson's organ is present and located in the ventral portion of the nasal passage. Both sight and olfaction (and taste in aquatic turtles) are used to detect the presence of food. Most turtles are omnivores, but some are completely herbivorous. Turtles lack teeth. Instead the jaws are covered with a horny sheath with which they cut, slice, or otherwise break up food and then swallow the pieces.

Hearing in turtles is usually well developed, but the tympanic membrane is often hard to detect in many species. Tests have shown they respond to sound frequencies up to 3,000 kHz. This frequency range is just audible to humans. On several occasions the junior author has attempted to maneuver his canoe close to a basking aquatic turtle whose eyes were tightly shut, only to have the slightest click of a paddle against the gunnel send the "sleeper" into the water. Land tortoises often vocalize when mating, but most other species are mute.

Land Tortoises (Family Testudinidae)

Tortoises currently occur on all continents except Australia. A number of species were once found there also, including one about the size of a car, up to around 50,000 years ago when the first humans arrived. These giant tortoises and similar forms were once widespread and abundant. Some 50 species have been described from Oligocene and Miocene fossils. Most died out in the Pliocene, but even in historic times giant tortoises occurred throughout the islands of the western Indian Ocean and the Galapagos Islands. As with the Australian giant tortoises, they disappeared from most islands with the arrival of humans who brought fire and competing herbivorous livestock, and introduced predators. Today only small numbers remain on the more inaccessible and remote islands or where features of terrain and vegetation provide special protection. The living remnants are *Dipsochelys gigantea* of the Seychelles and Aldabra islands in the Indian Ocean and the *Geochelone* species of the Galapagos Islands.

These two species and about 37 other living smaller forms are robust tortoises with high, domed shells, which range in diameter from 12.7 cm (5 in.) to nearly 1.5 m (5 ft). The hindlimbs are elephant-like, and the feet are stumpy with unwebbed, short digits. The anterior surface of the forelimbs and dorsal head surface are covered with armorlike horny scales. These limbs are folded in front of the withdrawn head when the tortoise is disturbed, sealing off the front of the shell. In addition to a concave plastron, males have a long projecting gular horn on the plastron beneath the throat area.

All tortoises are land-dwelling and herbivorous. They are also long-lived, especially the very large species. There is an authentic record of a *Dipsochelys* living over 150 years.

DESERT TORTOISE *Gopherus agassizii*

PROTECTIVE STATUS: FT, ST

IDENTIFICATION: The Desert Tortoise is the only naturally occurring living tortoise-type turtle in California. It has a dome-shaped carapace with an adult-diameter range of 15 to 38 cm (6 to 15 in.) and shields that show pronounced growth rings, except in very young or old individuals with worn shell surfaces. The number of rings increases with time, but a ring count is usually not a reliable indication of age in years. The carapace is dull brown or horn color, with individual shields usually light brown or yellowish in the center, especially in younger animals. Juveniles have a soft, almost round shell, with a diameter of about 4 cm (1.5 in.) at hatching. Shell ossification occurs slowly, with hardening becoming pronounced about the fifth or sixth year.

RANGE: In California its range coincides closely with that of the Mojave Desert and the part of the Colorado Desert south of the Transverse Range and east side of the Salton Trough. It is absent from the Coachella Valley except from the Boyd Deep Canyon Research Center area. The historical situation in this valley may be confusing because of escaped or released pet tortoises.

NATURAL HISTORY: The Desert Tortoise is an arid land species that requires firm but not hard ground for burrow construction. Like many other desert vertebrates, the ability to dig a burrow well into the cool subsurface layer of the desert floor is its key to existence. Where soil

Plate 117. Desert Tortoise on a foraging walk.

LAND TORTOISES (TESTUDINIDAE)

conditions are incompatible with this behavior, it usually cannot exist, although it may sometimes use deep rock shelters or large eroded crevices in the banks of washes as retreats. Burrows are often constructed at the base of bushes where the root system gives added support to the soil. The opening is round above and flat on the bottom, a shape that accommodates the cross-sectional contour of the shell. The burrow depth for adults ranges from 1 to 2.8 m (3 to 9 ft) in most areas, although shafts up to 9 m (about 30 ft) have been recorded. In some desert areas only the American Badger (*Taxidea taxus*) constructs a similar-shaped burrow entrance, but an active tortoise burrow usually has large fecal pellets laden with vegetative matter in the vicinity of the entrance and is never plugged as are some Badger entrances.

This species also requires adequate ground moisture to prevent the desiccation of buried eggs and sequestered young, plus an annual crop of desert grasses and forbs for food. Where these requirements are met, the Desert Tortoise can be found in washes, sand and gravel flats, rocky hillsides, canyons, oases, and areas of fine windblown sand.

The junior author and former graduate student Bill Voigt tracked and recorded Desert Tortoise activity and body temperature via telemetry in the western Mojave Desert during the late-spring period. At this time of year, a tortoise begins the day by emerging and basking at the mouth of its burrow. The common basking position is facing toward the sun with the neck and forelimbs fully extended in such a way as to expose bare skin in the neck region. Sub-carapace thermisters revealed that the shell is a relatively good insulator, and thus for rapid warming the exposure of skin capillaries in this manner is most efficient. Once the deep-body area is warmed to around 30 degrees C (86 degrees F), basking ceases and the tortoise begins a long, more-or-less circular foraging walk during which it grazes on small grasses and forbs, especially those with flowers or seed heads. The length and time of this walk depends on the warmth of the day, and its path eventually brings it back to the burrow as ground surface temperatures pass into the range of around 40 degrees C (104 degrees F). The tortoise then retreats until late afternoon, when decreasing surface temperature permits a second but usually shorter foraging walk. As the desert continues to cool near sundown, the tortoise retreats for the night. This activity pattern is varied according to the season. In early spring there is usually only one foraging walk during the midday period. During an exceptionally hot midsummer period, emergence may not occur for its duration or tortoises may be active at night. Hibernation normally occurs during the midwinter period, especially in the northern Mojave.

In one experiment deep-body temperatures were recorded from ingested miniature transmitters in free-living tortoises and also from two others that were maintained in a desert enclosure but were not allowed to retreat into a burrow at night. The core temperature of the latter group declined rapidly at dusk, while that of tortoises in burrows slowly decreased throughout the evening and never did reach the surface substrate lows. The insulating effect of the burrow, which permits this retention of body heat well into the night, may be important to the digestive process of this herbivorous ectotherm.

Although free water in the habitat is not essential, the Desert Tortoise will drink deeply when water is available. Drinking usually occurs at rain pools, some of which are shallow depressions constructed by the tortoise. It has been suggested that this behavior may be triggered by a rise in the normally low relative humidity of desert air that often occurs before a rainstorm. One captive individual increased its weight over 40 percent in one long drink. As you might assume from such an intake, the urinary bladder is large, and urinary wastes are stored for considerable periods. These are largely in the form of uric acid as in other reptiles. It is relatively insoluble and occurs as a granular deposit in nearly pure water. When first picked up in the field, tortoises commonly release urine from the bladder, especially if turned upside down. The fluid is nearly odorless and often a tea color. It has been suggested that a person dying of thirst on the desert might obtain a lifesaving drink if he or she could find a Desert Tortoise!

The male *G. agassizii* is somewhat larger than the female and has a longer gular horn, a longer tail, and more massive claws. Like other male turtles, the plastron is concave. Males also have larger chin glands than females. These glands probably serve as olfactory and visual recognition cues during courtship and/or combat. The male courts with his head extended and moves it up and down. He nips at the female's nose, front feet, and shell edge, and may lunge at her with his gular horn. After all this he mounts and copulates, a process accompanied by hissing, puffing, or grunting noises. The male's moderately prehensile, fingerlike tail is used to manipulate the female's vent region to ensure penetration and to pull back on the underside of her shell to prevent sideways movements that might dislodge him.

The nest cavity is dug by the female with her hindlimbs and may be up to 25 cm (10 in.) deep. Urine is often released while digging to help soften the soil. Tortoises often compact the nest with their feet and then urinate on top of it. Experiments have shown that the urine has a repelling effect on resident predators like the Coyote (*Canis latrans*) and the Kit Fox (*Vulpes macrotis*). The white, hard-shelled eggs are oval, sometimes flattened on one side, and about 45 mm (1.7 in.) long. Clutches range from one to 15 eggs, with an average of five, and two or three clutches may be laid in one year. June appears to be the peak egg-laying month, and hatching occurs from mid-August to October. During low-rainfall years when forage is scarce, nesting may not occur. Hatchling mortality is apparently very high, as indicated by the extreme rarity of encountering one in the field, even in areas where adults are relatively abundant. Their small size and soft shells make them easy prey for a wide range of avian, mammalian, and reptilian predators. This poor survival of young is offset by the long life of the adults, which may be capable of living a century or more.

CONSERVATION NOTE: The Desert Tortoise is the official state reptile of California, but like some other state animals, we have not given it the respect that this title deserves. Our official state mammal, the Grizzly Bear (*Ursus chelan*), has long since been exterminated in this state, and the state freshwater fish, the California Golden Trout (*Oncorhynchus mykiss aguabonita*), has had its original genetic makeup substantially altered by hatchery breeding and the introduction of hybridizing subspecies.

In the case of the Desert Tortoise, numerous human-caused impacts have greatly reduced its numbers during the past several decades. These include urban, suburban, and military reservation expansion, more on- and off-road vehicle traffic in desert areas, road development, mining, overgrazing, and the renewal of desert land grazing leases (sheep and cattle eat the same food as tortoises), the increase of a major hatchling predator, the Common Raven (*Corvus corax*) due to expanding garbage dumps and more electric pole perching sites, irresponsible shooters, and the spread of disease from released captive individuals. An additional threat is the spread of invasive exotic annuals, such as Red Brome (*Bromus madritensis rubens*), which have altered natural fire regimes in our deserts. A tortoise reserve has been set aside by the Bureau of Land Management in the western Mojave Desert, but its future is uncertain because of increasing human impacts adjacent to that area. More fully protected reserve land is greatly needed if we are to save this gentle reptile whose ancestral line predates the dinosaurs.

Pond Turtles (Family Emydidae)

WESTERN POND TURTLE	***Actinemys marmorata*** ***(Clemmys marmorata)***
NORTHWESTERN POND TURTLE	*A. m. marmorata*
SOUTHWESTERN POND TURTLE	*A. m. pallida*

PROTECTIVE STATUS FOR *A. M. MARMORATA* AND *A. M. PALLIDA*: CSSC
DESCRIPTION: The carapace of adults has a low crown and ranges in length from 8.9 to 21.6 cm (3.5 to 8.5 in.). It is a plain olive, brown, or black-

ish color with usually a network of dark spots, lines, or dashes that often radiate out from the centers of the shields. The plastron has six pairs of yellowish shields with dark blotches. The head and neck usually have dark spots or a network of irregular dark lines but do not contain fine yellow stripes or a broad red swath. Young have a solid brown or olive carapace with yellow at the edge of the marginal shields, and the plastron is yellow with a large central black figure.

RANGE: It occurs from Washington south through the Cascade Range, North and South Coastal ranges, Coastal Southern California, and the western Sierra Nevada foothills. Small, isolated populations occur near Susanville in Lassen County and in Afton Canyon and Camp Cody along the Mojave River in the Mojave Desert. There is a broad zone of intergradation between the northern subspecies (*A. m. marmorata*) and the southern one (*A. m. pallida*) in central California. The latter now borders on extinction in the southern San Joaquin Valley, the area where it was historically extremely abundant, and in Coastal Southern California.

NATURAL HISTORY: In contrast to many southeastern states such as Florida and Alabama, which have about a dozen species each of freshwater pond-type turtles, California has only one, the Western Pond Turtle. It is a fully aquatic species that centers its activities around relatively quiet water habitats such as ponds, lakes, marshes, broad rivers, and irrigation ditches with mud or rock basins and some aquatic vegetation. It may also occasionally be found in deep-pool areas of rapidly flowing streams and may even enter brackish or full seawater for brief periods. The terrestrial habitats surrounding these sites range from grassland and cropland to open forest.

Plate 118. Western Pond Turtle.

Basking sites are an essential part of this turtle's habitat and include open bank areas, partially sunken logs, and emergent vegetation mats. The junior author used telemetry to document the daily basking behavior of free-living *A. marmorata* adults in a cool water habitat in late spring. Basking commenced a couple of hours after sunrise and lasted until early afternoon. Each basking period lasted from 20 to 35 minutes, during which time the turtles' deep-core body temperature rose from near that of the pond water to the range of 30 degrees C (86 degrees F). In the absence of any disturbance the turtles would then slip off the log basking site and forage for about 10 minutes, at which time their body temperature had declined to near 20 degrees C (68 degrees F). They would then haul out again and repeat the basking sequence, usually with the neck and skin of one forelimb positioned toward the sun. This type of periodic warming most likely increases foraging efficiency in cool water and perhaps aids in digestion as well. For the field herpetologist it provides an opportunity to view this species. Indeed, the best way to determine if it is present or absent at an aquatic habitat is to view what appears to be good basking sites throughout the mid- to late-morning period on a sunny spring or early-summer day. Your observation point should be well removed from the potential basking area, and your movements kept to a minimum.

In many areas of the state this turtle hibernates during much of the cool winter season. Hibernation most often takes place in the soft bottom substrate of aquatic habitats, during which time it obtains oxygen through its highly vascularized pharyngeal and cloacal linings. At those sites where the aquatic substrate is not appropriate for hibernation, this turtle utilizes the burrows of the California Ground Squirrel (*Spermophilus beecheyi*) for winter retreats, and sometimes such sites are a considerable distance from their wetland habitat. The potential for short fall and spring inland migratory movements of the Western Pond Turtle to and from burrow hibernacula should be taken into consideration when defining the total habitat requirements for a specific population.

Like most pond-type turtles, this species is an omnivore, with foods ranging from aquatic vascular plants, earthworms, and insects to amphibian eggs and larvae, crayfish, and carrion. From three to 14 eggs are laid either along the aquatic habitat shore or some distance from water in open hillside or field sites, as long as appropriate soil moisture and warmth are present. Clutches are laid between April and August, depending on the latitude and elevation of a habitat. As with many other turtle species, hatchlings and very young juveniles are rarely observed in the field, which indicates that the mortality in this group is high. The length of a hatchling's shell is about 25 mm (1 in.), and it is soft and pliable. These features make young turtles vulnerable to nearly all vertebrate predators that forage in and near aquatic habitats.

One other explanation for the lack of juvenile turtle sightings is that the young may rarely leave the water to bask and therefore go unnoticed.

One summer while surveying for amphibian larvae with a seine net at a San Francisco Garter Snake (*Thamnophis sirtalis tetrataenia*) pond feeding site, the junior author captured several half-dollar-sized Western Pond Turtles in the first seine haul. During the prior months of daily observations of snakes and frogs at this pond, only adults turtles were seen. These young were apparently leading a completely aquatic existence, and their tiny heads went unnoticed when they occasionally protruded through the floating surface aquatic vegetation for breathing.

CONSERVATION NOTE: The Western Pond Turtle was very abundant throughout its range up to the late 1800s. It is estimated that in the southern part of the San Joaquin Valley alone the population was over 3.5 million. It was the only aquatic freshwater turtle in California, and as such faced no competition for that broad ecological niche. It thrived in habitats like the now-extinct Tulare Lake, which occupied an area some 100 km (60 mi) long and 50 km (30 mi) across at its greatest width. From the mid-nineteenth century through the 1920s this massive shallow lake supported a turtle "fishery" that supplied "terrapin" to San Francisco restaurants, and one could buy Western Pond Turtles for as little as 25 cents a dozen in that city's markets. For many centuries before this the Western Pond Turtle was a major food source for the Native American tribes, which depended heavily on the Great Valley wetlands for much of their food.

The initial reduction of this species by commercial harvesting was greatly expanded by the massive Great Valley wetland drainage projects of the first half of the twentieth century, which included the drainage of Tulare Lake. The most grandiose of these, the Central Valley Project, converted marshes and shallow lakes to crop fields and confined river channels within levees so that the newly drained agricultural land would not be reflooded. It resulted in major negative impacts to countless native, shallow-water animal species, including two endemic fishes that are now presumed extinct. In most areas it left only canalized rivers and open irrigation ditches as refuges for a limited number of surviving Western Pond Turtles. However, at this writing many of these are being replaced by concrete-lined channels and buried pipelines, both of which are useless to turtles.

During the past several decades, a third and equally devastating impact to the Western Pond Turtle has been developing. Two eastern species of pond turtle, the Red-eared Slider (*Trachemys scripta elegans*) and, to a lesser extent, the Western Painted Turtle (*Chrysemys picta*) continue to be randomly introduced by pet turtles released into many of the remaining Western Pond Turtle habitats in this state. They appear to be thriving in these habitats and seem to have an ecological niche quite similar to that of our native species. They are also nearly twice as fecund, with maximum clutch sizes of 25 eggs as compared to 14 for the Western Pond Turtle. They come to the pond turtle habitats of California from aquatic sites east of the Sierra Nevada that they share with one or more other pond turtle species. In these home habitats they have acquired a number of behavioral traits such as the ability to utilize a wide variety of foods and nesting sites, which have permitted successful competition with their neighbors. In contrast to this situation, our one native pond turtle has not faced competition with similar species and therefore has remained comparatively unspecialized and therefore vulnerable to competitive encounters with these introduced forms. In addition to such basic competitive superiority, there is also a further risk of introducing diseases to which the native turtle is not immune and therefore may succumb.

These nonnative species have been available to the public for many years in pet and aquarium stores. While preparing this portion of the book the authors visited several such stores in the Bay Area, and all offered one or both species for sale, with some carrying both juveniles and adults. Turtles are popular pets, partly because they are relatively easy to keep. This feature, coupled with their long life span sets the stage for the introduction problem. Many pet turtles begin their tenure with a family as a child's pet. Eventually the kids grow up and go off to college, or the turtle outgrows its aquarium and/or its initial welcome. In all cases the end result is almost always its release in the nearest aquatic habitat, as seen by the presence of both species at most freshwater sites in the heavily populated areas of Southern Coastal California.

In the northern part of the state some very unfortunate introductions have occurred in the Sacramento-San Joaquin Delta, and these nonnative turtles are now in the process of spreading throughout this massive, interconnected waterway. Several years ago the junior author was canoeing in the lower Stanislaus River near his farm in south San Joaquin County and decided to visit his favorite Western Pond Turtle viewing site in an oxbow pond off the main stream. As he glided the canoe into this area, he noted there were several distinct bumps on the exposed portion of a large half-submerged log some distance away, just where he had always found Western Pond Turtles basking in the past. However, as he began viewing them with field glasses, a disturbing difference was soon noted. One of the adult turtles had a wide red stripe across its ear area!

The most logical solution to this problem would be to ban the sale of live, nonnative turtles in California, a measure that has already been taken with many other potentially invasive animal species. If such an action would be viewed by state lawmakers as a denial of a citizen's rights to have a pet turtle, then an alternate plan would be to have a state-sanctioned Western Pond Turtle hatchery in much the same way that we have over two dozen state-operated trout and salmon hatcheries in California. The Western Pond Turtle would then be declared the only legal turtle to sell in this state, with the result that many citizens would eventually take part in re-introducing this protected species.

One further way to reduce and eventually end the sale of pet turtles in California is to make the buying public more aware of the health risk that such pets pose, especially to the children for whom they are usually purchased. Pet store turtles are often carriers of *Salmonella*, which then infects their owners through hand-to-mouth transmission after contact with the turtle or the water in which it is kept. Young children may also be directly infected by kissing "baby turtle." While *Salmonella* is rarely fatal, it does cause digestive disturbances, diarrhea, and fever, especially in children. A short but effective mass media campaign to inform Californians of this problem might well be the quickest way to curtail the foreign turtle introduction problem while there is still time to preserve some remaining pure Western Pond Turtle habitats.

POND SLIDER TURTLE   *Trachemys scripta*
INTRODUCED

IDENTIFICATION: This turtle has a large carapace length range of 8.9 to 36.8 cm (3.5 to 14.5 in.). It is the only pond turtle currently in California with vertical yellow streaking on the costal shields. The subspecies that has been introduced into California, the Red-eared Slider (*T. s. elegans*) has a broad red stripe behind the eye and "eyelike" spots, usually symmetrically arranged, on the plastron. The carapace of the young is green with yellow streaks, and there is usually a red or yellow streak behind the eye. This is the size and color form most often sold in pet stores.

RANGE: This species is widespread throughout the southern Midwest and eastern United States. In California it now occurs in most freshwater habitats in Coastal Southern California. It has also been introduced into several areas of Santa Barbara County, the San Francisco Bay Area, and the Sacramento-San Joaquin Delta and connecting drainages.

NATURAL HISTORY: Like several species of turtles of the southeastern United States, this is a habitat generalist. It is at home in a wide variety of freshwater aquatic habitats including lakes, ponds, sloughs, slow-flowing rivers, canals, and irrigation ditches, as long as there are good aquatic plant stands in relatively quite waters with good basking sites. This is a highly aquatic species, seldom venturing far from water, an observation that is supported by the frequent presence of algae on the shells of larger individuals. It is also an omnivorous feeding generalist, eating a wide variety of aquatic plants, aquatic insects and their larvae, crayfish, freshwater shrimps, snails, tadpoles, fish, and carrion. Young individuals tend to be more carnivorous than adults.

Nesting in this species occurs from early April to July throughout its range. The female constructs an urn-shaped chamber with her hind feet, softening the soil with urine as she digs. Nests are often placed some distance from the aquatic-feeding habitat, a behavioral trait that may reduce competition with other pond turtle species for prime shoreline nesting sites. One to three clutches are laid per season, with egg numbers ranging from two to 25, the latter number produced by larger females. Hatching occurs about two months later, from July to late September. Late-hatching young may overwinter in the nest.

As already summarized in the "conservation note" account for the Western Pond Turtle, the Red-eared Slider is now the prime introduced competitor for California's only native pond turtle species. For nearly two decades the junior author has conducted surveys for special-status amphibians and reptiles in the immediate watersheds of three large reservoirs that supply water to the City of San Francisco. When this work began in the mid-1980s, one or two Western Pond Turtles were usually sighted during each survey of a shallow bay complex at the end of one of these lakes. However, in the early 1990s turtle heads with a wide red patch on each side began appearing above the water surface in this area, and on one outing a few years later a flotilla of 22 Red-eared Sliders were seen here while our native pond turtle was not observed.

When writing natural accounts of the introduction of exotics into a native species' habitat, one is tempted to downgrade the introduced form and portray it as a villain bent on wiping out the endemic species. However, we must bear in mind that this introduced pond turtle is a successful freshwater reptile that has acquired through natural selection a wide variety of competitive behaviors that have allowed it to survive and flourish in its native multispecies habitat complex. It is certainly not the Red-eared Slider's fault that it is now in many California freshwater habitats, apparently successfully competing with this state's sole native pond turtle. Instead, the blame lies totally on the species that is responsible for its presence here.

WESTERN PAINTED TURTLE *Chrysemys picta*
INTRODUCED

IDENTIFICATION: The carapace is low-crowned and smooth, with an adult length range of 6.3 to 25.4 cm (2.5 to 10 in.). Its color ranges from black to brown or olive, with an olive, yellow, or red border along the front edge. The plastron of California introductions is usually marked with red, and a large dark, branched central figure extends along furrows between scutes. In young the plastron is orange or red with a well-developed central figure, and the carapace may have a weak dorsal keel.

RANGE: This is the most widely distributed freshwater turtle in North America. Its four subspecies extend from Vancouver Island to Nova Scotia, Canada, and south to Louisiana and the western tip of the Florida panhandle. It is found in all of the contiguous 48 states except Nevada, but its occurrence in several western states is due to introductions. In California it has been introduced into many freshwater habitats in the heavily populated areas of Southern Coastal California and is now also present in the vicinity of Goleta, Santa Barbara County, and Kaiser Meadows, Siskiyou County.

NATURAL HISTORY: Given the Western Painted Turtle's large geographic range, it is obviously a broad-spectrum aquatic-habitat generalist, although it tends to prefer sites with considerable aquatic vegetation with muddy or sandy bottoms and quiet or slow-moving water. This turtle appears to be more gregarious than other pond turtles and is often seen basking in groups of several to a dozen or more. Although it is usually closely associated with water, it has occasionally been found wandering considerable distances over land, sometimes in groups. Such behavior may be a response to overpopulation or food scarcity or both at a "home habitat," and by emigrating in groups the establishment of a new breeding colony would be assured if a

Plate 119. Western Painted Turtle in a California pond.

less or unpopulated aquatic habitat is located. In its home range it has also been observed to employ long-distance movements and homing ability to utilize two or more different aquatic habitats within a season. Homing behavior is also apparently used during travel to and from nesting sites that may be some distance from the home aquatic habitat. If such sites are inhabited by several pond turtle species, the ability to seek out desirable nesting areas away from a pond or stream shore area where predators consistently search for nests may be a significant competitive feature.

In the northern portions of its range the Western Painted Turtle usually hibernates in bottom mud and debris of aquatic habitats or in bank burrows such as those of the Muskrat (*Ondatra zibethicus*). However, this is a cold-adapted species and may be seen basking in early spring when part of its aquatic habitat is still ice-covered. The junior author vividly recalls making his way gingerly across the clear but relatively thin ice on a Wisconsin river one late-winter day and looking down to see an adult Western Painted Turtle swimming along beneath him in water that must have been only a few degrees above freezing.

Like most habitat generalists, this turtle is also a feeding generalist and a true omnivore. Plant foods include algae, duckweed, the common aquatic vascular plant *Elodea*, bulrush, wild celery, and water lilies. Animal food consists of spiders, aquatic insects and their larvae, earthworms, clams, snails, crayfish, fish, frogs and tadpoles, and carrion. This turtle has even been observed skimming fine food particles on the water surface.

Courtship and mating occur in the water from March to mid-June. At such times the male positions himself in front of and facing the female.

He then extends his forelimbs with palms turned outward and strokes the female's chin and sides of her face with the smooth backs of his claws. She may also reciprocate by stroking the male's outstretched forelimbs before copulation ensues. Nesting occurs from May through mid-July when the female seeks warm sites with moist sand or soil, which may be near water or as much as several hundred meters or yards from it. She digs a flask-shaped nest chamber with her hind feet, and if the soil is relatively hard and dry, she wets it with urine while digging. When the chamber is completed, a clutch of from three to 24 elliptical, flexible-shelled eggs, about 3 cm (1.2 in.) in length, are deposited within it. The drier excavated soil is then placed over the eggs, followed by the urine-moistened mud. The latter is kneaded with the hind feet and smoothed with the plastron. Soil surface debris is often dragged over the nest site to further hide it. Two clutches are usually produced each year, and up to five have been recorded in some southern areas. About two and a half months are required for incubation.

The Western Painted Turtle is one of a number of species that have temperature-dependent sex determination. Eggs incubated between 25 and 27 degrees C (77 and 80 degrees F) produce only males, while those experiencing a temperature range between 30 and 32 degrees C (86 to 90 degrees F) develop into females. A normal sex ratio is produced from incubation temperatures between these two ranges. This unusual bit of reproductive biology could have a bearing on the spread of this turtle at introduction sites in southern California, where late-spring and summer temperatures in open nest site areas might often be in the female determination range. This in turn could create a highly unbalanced female-to-male ratio, which, in a random polygamous breeding species such as this, would lead to its rapid increase at introduction sites where the native Western Pond Turtle (*Actinemys marmorata*) was once the only pond turtle species.

The Western Painted Turtle and the Pond Slider (*Trachemys scripta*) have been introduced into California from areas where they are usually just one member of a rich pond turtle guild in which each species has acquired traits that allow it to successfully compete with the others. Their high annual potential natality, the ability to utilize a wide variety of plant foods, and habit of traveling overland to distant nesting and alternate aquatic-habitat sites places our native Western Pond Turtle at an apparent disadvantage when encountering these "new kids on the block from the tough competitive neighborhoods back east." Hopefully the current tide of introductions can be stemmed before our only native California pond turtle is excluded from all of its habitats.

Softshell Turtles (Family Trionychidae)

The softshell or "pancake turtles" often appear to the first-time viewer as a cartoonist's creation. As the latter name implies, the shell is flat, round, and relatively flexible. Protruding from the anterior edge is a long neck and a narrow head with bulging eyes and a Pinocchio-like nose. Oversize paddlelike feet contribute to the exotic conformation. However, as with all of nature's designs, that of the softshell turtles has been carefully molded by natural selection to fill the niche of an aquatic, shallow-water, predatory turtle. The flattened body form represents a convergent evolution with fish groups like the flounders, rays, and skates. It permits these animals to partially or fully bury in soft aquatic substrate where they are protected from predators and at the same time may successfully pursue their own lie-in-wait predatory feeding strategy. Softshells usually assume this position in water depths approximately the length of the turtle's long neck and head so that the nostrils that open at the tip of the proboscis-like snout can be protruded above the surface in snorkel-like fashion for breathing.

Given these adaptations, softshells are the most aquatic freshwater turtles. They seldom move from water and quickly return to it when alarmed. They are powerful swimmers but are also capable of surprisingly rapid movement on land. Here, however, their relatively soft, leathery shell and small plastron does not afford the predation protection found in pond turtles, box turtles, and tortoises. This deficiency is compensated for by aggressive defense behavior. When annoyed they hiss violently and may strike and bite with great accuracy. The jaws are overlaid with fleshy lips and thus appear to be rather harmless. However, they are well adapted for crushing or cutting up prey and can inflict a most painful bite.

Seven genera and 25 species of softshell turtles are currently recognized. They occur in North America, Southeast Asia, the Indo-Australian archipelago, the East Indies, and Africa. Three species are found in the United States, and one of these has been introduced into California.

SPINY SOFTSHELL TURTLE   *Apalone spinifera*
INTRODUCED

IDENTIFICATION: The adult carapace length ranges from 12.7 to 53.3 cm (5 to 21 in.), which makes this the largest turtle in mainland California. The shell is flat with flexible edges and covered with a smooth, sandpaper-like skin instead of horny shields as in pond turtles. The long neck and head with its tubular proboscis along with the limbs are retractable within the margin of the carapace. Carapace color ranges from olive to brown or gray with dark-brown or black flecks. The front margin often contains small tubercles, and the entire edge is cream or yellow, as is the plastron. Males tend to be smaller than females, they have a more sandpaper-like texture to their shell, and their thick, fleshy tail extends beyond the carapace edge.

RANGE: This species is native to most of the central and eastern United States. A disjunct population also occurs in the upper reaches of the Missouri River in Montana. It was apparently introduced into the Colorado and Gila river drainages, and is now found throughout the irrigation systems of the Imperial Valley that drain into the Salton Sea. It has also been introduced into Lower Otay Reservoir and the San Diego River, San Diego County, the San Gabriel River, Los Angeles County, San Pablo Reservoir, Contra Costa County, and Lake Elizabeth, Alameda County. There are most likely other urban area aquatic sites that have yet to be reported. The subspecies introduced to California is the Texas Spiny Softshell (*Apalone spinifera emoryi*).

NATURAL HISTORY: The Spiny Softshell is primarily a river- and stream-dwelling turtle but also frequents permanent ponds, oxbows, lakes, reservoirs, canals, and deep irrigation ditches. It is especially attracted to quiet, relatively deep water with a mud, sand, or gravel substrate and where some aquatic vegetation is present. However, it generally avoids seasonal wetland sites. It often basks on shore near the waterline or in a floating position in shallow water. In cool climates it hibernates in the sand or mud of river bottoms, and during drought periods in the southwest it may estivate in mud.

It usually hunts from the lie-in-wait position described in the family introduction but also will actively seek bottom-dwelling prey by probing beneath objects or clumps of vegetation. Its primary foods include snails and clams, earthworms, insects and their aquatic larvae, crayfish, frogs and tadpoles, small fishes, and carrion. It has also been known to take some plant food.

Mating occurs in April and May, and nesting from May to August. Sandy or gravelly soils in open areas are preferred for nests, which are usually placed within 4.5 to 6 m (15 to 20 ft) of water. The nests are flask-shaped, about 15 cm (6 in.) deep, and dug with alternate movements of the hind feet. Between three and 39 round, hard-shelled eggs are deposited on the sloping nest sides and roll to the bottom. The entire nest construction and laying process usually takes less than an hour. Incubation time ranges from 55 to 85 days, with hatching occurring in August or September. Unlike the Western Painted Turtle, the sex of hatchlings is not affected by incubation temperature. Males reach sexual maturity by the age of four, but females are not ready to reproduce until eight to 10 years of age. The maximum longevity is unknown, but it is believed that it may be 50 years or more in the wild.

The Spiny Softshell is rarely found in pet stores, but in its native range it is occasionally harvested and sold for meat, and this is how it has come to be part of the Asian fish market live turtle trade in California. On a recent visit to two such markets in Oakland, the junior author was pleased to find signs above the turtle tanks informing customers that all turtles, which included some very large softshells, must be killed before taking them from the store. This is certainly an encouraging first step in curtailing future introductions of nonnative turtles in California.

Snapping Turtles (Family Chelydridae)

Snapping turtles occur from southern Canada south through the midwestern and eastern United States to the Gulf of Mexico. This family contains only two species, but one of them, the Alligator Snapping Turtle (*Macrochelys temminckii*), is one of the largest freshwater turtles in the world. Its shell length may exceed 61 cm (2 ft), and weights up to 91 kg (200 lb) have been recorded. In addition to its large size and formidable appearance, another unique feature is a small, pink, fingerlike process on the floor of the mouth. This turtle lies motionless on the bottom with mouth open and wiggles this structure that resembles a worm to attract small fish inside the strong, curved jaws. The Alligator Snapper has long been a favorite meat source for many people in the lower Mississippi drainage with the result that it is threatened by overharvesting in many areas. It smaller cousin, the Common Snapping Turtle (*Chelydra serpentina*), is also captured for food but not to the extent of the Alligator Snapper. It was for the fresh turtle meat market that it was imported into California and eventually released at select sites in the state.

COMMON SNAPPING TURTLE *Chelydra serpentina*
INTRODUCED

IDENTIFICATION: The carapace has an oval shape and an adult-length range of 20.3 to 47 cm (8 to 18.5 in.). It has three longitudinal ridges of moderate height and a sawtoothed posterior margin. The plastron is cross-shaped and much smaller than the carapace, which gives the appearance that this chunky turtle is far too big for its shell. The head is large with powerful hooked jaws, and the tail is long with a sawtoothed crest. Carapace color ranges from olive to brown or black.

RANGE: Its native range extends from southern Canada to the Gulf of Mexico, and from the eastern base of the Rocky Mountains to the Atlantic Coast. It is also found in tributaries of the Mississippi River in Montana, Wyoming, Colorado, and New Mexico. It has been introduced into a number of sites in California, including aquatic habitats in the Fresno area, Fresno County, the Andree Clark Bird Refuge, Santa Barbara County, and Lower Crystal Springs Reservoir, San Mateo County.

NATURAL HISTORY: The Common Snapping Turtle may be found in a wide variety of permanent aquatic habitats including lakes, rivers, streams, ponds, and marshes, but it seems to prefer more quiet, sluggish waters with an abundance of aquatic vegetation and soft substrate. It often lies partially buried in the muck and plant growth of pond or lake bottoms or may seek shelter beneath logs, brush, or other underwater debris. The small plastron makes full retraction of the head and limbs into the shell impossible, and thus it is potentially vulnerable to large predators during brief periods of overland travel. However, its pugnacious behavior when out of water appears to counter this anatomical deficiency. When encountered on land it may stand with legs fully extended and lash out by snapping large, powerful, hooked jaws, a trait from which this family acquired its common name. At such times this turtle may also secrete a foul-smelling musky fluid from glands along the lower edge of the carapace, which apparently repels some predators some of the time. This can be especially important to juveniles whose striking and biting threat is minimal. It may also be that a predator that is bitten in its initial attempt to kill a snapper will remember and associate this smell with that unfavorable experience and avoid the next one it encounters. Despite this formidable defense, small- to medium-size Common Snapping Turtles can be captured and carried safely by the tail or hind legs, as long as they are held with the head well away from the captor's body. However, a large individual may be able to reach one's legs even in this position, and it can inflict a severe bite. In water they are far less aggressive and simply retract the head as far as possible when disturbed.

Although highly aquatic, this species will occasionally venture onto land to bask and may move considerable distances overland, during which time it apparently employs sun compass orientation to return to its home waters. Basking may also occur in shallow water and when it is floating on the surface. Water is a good absorber of infrared radiation, and during the summer months surface waters, especially in shallow, calm areas, undergo considerable warming by midday. Thus, highly aquatic turtles like the snappers and the softshells probably gain body heat as much by conduction from the surrounding warm surface water as from shell exposure to direct solar radiation. In areas with cold-winter climates Common Snapping Turtles hibernate, either underwater in soft-bottom substrates or on land in large rodent burrows.

Throughout this book we have been assigning the title of "feeding specialist" to a number of amphibian and reptile species whose diet spans a relatively narrow range of animal and occasionally plant food. However, the Common Snapping Turtle must be put at the top of the feeding-generalist list. In the plant food column are algae, pondweeds, water lily leaves, and duckweed. Animal foods include leeches, insects and their aquatic larvae, crayfish, snails, clams, fish, frogs and their eggs and larvae, salamanders, reptiles (including other turtles), birds (including ducklings), and small mammals. Carrion tops off this broad-spectrum omnivore's dietary list. It captures animal prey either by overtaking it with its powerful swimming ability or by lying on the bottom, partly buried with only eyes and nostrils exposed, and seizing approaching prey with a quick thrust of its neck.

Mating can occur throughout the warm, active season, but it is most frequent in spring and fall. Nesting normally peaks in mid-June. Females will often travel over 1 km (0.62 mi) inland to locate a suitable nest site. These are usually moist but well-drained substrates in sunny, open areas. In some parts of the Midwest and East where lake cottages ring many Common Snapping Turtle lakes, cottage lawns have been selected as nest digging sites, because they also meet these criteria. Here or elsewhere the female scoops out a flask-shaped cavity in which she lays from 10 to over 100 round, golf ball-sized eggs, after which she covers the nest with soil and returns to the home water habitat. This is also a turtle whose hatchling sex ratio may be determined by egg incubation temperature. Eggs held in the 25 degrees C (77 degrees F) range produce nearly all males, while those incubated at 30 degrees C (86 degrees F) produce only females. However, this species will also produce only female hatchlings if eggs are kept near the lower thermal range for successful development.

Newly hatched snappers stay buried in the nest for several days during which the remaining nutrition in their yolk sacs is absorbed. They then emerge en masse and move quickly to the water habitat in a manner similar to that of hatchling sea turtles. These hatchlings are fleshy, small animals whose soft carapace and small plastron afford little defense against predators, and so the movement time on land must be kept to a minimum. The musk glands are functional in hatchlings, and they may provide the most effective protection against some predators. Young Snapping Turtles grow rapidly, but growth slows as they approach full adult status. The maximum age recorded for a captive individual is 38 years.

Most of the introduced Common Snapping Turtles in California have resulted from escapes or purposeful releases of specimens imported by Asian fish markets. As a result, wetland sites, some of which may be a considerable distance from such businesses, now contain small Common Snapping Turtle populations. A few years ago the junior author was conducting a visual survey for the California Red-legged Frog (*Rana draytonii*) in a shallow marsh at one end of a large reservoir in San Mateo County when a movement caught his eye in the dense aquatic vegetation ahead. To his total surprise it was the long, sawtooth-crested tail of a Common Snapping Turtle, and for a brief moment he was mentally transported back to the Kettle Moraine lakes of central Wisconsin where as a boy on family summer vacations he occasionally encountered this turtle in shallow bay vegetation. However, his youthful quickness had apparently slowed a bit, and an attempt to grab the retreating tail in the reservoir marsh fell short. He did, however, have a culprit to blame for the day's relatively low red-legged frog count. Hopefully the trend to ban the sale of live meat turtles in California will ensure that such sightings are few and far between in the future.

Sea Turtles (Families Cheloniidae and Dermochelyidae)

These are the largest living turtles, with carapace lengths ranging from 46 to 224 cm (18 to 96 in.) and weights from 13.6 to 727 kg (30 to 1,600 lb). Their distribution is circumpolar in warm seas, but some species occasionally range far north into temperate waters, especially in summer. The general body and shell design of sea turtles was already well established by similar-sized long-extinct reptiles such as *Henodus*, which lived in the Upper Triassic when dinosaur evolution was just getting underway, and little in the early sea turtle body plan has changed since. During that period there were other large marine reptile groups such as the fish-eating ichthyosaurs and plesiosaurs, but they have long since been replaced by large piscivorous fishes and toothed whales. However, modern-day sea turtles feed mainly on marine algae, vascular plants such as eel grass, and a variety of bottom-dwelling invertebrates, a food niche that the early sea turtle types was also thought to occupy. Their lineage may therefore have avoided the progressive evolutionary selection for the large marine piscivore niche, and this along with the universal mode of protection afforded by the shell has most likely accounted for their continued presence.

Today sea turtles are represented by seven living species. One is an Atlantic Ocean and Gulf of Mexico species (Atlantic Ridley, *Lepidochelys kempii*) and four others occur in the Pacific Ocean and rarely move into California's inshore waters. However, two other Pacific residents, the Green Sea Turtle (*Chelonia mydas*) and the Black Sea Turtle (*Chelonia agassizii*) are regular visitors to the southern California coast and thus have been accorded a full natural history account in this book. For those readers who occasionally cruise or sail off the California coast or are avid beachcombers, there is a possibility of seeing one or more of the other four Pacific species, either alive or dead. Therefore, brief natural history notes for these are also presented along with the following general biology of these two families.

BODY FORM: Sea turtle conformation is designed for a life at sea. The shell is low and streamlined, the paddlelike hindlimbs have one or two claws and serve primarily as rudders, and the forelimbs are elongated, flattened flippers that provide the swimming power much like those of sea lions and fur seals. Their pelagic marine existence demands that these turtles must be able to engage in long, sustained periods of rigorous, aerobic muscular activity. This is quite different from nearly all other reptiles whose white nonmyoglobic muscles permit only short bouts of activity. The red myoglobic muscles of sea turtles have the same sustained contractile ability as those of birds and mammals, and in the process generate a moderate amount of body heat. Using biotelemetry, the junior author recorded deep-body temperature from free-living Green Turtles in a large island lagoon in the Sea of Cortez. When resting at or beneath the surface, their temperature was that of the seawater. However, shortly after sustained swimming began it rose to from 2 to 3 degrees C (3.5 to 5.5 degrees F) above ambient temperature and remained elevated throughout this activity. The turtles weighed 32 kg (70 lb) each, and their relatively small surface-to-mass ratio retarded the loss of heat generated by swimming. In the Leatherback Turtle (*Dermochelys coriacea*), the largest of all turtles, the surface-to-mass ratio is even smaller, and this along with insulating subepidermal fat and a countercurrent heat exchange system in the limbs combine to produce the same sort of endothermy found in very large, active marine fishes. Leatherback Turtle deep-body temperatures of at least 18 degrees C (32 degrees F) above cool northern-latitude water temperatures have been reported.

REPRODUCTION: Given this long-term swimming ability, these turtles spend nearly their entire life at sea. However, unlike the ovoviviparous sea snakes, the oviparous female sea turtles must come ashore to lay eggs. After breeding in shallow offshore waters, they haul out on remote beaches and laboriously make their way to sites above the high-tide zone, where the nest is excavated with the flippers and hindlimbs. First a large, shallow pit is dug with all four limbs, and then an egg chamber is constructed in the bottom with the hind feet. Next dozens to hundreds of pink to white globular, leather-shelled eggs are laid, and then the nest hole is filled with the excavated sand. Females usually nest two or more times a season.

Hatchlings must make their way through 60 to 120 cm (2 to 4 ft) of sand to reach the surface. This usually occurs at night so as to avoid predation by diurnal birds and mammals. Despite this precaution, predation by nocturnal predators is still high during the trip from nest to water and continues as they are attacked by fish and large marine invertebrates once in the surf. The resulting high mortality is offset by the large number of eggs produced and by the en mass nest-to-sea migrations that most often occur. Thus, although those predators present may gorge themselves on hatchlings, there are still enough to ensure that a good number do make it to the relative safety of the offshore waters.

This sort of hatchling mortality has probably been a constant part of the biology of sea turtles since the late Jurassic Period. However, during the past few centuries a whole new spectrum of negative impacts have been directed toward these ancient mariners by our own species. These include the continued overexploitation of adults for meat, shells, leather, and oil for cosmetics. In addition to the purposeful killing of these turtles for such products, high mortality continues to result from drowning in shrimp and fish nets and trawls, and from oil spills. As human populations expand in nesting-beach areas there has also been an overharvesting of eggs. Long-established nesting beaches are also some of the most valuable pieces of human real estate in the world, and beachside homes, hotels, and off-road vehicle traffic have destroyed nests. Even the sounds of vehicles and jet aircraft have been shown to significantly reduce hatching success. The sand vibrations that they cause stimulates continuous activity of near-term hatchlings within the eggs, which then fail to hatch, presumably because of the exhaustion of these unhatched young.

CONSERVATION NOTE: Although conservation efforts have been mounted to combat many of these problems, the preservation of any wide-ranging marine species whose survival depends on close cooperation of numerous countries and ethnic groups within each is a difficult task. An even more challenging task will be to convince all countries of the world that the human-induced component of global warming must be reduced as soon as possible, because the sea level rises predicted for this century alone could negate many nursery beaches. One result of such warming, an increase of ocean surface temperatures, is now being viewed as the cause of more frequent and ferocious hurricanes, the surges from which frequently destroy sea turtle nests. During the recent high hurricane decade (1995-2004), Green Sea Turtle and Loggerhead Sea Turtle nests on the beaches of Dry Tortugas National Park off the Florida coast dropped by more than half. At this writing the world's sea turtle populations still appear to be declining, and the outlook for the survival of these remnants of the period when reptiles were the dominant large marine vertebrates is not good.

Sea Turtles (Family Cheloniidae)

GREEN SEA TURTLE *Chelonia mydas*
BLACK SEA TURTLE *Chelonia agassizii*

Green Sea Turtle

PROTECTIVE STATUS OF *C. MYDAS* AND *C. AGASSIZII*: FT

TAXONOMIC NOTE: The Black Sea Turtle was formerly viewed as a subspecies of the Green Sea Turtle but has recently been recognized by many as a new species, although its full taxonomic state is still being debated. As with several other recent subspecies-to-species elevations, we discuss both together because of the great similarity in their respective natural histories.

IDENTIFICATION: The Green Sea Turtle has an adult carapace-length range of 76 to 152 cm (30 to more than 60 in.) and a weight range from 55 to more than 275 kg (120 to more than 600 lb). The Black Sea Turtle is smaller, with a total carapace length up to 90 cm (35 in.) and weights up to 120 kg (264 lb). The carapace of both species is smooth with four large costal shields on each side. The color ranges from green or olive to brown in the Green Sea Turtle and from gray to black in The Black Sea Turtle.

The original common name of the Green Sea Turtle is not derived from carapace color but instead from its green subepidermal fat. Both have a single pair of large prefrontal scales between the upper eyelids that aid in recognizing them from other sea turtle species when only the head is seen above water. Males have a very long, prehensile tail with a nail at the tip and an enlarged curved claw on the front flipper that aids in grasping the female during breeding.

RANGE: Like other sea turtles, the Green Sea Turtle ranges throughout the warm seas of the world. The Black Sea Turtle is restricted to the eastern Pacific from Southern California to Chile and west to the Galapagos Islands, Hawaii, and Papua New Guinea. Both species have been found together in a warm-water electric plant discharge channel in San Diego Harbor and also share nesting beaches along the coast of Michoacan, Mexico. Individual sightings of what are presumed to be Black Sea Turtles have been reported as far north as Prince William Sound, Alaska.

NATURAL HISTORY: These highly aquatic turtles seldom venture onto land except to nest and occasionally bask and rest on the sandy or rocky edge of remote islands. More often they will bask while floating, either asleep or in a state of tonic immobility. A seasoned Mexican sea turtle fisherman from whom the junior author purchased specimens for his thermoregulation studies in the Sea of Cortez bragged of his ability to sneak up on a sleeping Green Sea Turtle in a small skiff and capture it by hand. In experiments with large juveniles in a temperature-controlled tank, this author was able to induce the surface-floating, inactive state by raising water temperature into the 28 to 30 degrees C (82 to 86 degrees F) range, and the turtles would remain in this condition for hours as long as that range was maintained. However, at any time this "floating sleep" could be terminated by slowly lowering the water temperature back to the normal low 20 degrees C (68 degrees F) range.

These are the most herbivorous of all marine turtles. They are grazers whose primary foods are algae, eel grass, and mangrove roots and leaves. Given this plant-oriented diet, they restrict much of their activity to bays, estuaries, and reefs as opposed to the open-sea orientation of most other species. However, adults will also take sponges, jellyfish, ctenophores, sea cucumbers, mollusks, crustaceans, and tunicates, and the young feed almost exclusively on these protein-rich foods. All of these foods have a salt concentration over twice that in land and freshwater food items. This, combined with the sea being an "aquatic desert" in which no inhabitant can obtain freshwater, places sea turtles in a demanding osmotic situation. They cope by secreting a highly saline solution from glands in the corner of each eye (sometimes referred to as "turtle tears"), and this along with additional salt excretion by the kidney allows them to exist on the preformed and metabolic water gleaned from their food.

Each year extensive migrations occur between feeding grounds and nesting beaches. Remote areas are sought, presumably to reduce predation

Plate 120. Green Sea Turtles captured for meat in Baja California.

on eggs and young. Migrations of from around 800 to 1,280 km (500 to 800 mi) for the Green Sea Turtle are not uncommon and can be completed in about two months. Mating occurs during the egg-laying season in waters off the nesting beaches. A single female may be sought by a number of males, all of which attempt to copulate with her. Nonreceptive females reject males by biting and assuming a vertical position in the water with limbs extended and plastron turned toward the male, all of which prevents him from attaining the dorsal copulatory position. Before copulation, a receptive female will allow the male to gently nip her throat region, and she may sometimes nip his in return. Copulating pairs may be followed by an escort of one or more males seeking an opportunity to mate. They have to be patient, however, since a single copulation may last as much as six hours. After copulation and/or egg laying, females cover their vent region by placing their hind flippers together with the plantar surfaces in contact, thereby thwarting further attempts by males to breed. Nest construction, egg laying, and hatching follows the classic sea turtle pattern described in this section's introduction.

CONSERVATION NOTE: The ability to survive long periods without food and water has made turtles in general and sea turtles in particular a long sought-after source of fresh meat, especially on long sea voyages, and the ongoing harvesting of both eggs and adults in many parts of the world continues to decimate all species. For several centuries these two species have been particularly hard hit because of their good-tasting red meat and cartilaginous parts of the shell, which when cooked produce a gelatinous base for "Green Turtle Soup." As a result, these species are in serious decline and have undergone an estimated 90 percent population decrease over the past half century. The placement of both sea turtles on the Federal Endangered species list now prohibits the importation and sale of sea turtle meat and shells in the United States. This is very important, because until recently, a common scene in several Sea of Cortez coastal towns was the one seen here of Green Sea Turtles in holding pens, from which they were loaded upside down and six or eight deep in trucks and taken to Mexican border towns and the San Diego area for sale. With additional restrictions by the Mexican government on sea turtle harvesting and the protection of nesting beaches, both species may hopefully remain regular visitors to the southern California coast. Supporting this hope are recent (2006) reports of recoveries of several Atlantic Ocean populations of Green Sea Turtles where six of eight recently protected nesting beaches have exhibited a doubling of their respective breeding populations in the past 30 years, and similar results from protected rookery sites have been documented in Hawaii and Australia. One current estimate places total world population of both species at 2.6 million.

LOGGERHEAD SEA TURTLE — *Caretta caretta*

PROTECTIVE STATUS: FT

IDENTIFICATION: This is a moderate-sized sea turtle with an adult carapace-length range of 20.3 to 91 cm (8 to 36 in.) and weights from 136 to more than 409 kg (300 to more than 900 lb). Those specimens that are occasionally seen off the southern California coast are juveniles that have traveled across the Pacific from nesting beaches in Japan and are therefore much smaller. These young have a prominent mid-dorsal ridge on the carapace that is less conspicuous in the adult, and there are five or more costal shields on each side. As the common name implies, the head is large and broad. Carapace color ranges from reddish or orange-brown in adults to a yellow buff, brown, or gray-black in young.

RANGE: Its range is circumpolar in temperate-sea areas. It is sporadically seen off the southern California coast as far north as Santa Cruz Island. One sighting has been reported from the beach at Fort Canby State Park, Washington.

NATURAL HISTORY: The Loggerhead Sea Turtle is a long-distance swimmer, as the sightings off the California coast of young turtles from Japan nesting beaches illustrate. It is primarily carnivorous, feeding on sponges, mollusks (including squid), crustaceans, sea urchins, jellyfish, and fish (some of which may be obtained by scavenging), although it will also eat eel grass and marine algae. The United States has designated it as a Federally Threatened species, and legal protection may hopefully reverse its decline.

OLIVE RIDLEY SEA TURTLE *Lepidochelys olivacea*

PROTECTIVE STATUS: FT

IDENTIFICATION: This is our smallest Pacific Ocean sea turtle with an adult carapace-length range of 51 to 74 cm (20 to 29 in.) and weights up to only about 45 kg (100 lb). As its name implies, this turtle has a uniform olive-brown to gray-green carapace and skin color to match. Six to eight costal shields are usually present.

RANGE: This species of *Lepidochelys* occurs in the warm waters of the Pacific and Indian Oceans. It has been seen as far north along the

California coast as Monterey Bay, and strandings have been reported on beaches in Mendocino and Humboldt counties.

NATURAL HISTORY: The Olive Ridley Sea Turtle appears to be more of a bottom dweller than other marine turtles. There it feeds mainly on sea urchins, crustaceans, mollusks, and some algae, but it will also take jellyfish and fishes in the water column. Its designation as a U.S. Federally Threatened species along with its listing as endangered by Mexico is a first sound step toward saving this species.

HAWKSBILL SEA TURTLE *Eretmochelys imbricata*

PROTECTIVE STATUS: FT

IDENTIFICATION: This a relatively small sea turtle with a carapace-length range of 46 to 91 cm (18 to 36 in.) and weights rarely exceeding 114 kg (250 lb). The carapace has strongly overlapping, shinglelike dark-brown shields with a yellow marbled pattern, and the jaws resemble a hawk's bill as its name implies. The young have a midline keel on the carapace, and the shields are dark brown or black.

RANGE: It occurs in the warm water regions of the Pacific, Atlantic, and Indian Oceans, where it frequents rocky coastal waters, coral reefs, bays, shallow lagoons, and even the lower reaches of small creeks that flow into marine waters. However, it also ventures well out to sea. It may occasionally be seen off the southern California coast.

NATURAL HISTORY: This omnivorous marine turtle feeds on algae, mangrove leaves and roots, sponges, coral, barnacles and other crustaceans, mollusks, sea urchins, and coral and other coelenterates. The sharp, curved upper jaw may be important in dislodging a number of these prey species from the substrate. It also uses it in defense when disturbed.

The Hawksbill has been exploited not only for its eggs and flesh; for centuries its shell has been the source of manufactured "tortoise-shell" items, and despite the invention of plastic in the early twentieth century, natural shell is still sought after. As a result, its prospects for survival are not bright, despite its status as a Federally Threatened species.

Leatherback Sea Turtle (Family Dermochelyidae)

LEATHERBACK SEA TURTLE *Dermochelys coriacea*

PROTECTIVE STATUS: FE

IDENTIFICATION: This is the largest living turtle, with an adult carapace-length range of 122 to 244 cm (48 to 96 in.) and weights from 273 to 916 kg (600 lb to over a ton). The carapace and plastron are covered with smooth leathery skin instead of horny shields, and there are prominent tubricate lengthwise ridges on both. The upper jaw is highly serrated. The carapace and skin ranges from dark brown to slate or black and may be marked with white or pale-yellow blotches. The plastron is white, which imparts the classic pelagic marine countershading effect.

RANGE: It has a circumpolar distribution in tropical and subtropical seas and also enters cooler northern and southern waters in the summers and autumns of each hemisphere. It has been caught off the Monterey coast and is occasionally seen in San Francisco Bay. There are also records of sighting off the coasts of British Columbia, Canada, and Alaska.

NATURAL HISTORY: It seems appropriate that we should close our natural history accounts with a few words about one of the largest living reptiles, the

Leatherback Sea Turtle. From stem to stern, its body is constructed for life in the open sea and differs from that of all other turtles. As mentioned in the identification comments, both its carapace and plastron are covered with smooth, leathery skin instead of horny shields, and there are seven longitudinal ridges on the dorsal side and five more on the ventral. Its shell consists of polygons of bony plate imbedded in the skin, with no attachment to either the vertebrae or ribs. The shell is low and does not extend beyond the limb insertions or neck base, which creates a more hydrodynamic body form. All these adaptations plus a muscle and circulatory physiology that allows it to engage in endothermic thermoregulation when active adapt this ancient mariner well to a pelagic life.

Its food spans the normal array for sea turtles, except that it has a special preference for jellyfish, a little-utilized marine invertebrate food source. Leatherbacks have been known to gather in groups to feed on a large jellyfish school. The deeply serrated upper-jaw edge may help to hold on to this slippery prey, and backward-projecting spines in the mouth and esophagus aid in swallowing these gelatinous morsels.

Despite its designation as a federally endangered species, the Leatherback Sea Turtle continues to decline, most likely due to the increased impact of modern long-line, drift, and gill net fishing. Satellite-assisted tracking studies are needed to determine if estabished migratory pathways exist, so that if such pathways are present, attempts can be made to restrict the use of such fishing methods in major movement areas.

OBSERVING AND PHOTOGRAPHING AMPHIBIANS AND REPTILES

Recording Your Observations

The two topics of observing and photographing are discussed together, because it is usually good viewing methods that produce good photos or videos of amphibians and reptiles in the field. However, we do not mean to imply that enjoyment of field observations must be accompanied by some sort of film or electronic image record. Indeed, the vast majority of natural history information in this book was originally documented by handwritten field notes during or immediately after observations of species in the field. Your own future enjoyment in observing wild "herps" will be increased greatly if you document your field experiences. The long-standing method for such records has been to write them in a loose-leaf notebook with waterproof ink on high-quality paper. The size of the notebook will vary with personal taste, the most important consideration being that it should fit into whatever carrying space you always take into the field (shoulder bag, large field pants pocket, backpack, fanny pack, etc.). If it is so large that it must always be carried in the hand, human nature dictates that it will often be deemed too cumbersome and be left behind.

TAKING FIELD NOTES

When the first edition of this book was published in 1972, the only alternatives to writing field notes by hand were to carry along a portable mechanical typewriter or an equally bulky (by today's standards) tape recorder. At present the former is often replaced by the electronic notebook in one's pants pocket or the laptop computer in a backpack and the miniature digital recorders that now fit easily into a shirt pocket. Indeed, the present and future generation of readers may find it more appealing to always record field observations on multiuse miniature digital units that they carry at all times for a variety of purposes. When combined with a simultaneous video record, auditory comments become even more meaningful. Speaking softly into the microphone of a camcorder does not appear to disturb most amphibians and reptiles, even at close range, and often the electronic eye catches fine details that the human eye has missed. A potential problem with auditory field notes if they are later handwritten or typed for a permanent visual record is the tendency to eliminate a word or comment here and there that at the time of transcribing may seem unnecessary. However, it is such little comments on seemingly insignificant bits of behavior that, if seen and recorded repeatedly, may lead to important new facts about a species' biology.

Whether the field note recording method is by pen or pixel, a standard format is advisable to ensure consistency and accuracy. The one presented in figure 37 was established by the Museum of Vertebrate Zoology, UC Berkeley, nearly a century ago, and has been adopted by many similar institutions. When using a laptop or notebook-type computer, a template for this format can be brought up for each new species account. Note that additional species observations are listed one after the other with accompanying new date and locality. When using a field notebook, additional accounts follow the original one and continue onto the following page(s). The recommended loose-leaf notebook permits the arrangement of all pages of each species account in this manner, and the grouping of all field records of each species is even easier when taking computer notes.

Species Accounts

The species' common and Latin name should head each paper or electronic page. The underlined locality of each observation is very important, and thus it, along with the date in the left margin, begin each new entry for a species. Each new account should start with a reference to some nearby natural or relatively permanent human-made feature or the approximate directional mileage from same. The elevation, which can be obtained from a U.S. Geological Survey topographic map, a detailed road map, or a pocket or auto GPS (global positioning system), is also an important entry. The behavior of a heliothermic lizard species observed at a high elevation may not be the same as that seen near sea level. The name of the major geographic area, county, and state should complete this entry. The time of each observation comes next at the beginning of each observation and may be followed by a more precise description of the exact location. This latter information can also be presented in the journal account to be described next. The journal is usually the best place for details of the weather and physical/botanical habitat, because this negates the need to repeat such information in several species accounts that may be entered for the same locality on the same day.

In recording your observations of a species' behavior, be careful of the descriptive terms that you use. We all have a tendency to assign to other animal species our own particular species' traits, feelings, and motives (i.e., anthropomorphism). Phrases such as "the lizard is nervous," "the large frog got mad and jumped at the smaller one," and so on, can often mask what may really be determining certain activities. For instance, large ranid frogs often eat smaller ones, and thus the latter observation may have been simply an incident of attempted feeding and not some imagined form of anger. During some future observation sequence, a similar encounter might end with a large frog actually catching and eating a smaller one, in which case one might be tempted to assume that it was now "happy" instead of "mad." Yet it is very doubtful

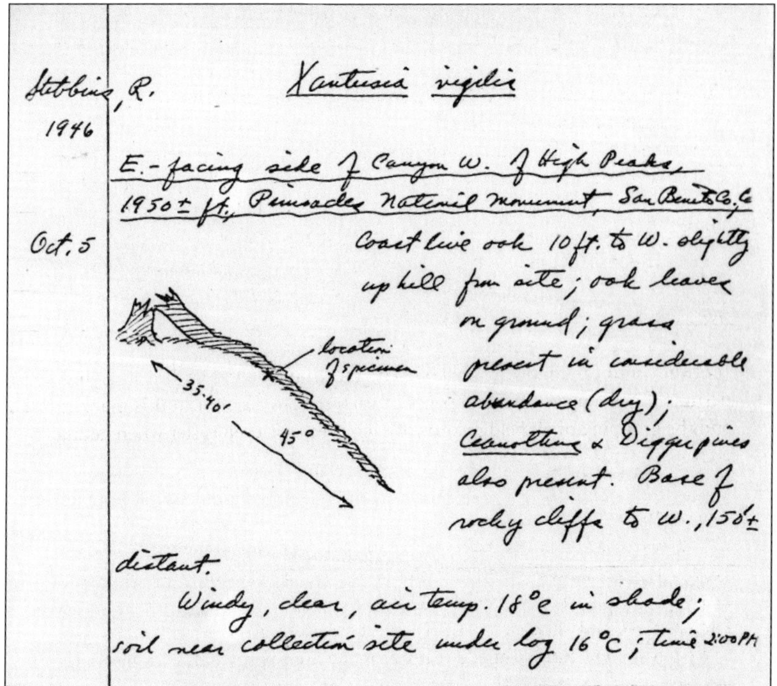

Stebbins, R.
1946

Xantusia vigilis

E.-facing side of Canyon W. of High Peaks,
1950± ft., Pinnacles National Monument, San Benito Co., Ca.

Oct. 5

Coast live oak 10 ft. to W. slightly up hill from site; oak leaves on ground; grass present in considerable abundance (dry); *Ceanothus* & Digger pines also present. Base of rocky cliffs to W., 150± distant.

(sketch: location of specimen, 35–40', 45°)

Windy, clear, air temp. 18°C in shade; soil near collection site under log 16°C; time 2:00 PM.

Adult found 100 feet diagonally upslope toward SW from above individual; 20-25 ft. from base of 200 ft.± cliff. Sparse stand of Digger pine and Coast live oak here. Slope with dry grass, buckwheat and deer weed (*Lotus*). Some *Ceanothus* present. Lizard under much fissured log with bark absent above; log 35' x 1½' lying at rt. ∠s to slope. Specimen midway on log opposite Deer weed plant. Soil with many small ash-gray rocks, some with reddish tinge, little fine material; bits of decayed wood & bark present; slightly damp; temp. 17°C. Pine needles on ground.
Time 2:15; air temp. in shade 18.5°C; lizard sluggish

Figure 37. A species account from the senior author's 1946 field notebook. The information here remains a useful guide to finding the Desert Night Lizard in Pinnacles National Monument.

that an amphibian experiences either of these emotions as we know them, and so you should avoid the temptation to attribute a human state of mind to amphibians and reptiles.

The Journal

The second part of a field paper or electronic notebook is the journal, an account of exact locations and characteristics of the sites where species accounts were recorded for each day. The heading for each page should be "Journal," with the date indicated in the lefthand column as in the species accounts. Here the location of each site visited on that day is the underlined heading and should be described as completely as possible. A sketch map showing the route taken to a site with distances along hard surface and dirt roads should be given whenever possible. Such information can now be greatly enhanced by a "fix" with a GPS unit. These become smaller, more accurate, and more reasonably priced each year, and currently can guide you to the very rock on which a lizard was displaying or under which a salamander was discovered in years past. However, batteries do discharge and electronic circuits occasionally fail, so the journal description of a site location should still accompany such digital records.

The site description for each area visited should include the kinds of plants present, characteristics of the substrate (sandy, rocky, type and amount of organic ground cover, amount of moisture under it), and the weather and seasonal climate conditions. For amphibians, the relative amount of soil moisture and the presence or absence of standing-water pools are usually the most important weather features. In notes concerning heliothermic anurans and most reptiles, available sunlight and the temperature regime near the ground are key items. For the latter a small spirit or electronic thermometer is a handy item for one's field bag. Some key temperatures are that of the ground surface under a rock or fallen log, at the exposed substrate surface in both sun and shade, and the air temperature about 2.5 cm (1 in.) above the ground in both sunlight and shade. These constitute that major spectrum of temperatures that an amphibian or reptile may experience on a given day, and marked variations between readings on different days can explain why the same complement of species were not seen on each trip to a site.

When possible, the climatic conditions for the period of your observations should be described. Here one should focus on indicators of the seasonal climatic that are present on that particular day and year. A cool, wet spring season at lower elevations in California will often produce an amphibian or reptile activity pattern that differs markedly from that found during a warm, relatively dry spring. New plant growth and flower production is usually a good indicator of how advanced the spring climatic period may be. Are common plants such as Wild Mustard (*Brassica nigra*) or California Poppy (*Eschscholtzia californica*) in full bloom or only in the flower bud stage? The activities of other animal groups are also good indicators. Is bird nesting well under way, or are male songbirds just beginning to display on their territories? Insect availability and abundance are usually the best of animal-based climate indicators. Are common butterflies abundant yet? Have you experienced your first

Plate 121. Herpetology students photographing a horned lizard.

mosquito bite of the season? Have grasshoppers undergone their final-instar molt? The recording of these and many more seasonal indicators in your journal can be increasingly helpful in understanding variations in amphibian and reptile activity and behavior from year to year.

Photography and Videography of Amphibians and Reptiles

Like many readers, both authors have experienced the thrill of African big-game photography. Slowly maneuvering into the right position for that "perfect" telephoto shot is an exciting wildlife experience. However, carefully stalking a wild reptile or amphibian so as to record those poses and behaviors that best represent a species can be equally rewarding. Given the current availability of reasonably priced, high-quality digital cameras and camcorders, and the former costs of film and developing that they eliminate, the equipment for high-quality photos and "movies" of amphibians and reptiles is now within the economic reach of many people. In addition, the instant viewing of pictures and video in the field that this new equipment offers allows you to judge whether a "re-take" is necessary to obtain that perfect record of what may be a once-ever personal observation.

Whether you use the new line of digital equipment or prefer the 35-mm film camera, both of which produce high-quality color images for publication, a few basic techniques deserve mention. One concerns the use of a flash. Most new digital and film cameras have built-in good-quality flash units. Top-quality systems that automatically adjust the strength of the flash to the zoom lens setting for each picture may also be added. Given this availability, the question remains whether to rely on natural light or use a flash for an amphibian or reptile portrait. The determining factor here is usually whether natural illumination of the subject is adequate to allow enough depth-of-field focus to produce a sharp image of the entire amphibian or reptile. When photographing basking lizards or ranid frogs in sunlight, the use of a high aperture setting (f/12 to f/22) will usually produce a sharp image of the entire animal, especially when using the relatively fine-grain, high-speed films (400 ASA or greater) now available. However, close-up photography of specimens in shade or subdued light will not permit the use of a small aperture (high f settings) unless a flash unit is employed. Many new cameras now make this decision of flash or no flash for you, but you should still consult the camera's digital display screen to make sure that such automation incorporates the highest possible f setting. If it does not, take the designated steps to override the "automatic" setting of the camera and manually select a small-aperture setting to ensure large depth-of-field focus.

One other consideration concerns the use of standard or zoom telephoto lenses. These are a must for the field photography of most California reptiles and anurans. However, they are really designed for "people photography," and although they are capable of producing a frame-filling picture of a person at some distance, the size of a lizard or frog on the view display screen is often smaller than desired, even at the closest possible focusing distance. This problem can easily be corrected by inserting a short- to medium-length extension tube between the telephoto lens and the camera body. This tube, when coupled with a 200- to 400-mm telephoto lens, can permit you to fill the view frame with the complete image of a basking lizard or frog at a distance of only several meters (about a dozen feet).

Even with this new spectrum of photographic equipment and associated techniques, it is still your skill in locating and approaching amphibians and reptiles in the field that determines your success in this exciting and rewarding pastime. We next briefly review several photo/video situations that you can explore and suggest methods for each.

NATURAL FIELD PHOTO OR VIDEO

There is nothing more satisfying in the photo/video area than recording images of wild animals in their natural habitats.

Salamanders

Before the advent of the self-focusing and automatic-aperture cameras, a photographer would literally have to bite the dust (or mud in this case) to obtain a ground-level profile picture. Today the camera or camcorder has only to be placed on the substrate and pointed at the subject. Many of these units also have an adjustable image display screen that allows one to frame a horizontal, ground-level scene while looking down at it. The result is usually a far more dynamic salamander profile portrait than the standard dorsal view.

In the area of herpetology, this is most easily accomplished with salamanders. Migrating newts, for example, may often be encountered during daytime and will usually continue on their way with occasional brief pauses despite the close presence of a quiet, slowly approaching photographer. Given such an encounter, it is often one's first inclination to photograph such animals from above, as usually observed. However, this is somewhat comparable to taking a picture of a friend from an upper-story window. Although your friend can usually be recognized in the resulting photo or video, many of the facial and body features unique to that person are lost. The same is true for most amphibians and reptiles, where profiles or views at oblique angles are far more revealing, for this is how the animals view each other.

Other salamander groups are not usually as "photographer friendly" as newts. Ambystomids such as the California Tiger Salamander (*Ambystoma californiense*) are normally seen aboveground and out of water only during their nocturnal migrations between aquatic-breeding sites and upland burrow retreats. However, these movement pathways are usually predictable, and the devoted herp photographer with good rain gear and a camera in a clear plastic bag with a hole for the lens to "see through" may obtain some dynamic field pictures. A clear filter such as a haze or UV filter should protect the delicate lens surface, and the edges of the hole cut in the bag should be taped to the filter ring so that no moisture can reach the camera body. The camera's built-in flash will project its light through the bag, and the photographer can see a moderately clear image of the subject through the plastic so as to frame it while the auto focus does the rest. The pliability of the plastic bag allows one to press the shutter button when all is ready.

Although most plethodontid salamanders spend much of the wet season aboveground, they remain sequestered under logs, loose rock, or dense leaf litter during daylight hours. Thus, an initial disturbance such as lifting a ground surface object is necessary to locate them. Once exposed, most exhibit normal posture and movement as they seek to regain a retreat site at a very slow pace, which allows for profile photos and videos as described for migrating newts. However, some salamanders such as the subspecies of *Ensatina* may assume defensive postures as described in their respective natural history accounts, and for the next several minutes one is rewarded with subjects that "sit" motionless for their portrait much as humans once did in the early days of photography.

Frogs and Toads

Anurans, for the most part, are a more challenging group of amphibians to photograph. Toads, which rely heavily on poison gland secretions and the release of urine for defense, are usually not prone to rapid retreat when approached by careful, slow stalking. The same usually holds true for species that rely heavily on cryptic coloration for defense such as the Pacific Chorus Frog (*Pseudacris regilla*). The photo of this species in the natural history accounts (see Plate 50) was taken with a standard 50-mm lens and short extension tube as the photographer leaned forward to within an arm's length of the subject.

Most ranid frogs, on the other hand, pose far more of a challenge. Their presence along a pond or creek bank is often hard to detect because of the usual dense plant or rock cover in such areas, and the first indication of the frog's presence is usually a loud plop as it jumps into the safety of the water ahead of the photographer. Here the best method is to move along a shoreline with the greatest caution possible while continuously scanning the area ahead with short-focal-length binoculars. Once a basking frog is spotted in this manner, the battle is half-won, and now a very slow, cautious stalk should bring you within the desired focusing distance. The one variable that may still cause a problem is the position of vegetation and other cover items between you and the frog, and sometimes this is so dense that a good photo or video sequence is simply not possible. Here is where the determined amphibian photographers must get their feet wet to obtain their prize. Frogs along shorelines often are less disturbed by a slowly moving object coming toward them from the water than from land. Such frogs also select basking sites with some open space between them and the water, and with a careful aquatic stalk this now becomes your focal pathway.

Turtles and Tortoises

The photographing and videotaping of turtles in the field spans the range of challenges that wild reptiles present. On one hand, there is no better California reptilian subject than the Desert Tortoise (*Gopherus agassizii*). All you need is a medium-length telephoto or zoom lens and patience equal to that of your subject, and you can record a spectrum of activities including morning basking and grazing. A word of caution concerning the photographing of this and other protected species is appropriate here. You must keep in mind that any handling or otherwise manipulating an endangered, threatened, or otherwise protected species, in this case to perhaps produce a better pose, is a violation of state and/or federal law unless one has the appropriate permits to do so. It is also illegal to interfere with the normal activities of such species, and hence the recommendation of a telephoto or zoom lens that enables you to obtain good photos and video of your subject without its reacting to your presence. This distance varies greatly with species. The junior author and his students have accompanied a number of Desert Tortoises on their morning foraging walk at a distance of approximately four meters (13 ft) with no observed disturbance. Such cautious, distant viewing does not violate any provision of the state or federal endangered species acts. Also, if you crouch and remain still some tortoises may approach, perhaps to seek the shade that this new large object in its habitat has produced. Photographing and videotaping from a safe distance is also to the viewer's advantage, because one reaction of a tortoise to a closely approaching human is to pull in its head and neck, cover the front carapace-plastron opening with its broad forelimbs, and remain in that position for some time. Many novice snapshots of this species show only that defensive pose.

In contrast to this situation, approaching a basking Western Pond Turtle (*Actinemys marmorata*) at a distance several times the tolerance limit for the Desert Tortoise will usually send it quickly into the water. Here the larger telephoto lens usually produces a better image. It is also helpful to recall the basking behavior for this species as outlined in its natural history account. Unlike most heliothermic lizards that normally exhibit only one morning basking period, the Western Pond Turtle often engages in a series of basking bouts interspersed with shorter aquatic foraging periods. Given this behavior, your anticipated morning of pond turtle photography is not lost if you happen to initially disturb your intended subject(s). Instead, take advantage of their temporary absence on a basking site to position yourself a safe distance away, preferably within some sort of vegetative cover, focus in on the vacated basking log or bank segment, and quietly wait.

We stress the word quiet here, because an unusual sound will often send them into the water as fast as a person's movement. Here is another advantage of the digital camera, in that the click of its "shutter" is silent, and even if the camera is set to "shutter sound" the electronic sound is far less disturbing than the noise made by the physical focal-plane shutter of a 35-mm film camera, and the near-silent operation of camcorders is a great improvement over the whirl of the old wind-up movie cameras.

Lizards

Many lizard species are ideal subjects for field reptile photography because of the territorial behavior of the males. Besides being usually the more colorful sex, territorial males normally perform their advertisement displays on elevated, well-illuminated sites that they are reluctant to vacate so long as the photographer's approach is slow and stops short of the tolerance distance of the species at hand. Such distances vary considerably with species. For example, a Western Fence Lizard (*Sceloporus occidentalis*) will normally allow human approach to within about 2.5 m (8 ft), whereas a male Chuckwalla (*Sauromalus ater*) will often vacate his rocky display post and squeeze into a nearby crevice before the photographer approaches by twice that distance. It will often take the first such retreat of an intended subject for you to "get a feel" for the approach limit for that species, but once that distance is identified and honored, the results from both still and video photography can be most rewarding. This is because most male heliothermic lizards readily display their species-specific head bob and/or pushup sequences, especially during spring. Recall also that many exhibit a form of displacement behavior in which the territorial display, normally reserved for approaching lizards, is often presented to other intruders, including humans. In short, these territorial males are real "hams" and will usually give a top performance to any patient photographer. An added attraction here is when another male of the same species advances and challenges the territorial male, and if the latter does not retreat, your photo or video sequences may equal or surpass Hollywood-generated scenes of battling dinosaurs.

Snakes

As with turtles, snake field photography spans the range from relatively easy to quite challenging. Among the best subjects are the rattlesnakes, because most usually assume a coiled, rattling defense pose when encountered instead of making a fast retreat as do most racers. Several of the rattlesnake photos in this book depict individuals that were exhibiting this defense, and in such situations you can keep pressing the shutter button until that perfect pose with raised rattle and extended tongue is achieved. The one obvious precaution here is to be sure that you stop your approach to the snake at a distance that is significantly greater than its striking distance (usually about one third of its length). Also be aware that the excitement of acquiring excellent photos or video footage may sometimes take one's attention off what the rattlesnake is doing. The Sidewinder (*Crotalus cerastes*) is especially skillful in moving slowly toward you without being noticed, and you may suddenly find your subject far too close for comfort. A few of the pictures in this book were taken with the junior author's now antique-grade 35-mm camera, the viewing field of which goes dark after each shot until the film is advanced by hand. To hold his focus field, he acquired the habit of keeping his eye glued to the viewfinder, even during this relatively short blackout period. He quickly broke that habit, however, after one photo sequence of a Western Diamond-backed Rattlesnake (*Crotalus atrox*), in which he suddenly found that the snake had moved in dangerously close once the view window was reopened.

Other species such as the Gopher Snake (*Pituophis catenifer*) will sometimes also coil, vibrate its tail, and occasionally strike when approached, but on other occasions simply retreat. Such retreats are usually not as fast as those of racers, however, and one can often get ahead of the snake's intended travel path and make a good photograph or action video as it passes by. Species such as the Rubber Boa (*Charina bottae*) and the Ring-necked Snake (*Diadophis punctatus*) are among a photographer's favorite subjects, because they rarely retreat when encountered under ground-cover objects and often execute their respective defensive tail displays (see natural history accounts). In great contrast to such cooperation are the actions of racers and most garter snake species, which usually make a speedy retreat when discovered. Indeed, one of the most abundant snakes in coastal central California, the Western Yellow-bellied Racer (*Coluber constrictor mormon*), is usually one of the least successfully photographed in the field for just this reason. Whatever the degree of difficulty may be in obtaining a good photo or video sequence of a wild amphibian or reptile in its natural habitat, the satisfaction that comes with capturing such images is usually well worth the effort.

PHOTOGRAPHING CAPTIVE OR TEMPORARILY RESTRAINED SPECIMENS

Given the aforementioned rewards of wild amphibian and reptile photography, there are some situations in which photographing specimens in terrariums or under temporary restraint in the field is the only way to obtain the desired results for a given situation. Instructional photos whose purpose is to show specific features normally hidden from view, such as the "spade" on the hind foot of a spadefoot toad or the ventral display coloration of a lizard, can be best shown in live specimens by the use of hand restraint. Even though the subject is in hand, one must still be aware of certain basic factors to ensure good results. With hand-held subjects you can control the amount and angle of light striking it, and it often takes several positionings of the subject with respect to the sun or artificial-illumination source to achieve the best result. Backgrounds can also make or break a specimen-in-hand picture or a "show and tell" video sequence. Because most such pictures are taken as the person with the animal holds it in front of his or her body, the color of the garments is paramount. A small lizard held in front of a white shirt front will usually cause the automatic-aperture system of the camera to set the exposure for the shirt, not the animal, with the result that the latter is quite underexposed. The impact of a photo whose intent is to show the color pattern of a subject will be diminished if it is held in front of a gaudy, multicolored garment. A solid, medium-gray background is a good standard for such pictures, and if several are anticipated, a piece of cloth of that color can be carried in your camera case for just that purpose. A good alternative to using a neutral-gray backdrop is to have an assistant hold the animal to one side of his or her body so that distant, out-of-focus vegetation composes the background.

The use of photo terrariums or the short-term restraint of animal subjects, a process sometimes referred to as "nature-faking" may be both useful and justified under certain circumstances. For example, during family or class field trips on which a number of people disperse and catch various amphibians and reptiles for everyone to see before release at their respective capture site, participants often want to take photos, and not just of a hand-held animal. With a little advance planning, one or more terrariums can be set up at the "base camp" and landscaped with local substrate and plants. Captured specimens are then brought to this site and held in an appropriately decorated terrarium until all have a chance to photograph them.

One problem with this method is the reflection from the glass through which the pictures are taken. This can sometimes be eliminated by simply adjusting the angle of the glass to the sunlight path. A more reliable method is to construct or modify a photo terrarium with the front piece of glass loosely held by vertically grooved corner posts so that it can be easily raised. Once most amphibian or reptile species settle down in such a completely enclosed unit, one person can slowly raise the upper front glass panel for a short time while others take their pictures. If the animal begins moving toward the momentarily opened front side, the glass can be quickly lowered to prevent the subject from escaping.

When employing this central photo station technique, it is important that all subjects be returned to their exact site of capture, because reintroduction into an unfamiliar area could result in rapid predation due to the lack of familiar retreats and, in the case of males, the loss of their territorial status, because he is now probably in another male's realm. This can be avoided by clearly marking the capture site with some sort of flagging placed as high as possible on an appropriate bush or rock. If the site is not within easy view of the base camp, a sketch map and/or coordinates from a good-quality GPS unit will also be needed. Direction and distance can be very deceiving in the field, and it's best not to trust just memory to relocate capture sites.

When photo terrariums are not available, more blatant forms of nature fakery must sometimes be resorted to. One is simply to hold one hindlimb of a legged amphibian or reptile or the tail portion of a snake so that it and the holder's thumb and forefinger are hidden from view behind the rock or tree limb on which the subject is posed. This can also be done in soft substrate by burying the holders hand and arm along with the animal's held appendage. The problem with this technique is that the held specimen is usually in a full escape mode and rarely exhibits the limb placement, head carriage, and eye direction that one who is familiar with that species normally sees and expects.

Another form of photographic fibbing entails the chilling of specimens to or slightly below their low ecological lethal thermal threshold, then quickly positioning them in a natural setting for the picture. In the field this can be done by placing the container holding the animal in a portable cooler with care not to allow it to contact the ice or ice water. However, some groups such as lizards never seem to assume an alert posture and stance while in this state. Indeed, there is usually a tendency for them to close their eyes, for indeed they are really on the threshold of exothermic sleep. Given this situation, it is best to focus in and then wait for the lizard's body temperature to rise back to the emergence temperature range. At that point there usually occurs a moment when the animal finally does appear alert, and you can get a few quality pictures or video frames before it dashes off. Because the final part of this technique entails the release of the subject, it is important to employ this method at or very near the site of capture.

Among reptiles this technique actually works best with snakes, because they lack eyelids and thus always appear awake. This is also true for dead snakes, and occasionally a near-perfect road-killed specimen is recovered that, when propped up in a characteristic pose within a rock or branch network, can look rather convincing. The authors know a few photographers who take dead-snake photography one step further by "wiring" the deceased specimen. Here a flexible wire is passed through the mouth and down the length of the body, which enables them to bend the snake into an alert-appearing pose without the aid of rock or branch props, and once again the eyes remain wide open. We assure the reader, however, that no wired snakes are pictured in this book.

CAPTURING AMPHIBIANS AND REPTILES

The capture and short-term restraint of amphibians and reptiles is needed for several of the photographic techniques previously mentioned and also for meaningful field instruction. The experience of seeing at close range and touching the various reptile scale types and viewing the nasolabial groove on a plethodontid salamander's snout with a hand lens will usually be remembered far longer than if the only exposure was through verbiage in a classroom or print in a book. The following are several methods by which these animals can be safely captured, momentarily held, and then released in the field.

Before setting out to capture one or more species for the above purposes, you need to be aware of the rules and regulations that govern this activity. We have mentioned several times in this book that at present (2011) you may legally capture the majority of California amphibian and reptile species by purchasing a freshwater fishing license. For those species not listed in the regulation booklet that is supplied with the license, you need to consult a reliable and current resource as to their protective status. This book will serve that purpose for awhile after publication, but because new species are occasionally added to the protective rolls, it's best to consult one of the CDFG web sites for current listings. To capture any such listed species, you will then need to apply for a scientific collecting permit with attached memorandum of understanding for the species involved. Even with the proper state-issued license or permit, it still is illegal to collect in a regional, state, or national park without obtaining written permission to do so from the park authorities.

Once these important precapture arrangements have been made, the next step is to select a potentially productive capture site. Familiarity with a species' preferred habitat(s) is paramount for success, especially when searching for species that are habitat specialists. Looking for fringe-toed lizards on hardpan desert, night lizards where there are no fallen branches of the Joshua tree (*Yucca brevifolia*) or other large yucca species, slender salamanders along the shores of cold mountain streams, and so on, will be a complete waste of time. Often the location of a habitat generalist can be equally challenging because of the variety of situations in which they may be found. Here a knowledge of seasonal habitat use for each species can be very helpful. For instance, some habitat- and feeding-generalist garter snake species often forage in upland habitats

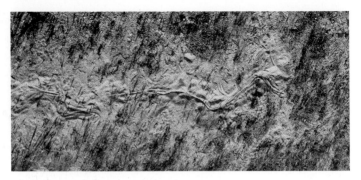
Plate 122. Tracks and tail drag of a whiptail lizard.

during the wet spring season but then concentrate their activities near pond and stream shoreline sites during the dry summer and early-fall seasons.

In desert areas, much field time can be saved by looking for signs of reptile activity. The presence on a rock of small dried pellets of dark fecal material with small white uric acid caps at one end indicates that a lizard uses this site, possibly for display or as a feeding perch, and therefore may return to it soon. In soft desert substrate, the tracks of lizards and snakes are one of the best indicators that you are in a good collecting area.

Your particular hunting strategy is also important. Some herpetologists prefer to take long hikes through a potentially productive area on the assumption that the more ground they cover, the more herps they will see. An alternate method entails a more sedentary approach in which one remains in a relatively small area of good habitat and simply waits for herps to appear. On the junior author's annual herpetology course trips to the Mojave Desert, it was often the group of students whose turn it was to cook breakfast that saw and caught more lizards and snakes for the morning's "show and tell" session as they emerged and/or moved through the field camp area while the rest of the class was on a long and less productive morning "herp walk." The senior author and his longtime friend and fellow herpetologist, Dr. Nate Cohen, are prime examples of these contrasting approaches to collecting. The former is of the "long walk" school and on one of their past desert trips embarked on an extensive, hot search while Dr. Cohen chose to remain at a roadside stop area to see what might be lurking there. The result of this contrasting methodology, in somewhat exaggerated form, may be seen in figure 38.

Capture Techniques

HAND CAPTURE

Once these important pre-capture decisions have been made, a capture technique must be chosen. For slow-moving species such as most salamanders and toads, capture by hand is usually the most practical. Often this follows the lifting of ground-cover objects such as rotting logs, which should always be replaced in their exact position so as to preserve the microhabitat created by

Figure 38. The senior author's illustration of the comparative results between the "sit and wait" reptile-collecting method and the "hike all day" approach.

their presence on the substrate. When making hand captures of such species, you must be aware that the human hand can be a lethal weapon if not used correctly. The skin temperature of our palms is often at or above the high thermal limit of many small ectotherms, especially plethodontid salamanders, and their large ratio of body surface area to body mass permits rapid heat exchange with the immediate environment—in this case the closed human hand. Thermal injury to these amphibians can be avoided by holding a specimen by thumb and forefinger only and quickly transferring it to a cool container.

The hand capture of relatively fast-moving species is a different matter. Here a successful stalk is necessary to attain a position within an arm's length of the animal. However, many heliothermic lizards, snakes, and ranid frogs have a tolerance limit to human approach that is far greater than the length of one's arm, and thus attempts at hand capture are usually unsuccessful. For those species that do permit such a close approach, the problem now becomes how to execute a quick hand capture without injuring the specimen as you bring your hand down upon it. Many "hand grabbers" develop the technique of allowing their palm base to strike the substrate just short of the animal, so that the force of the strike is not brought to bear on the specimen. At the moment of impact, the cupped hand should lightly cover the animal and one's fingers close around it.

Even if a hand capture is successful up to this point, the animal still may escape. The mucus coating of most anurans makes them hard to hold on to, and highly active lizards and snakes have the uncanny ability of finding any small opening through which to escape as you attempt to close your fingers around them. To prevent this, bring your free hand into play. If a snake's head or frog or lizard forelimb is protruding from under your capture hand, grasp one of these items with the other. Hold a snake on both sides of the neck just behind the head with your thumb and forefinger. Moderate pressure can be applied without risking injury, but once you begin to lift the snake from the substrate be sure to support its body with your free hand. Pressure on a frog or lizard limb can be greater, about the amount you can apply to your little finger without inducing pain. If no head or limb is protruding from beneath your hand, you will need to search with your free thumb and forefinger under your cupped hand and feel around until you find and grasp a neck or appendage.

If a wary species cannot be successfully approached to within an arm's length, a long-handled net may sometimes be used successfully for a capture, although some often are able to escape through spaces between the net rim and the ground surface. In addition, a metal net rim brought down hard upon the substrate may injure or kill fast-moving anurans, lizards, and snakes, which usually begin their retreat at your first movement, and the trajectory of a net is much harder to control and adjust than just that of your arm and hand. Long-handled dip nets are a preferred item for collecting aquatic amphibian larvae and slow-swimming adults, because injury due to the pinning of an animal between the net rim and the substrate is usually not a factor. However, the use of any type of net, jars, rubber boots, etc., employed for aquatic collecting runs the risk of transporting chytrid fungus, a major killer of aquatic amphibians, from one capture site to another. The saturation of any such item that contacts the water or moist shoreline in a 10 percent bleach solution before use at each collecting site should be a standard part of this capture method.

NET AND NOOSE CAPTURE

In the late 1950s a colleague of the senior author, Dr. Richard Eakin, published a note on a lizard capture method that has become the standard technique for a variety of field studies (home range, growth, etc.) in which the capture, marking, release, and often recapture of individuals is necessary. It entails the use of a slip noose constructed from a single strand of lamp cord (zip cord) wire that is attached to the tip of a long, thin pole. A light bamboo fishing pole or fly rod with the terminal grommet removed works very well. Lizards are usually not disturbed when such a twiglike item is slowly maneuvered close to them so that the noose can be lowered over the animal's head. Once it encircles the neck area the pole tip is raised, lifting the lizard quickly off the substrate. While dangling from the noose it is essentially helpless, but the battle is still only half won, because while you are attempting to grab it, it may gain sufficient purchase on your hand to slip the noose and escape. To avoid this, let the pole slide through your hand as you move rapidly toward the dangling noosed lizard. When you reach the tip of the pole allow the lizard to grasp your clothing in your chest area and then with your free arm quickly perform

Plate 123. Herpetology students trying their hand at lizard noosing.

the cupped-hand capture technique previously described. Even with the lizard now in hand, a novice nooser often has difficulty loosening and removing a noose from a lizard's neck, especially if tension from the weight of the noosing pole is still being applied. At this point it is best to have a small scissors ready to cut the noose material at its attachment point to the pole tip. With all tension released, one can now easily loosen and release and later re-attach the severed noose to the pole.

During countless past lizard-noosing instruction sessions by both authors, a commonly asked question is if this capture method hurts or injures these reptiles. We and our many colleagues and past students who employ this technique have never seen any indication that this is so. The neck anatomy of lizards appears to be far more resistant to stress than ours, and thus the comparison of a noosed lizard and a noosed human is not a valid one.

Of course, one must use good common sense during the noosing process. If a lizard is being noosed in dense brush or well back in a rock crevice where its body may become held or wedged within these retreats, serious injury could result if the noose is harshly jerked. In such cases the lizard should beheld in place through constant, light tension on the noose while one makes a slow, deliberate hand capture.

We have also found that once exposed to this form of capture, many people become "hooked" on lizard noosing, and on extended field trips some are often reluctant to return to camp until they have succeeded in capturing and releasing that one lizard that continues to elude them. One probable reason for such attraction to this technique is that its challenges are similar to those of fly fishing, where a careful approach to a pool where fish have been sighted followed by the proper presentation of a fly or lure to them is usually the key to success.

There have also developed several "schools of noosing" among converts to this pastime. While the authors believe that a thin strand of copper lamp wire is the best noose construction material for small to medium-sized lizards, others prefer more flexible and stronger materials such as dental floss, because it always tightens securely around a lizard's neck no matter at what the angle the pole tip is pulled. On the other hand, a floss noose does not keep its shape well under wind conditions and unlike copper wire is hard to maneuver through brush and into crevices to extract a sequestered species. There is yet a third group of noosers who prefer fine monofilament fishing line to either of these materials. It holds its circular shape when taken off a spool and, like copper wire, is fairly maneuverable but far stronger. This latter point is important when attempting to noose very large species such as the Chuckwalla (*Sauromalus ater*). However, thin copper wire can be strengthened by twisting two strands tightly together.

CAPTURING SNAKES

As efficient a technique as noosing is for lizards, it does not work for most kinds of snakes because the head blends smoothly with the neck, and thus there is no indentation behind the skull where the noose loop may be secured. One exception among California snakes are the rattlesnakes, which do have a relatively narrow neck compared with their broad head and thus could be successfully noosed. However, to attempt a capture via this technique would be hazardous to both the rattler and its capturer. Once noosed, the latter could not rush up and grasp the snake as is done with noosed lizards without running a very high risk of being bitten. The rattlesnake may also suffer, because its very heavy body would now be supported entirely by its neck, and these snakes often engage in extensive body twisting when noosed.

Given these problems, the only safe and humane way to capture rattlesnakes other than by trapping is to use a mechanical snake stick with grasping tongs that are controlled by a lever on the handle. These are about one meter (one yard plus) long, and with one's arm extended a rattlesnake can be kept well away from the captor's body while holding the snake in its mid-body region. Once secured in this manner, a rattler can easily be manipulated as desired. However, to use this method successfully, you must acquire a "feel" for holding a snake with a mechanical extension of the hand. The best way to acquire this is through practice, either with a relatively large nonvenomous species such as a Gopher Snake or, if not available, an appropriate-sized section of flexible plastic tubing.

Trapping

For field surveys whose intent is to determine what species complement may be present at a given site and/or to estimate the population size of one or more of its members, the system of a drift fence with a funnel trap at either end, developed by Dr. Henry Fitch in the early 1950s, remains the most efficient method. However, like most other trapping methods it is relatively nonselective and thus has the potential of catching protected as well as unprotected species. It is also not one of methods that the CDFG lists as a legal means of

Plate 124. A low drift fence with pit trap for salamanders.

catching the many species for which it grants a limited take for fishing license holders, and therefore special permits from both the CDFG and USFWS are usually required before using this capture method.

As for the method itself, drift fences may be constructed of any moderately sturdy material supported by wooden stakes at approximately four-foot intervals. The base of the fence should be buried about 7.5 cm (3 in.) below grade to discourage animals from burrowing underneath. The height varies with the species or species complement sought but need not be so high that it is impossible for these animals to climb or jump over. A height of 15 cm (6 in.) above grade is usually effective for ambystomid salamanders, whereas 30 to 40 cm (12 to 16 in.) is adequate for most snakes and lizards. These relatively low heights are normally effective, because most species tend to move or "drift" along the fence for some distance before either moving away or attempting to span it. The junior author has found that for many California amphibians and reptiles about 9 m (30 ft) is an effective drift fence length.

Plate 125 shows a funnel trap system in which a flexible thin hardboard drift fence is used to conform to the terrain surrounding a rock outcrop where the presence of the Alameda Striped Racer (*Masticophis lateralis euryxanthus*) is suspected. Wooden traps with large entrance funnels are positioned at each end of the fence. Funnels and trap backs are constructed of 3-mm (1/8-in.) mesh hardware cloth, because larger mesh sizes may restrict and kill small snakes if they attempt to squeeze through. The large end of the funnel should be at least 30 cm (12 in.) in diameter and positioned so that it bisects the end of the drift fence. To ensure that an active trapped snake does not randomly locate the small end of the funnel and escape, an inward-swinging clear plastic flap is suspended over that opening within the trap body.

Plate 125. A high snake drift fence winding through rock outcrops.

When employing any of these trap designs, several precautions must be taken to protect the trapped animals. Perhaps the most important is that traps should be checked at least once a day or night, and in areas of high animal abundance, several checks over a 24-hour period may be necessary. When trapping in warm weather an elevated reflecting sun roof should cover each trap to prevent captives from suffering hyperthermia. A hardware cloth trap back instead of a solid one will also help to cool the trap by allowing cross ventilation. When trapping in areas where amphibians are present, the possibility of desiccation is a serious problem. A seemingly reasonable solution would be to place a small, shallow container of water in the trap in which a frog or salamander would then presumably go to rehydrate. Unfortunately, most amphibians don't share this reasoning and instead often remain in the dry areas of the trap where they sometimes desiccate past their lethal limits. The way to prevent this is to provide a saturated substrate for the entire trap floor. Saturated water crystals of the type used to keep plant potting soil moist are ideal for this purpose and, once saturated by soaking in water, can be kept moist by spraying with each trap check.

When using traps it is necessary to protect against trap robbers. During the interval between the time a specimen is trapped and when you check the trap, captured animal(s) are in an essentially helpless position with only the trap body to protect them. The once-popular open-pit or bucket traps are especially vulnerable to a variety of mammalian, avian, and even larger reptilian predators, and we do not recommend their use.

Two of the most dangerous trap robbers are the Raccoon (*Procyon lotor*) and the Coyote (*Canis latrans*). These species are abundant in most areas in California and are powerful enough to rip apart a trap if it is not solidly built.

Special attention should be given to securing the back of the trap, which is usually the side that is opened to retrieve captured specimens. It is also a good idea to stake traps down, because both of these thieves are adept at rolling or actually carrying them some distance away. However, when trapping in areas that people frequent, the most severe trap predation may be from our own species. Offenders range from curious youths or young vandals to adult amphibian and reptile collectors. Here common sense in the choice of trapping sites is usually the best protective measure. Locations adjacent to road or hiking paths should be avoided, even if a very attractive habitat segment is situated in such a location. Drift fence height should also be kept as low as possible, and fence color should blend with the surrounding grass and brush. Even your presence while installing or checking traplines may advertise your survey effort, and when possible such activities should be conducted during periods of low human presence near the trapping site.

Using Cover Boards

An alternate method for surveying the amphibian and reptile complement of a given area is through the use of cover boards (figure 39). This approach is based on the attraction of most species to any ground-cover object that offers the appropriate thermal and moisture features for daily, nightly, or extended retreat periods. In some habitats such as a woodland with numerous fallen, rotting logs and tree trunks, it is usually better to carefully turn and then replace these natural ground-cover objects than to supply additional ones. However, in most other habitats, relatively large ground-cover items beneath which a satisfactory retreat microclimate exists are often very scarce, and thus if such cover is provided it may attract one or more wandering species, which then remain sequestered for varying periods of time. Indeed, it is not unusual to find a mix of both amphibian and reptile species under one large cover board.

Cover objects may consist of a variety of materials. Readers may recall that the junior author's first reptile capture resulted from the lifting of discarded winter "long johns" in a vacant lot, and less elaborate fabric items such as large pieces of canvas or even heavy, opaque sheet plastic work well, though they may be blown away by heavy winds if not weighted down. Corrugated-metal roofing sheets also work well, but the most popular cover objects for herp surveys are pieces of plywood or similar wood panel products. Boards should cover a minimum area of about one square meter (square yard) but need not be in the basic square form, because rectangular boards are easier to carry about in the field. Like drift fence trap lines, cover boards should be placed

Figure 39. A basic cover board design with one end slightly elevated to allow wandering amphibians and reptiles to seek shelter beneath.

at sites within the survey area where movement and activity of the species desired would most often occur. Ecotone situations such as the upper shorelines of ponds, streams, and marshes or along the border between woodland and grassland are usually productive. Position the boards in areas of scant or no plant cover, because dense grass and forb cover will prevent a board from settling near enough to the ground to produce and maintain an acceptable thermal/moisture microclimate beneath. A space of about one index-finger width between the board and substrate is usually necessary to ensure that this condition is achieved. The cover board design shown in figure 39 provides a space gradient beneath the board to accommodate the thigmotactile tendencies of individuals of species that vary in size.

Board check visits to the site should be scheduled with the activity of the species that one wishes to capture or observe in mind. Most heliotherms will use this cover as a nightly retreat and may still be present in a suboptimal activity state in the early-morning hours. Conversely, nocturnal snakes and amphibians would be expected under the boards during daylight hours, especially in warm, sunny weather. When checking for animals beneath a board, do not lift it completely off the ground, because this will not only startle any animals beneath but will also temporarily destroy the microclimate that has developed there. Instead, lift one edge just enough so that you can see beneath with your eyes near ground level. A small flashlight can be very helpful for such viewing. If one or more herps are detected, it is usually best to carefully lower the board and then plan your next move. If you desire only a photo or video record of a specimen, secure your camera or camcorder in one hand so that the other is free to again lift the board or, best of all, have a helper slowly lift it for you.

Most species tend to remain in a sequestered position for a short time when their retreat roof is slowly lifted. One or more close-up pictures at this time can often be sufficient to obtain a record of the individual markings of a specimen, so that it can be identified if seen and photographed again. When sequestered animals eventually leave their now-disturbed retreat, they often do so in a slow, deliberate manner that provides good "photo-ops" of these species in their natural habitat. If instead your goal is to actually capture a specimen, the hand-capture method previously described can now be employed, and if a lizard or snake is at or below its low ecological lethal temperature, a fast hand capture may not even be necessary.

The advantages of this survey technique over trapping are several. One is that the risks associated with trapping (hyperthermia, desiccation, trap predation) are eliminated, because the animals are free to choose and vacate these retreats whenever they desire. Cover boards or similar cover objects are also far easier to construct and install than the units of a drift fence-funnel or bucket trap system. Finally, if you employ this technique to simply observe and/or photograph an amphibian or reptile by the method just described, then you are technically not violating any of the tenets of the various laws and regulations that protect some species that may appear under cover boards. For all of these reasons, this approach to amphibian and reptile field studies and surveys is by far the best suited for supervised individual student and biology field course projects.

Recognizing Individual Amphibians and Reptiles in the Field

In field capture-release studies directed toward estimating the population size of a species or in visual surveys whose goal is to determine the territorial limits or home range of animals, the ability to recognize individuals either recaptured or seen naturally again is paramount. However, many of the standard marking methods used for bird and mammal studies such as colored leg bands, ear tags, or collars are not appropriate for amphibians and reptiles. Other techniques such as the application of paint or ink to the skin also present problems, because lizards and snakes periodically shed their outer skin layer, and the mucus-laden skin of most amphibians is not conducive to the use of most marking materials.

During the past few decades the method of permanently marking individuals of nearly all vertebrate species through the implantation of microtransponders has been widely used in professionally based scientific studies and is now being used to identify lost pets and select members of our own species. These transponders or "pit tags" are minute cylinders, each containing a specific electromagnetic sequence that can be read like the bar code on a product when held within a few inches of a decoding unit. They are implanted with a hypodermic needle, usually in the subcutaneous area. However, this equipment is rather expensive, and its use often requires appropriate permits.

Given the preceding review, our recommendation for identifying individuals of most amphibian and reptile species in the field entails the use of the individual marking system that genetic variability has already provided. With relatively few exceptions, California species are not monochromatic but instead have two or more skin colors that are usually arranged in a pattern that varies between individuals, in much the same way as human faces vary. Bold color pattern variation in species like the California Tiger Salamander (*Ambystoma californiense*) can be easily sketched in one's field notebook. Most others are less obvious but still distinct, and here is where photography, especially with the digital format, can provide precise individual-recognition information. During the first years of this century the digital camera business has expanded to such a degree that essentially anyone interested in the natural world has a fairly high-quality one, and if you happen to forget it on a particular outing, the ever-present cell phone can now take a very acceptable picture.

This method requires that you take a picture of either the whole animal or that area of its body where variations in the color pattern are most prominent. Such pictures may be of hand-held specimens that have been noosed or hand-captured, or they may be of totally undisturbed animals as they are observed under a cover board or abroad in their natural habitat. Pictures of each specimen seen or captured in a field study are then organized into species groups and stored on a reference disk for future comparison and analysis. Any individual with bold marking differences permits easy determination as to whether it is a new capture or a recapture. Even very subtle pattern differences or acquired skin blemishes that may go undetected when viewed with the naked eye will become apparent when pictures are greatly enlarged and then compared to those of previously photographed individuals on a split computer screen. Laptop units permit such comparisons to be made in the field as well as in the laboratory.

One final feature of this completely noninvasive individual-recognition technique is that it can be used with free-living special status species without the need for special permits, because the state and federal endangered species acts stop short of prohibiting the photographing of undisturbed wildlife. The techniques presented in the photography and videography section describe how, with the proper lens, this can be done at a safe distance.

AMPHIBIAN AND REPTILE HUSBANDRY

The successful husbandry of wild amphibians and reptiles is an art in itself that, unfortunately, is not often successfully mastered. Far too often the thermal and food requirements of captive amphibians and reptiles are not met, with the result that caged specimens slowly deteriorate and die. We therefore begin this section with yet another reminder of the great satisfaction to be gained from recording your observations of wild herps on film, digital camera cards, or magnetic tape to produce images that will never die.

However, if limited husbandry of specimens for display or research is appropriate, the decision as to how to obtain such species becomes paramount. If the purpose of such husbandry is simply to have one or two specimens for observation in the classroom, home, or for experiments in the laboratory, then by far the best decision is to purchase such animals from a pet store that specializes in amphibians and reptiles. Many of these now offer captive-raised animals, and therefore you will not be affecting a wild population through such a purchase. Animals from a reputable store are usually adapted to captive feeding and in good physical condition, because emaciated, lethargic animals usually don't sell. For short-term use in displays or laboratory behavioral studies, animals maintained in a good physical state could even be sold back to a pet store or obtained on some sort of "rent a herp" agreement. A business-wise store owner may even agree to a short-term free loan for classroom display in the hopes that parents of some of the students may soon be dragged into the store by their offspring.

Given the above possibilities, one still must make a wise buy. Healthy lizards have a filled-out tail base, and their pelvic girdle bones should not be detectable beneath the skin. The body of both amphibians and reptiles should also be filled out, not emaciated, and the animals should appear alert to their surroundings. The scales of snakes and lizards should not be elevated, a sign that usually indicates the presence of mites beneath. Large-scale mite infestations are one of the most common killers of captive snakes, and regular checks with a good magnifying glass should be made to detect these minute creatures.

Other major problems that can arise in captive amphibians and reptiles are bacterial or fungal infections. As long as the protective skin and/or mucus body covering remains intact, the former is usually not a serious problem in short-term captivity. However, if even a small portion of these outer coverings is destroyed, serious infections may occur. One major cause of this in captive anurans is the abrasion of head and back skin that occurs while attempting to escape by jumping and hitting a hard screen or hardware cloth terrarium cover. Reptile skin is also open to bacterial and fungal infection if abraded. This is especially true of snakes held in aluminum screen or hardware cloth cages. The natural escape mode of a snake is to probe with its snout, a behavior that in nature would allow it to enter dense ground cover or a soft soil retreat site. However, the persistence of this behavior in such cages will soon produce an open wound in the snout area that can lead to an infection of the entire mouth lining, which in turn negates feeding. Internal parasites may also be a problem in piscivorous species such as garter snakes when fed commercially raised food such as Mosquitofish (*Gambusia affinis*). Before feeding live fish, one or more should be sacrificed and the gut and body cavities inspected for the presence of nematodes that may then be passed on to a new host.

The more questions concerning the husbandry practices of the specimen in question one can ask of a knowledgeable salesperson in the store, the better. Of special importance is information on the thermal and light regime of the terrarium in which it is housed. For most lizards and snakes, there should be one area in the cage where either a 75- or 100-watt lamp bulb with reflector or a buried heating coil creates a substrate surface temperature similar to that preferred by the captive species. For most heliothermic lizards, this is around 35 to 40 degrees C (95 to 104 degrees F), and for anurans and nonbasking reptiles, a range of about 30 to 34 degrees C (86 to 93 degrees F) at the warm end of the terrarium is appropriate. Conversely, temperatures above about 25 degrees C (77 degrees F) should be avoided when keeping most salamander species. These amphibians are best held in a "warm" refrigerator and brought out only at display time. The daily heat and light cycle is also important. The warm, lighted area in a terrarium should not persist for more than about eight hours each day and should approximate the current daily sunlight period of a species' native habitat. Even then there should be shaded areas for the animals to retreat to. As discussed in the reptile physiology section, sustained warm conditions can cause metabolic "burnout" in these animals. Thus, for the long-term husbandry of most temperate latitude species the terrarium temperature should be markedly reduced during the winter months so as to stimulate a period of retreat or hibernation.

When housing amphibians, the prevention of dehydration is usually the most important factor to be considered. For this reason, covered-glass terrariums with a moist substrate that maintain a humidity of near 100 percent are greatly preferred over screened cages that permit the circulation of relatively dry room air. If in addition to a saturated substrate, a "pond" in the form of a shallow bowl or pan is also provided, such containers should be removed, rinsed, and refilled often to prevent a buildup of contaminants.

Of equal importance to these physical factors is the type of food that your intended purchase is being fed in the store and presumably eating. If possible, try to be there at feeding time so that perhaps the most important question concerning such a purchase can be answered: Is this animal eating? Captive herps, especially snakes, sometimes simply stop feeding in captivity and never begin again. Treat your pending purchase as you would other rather expensive ones that you make, and insist on seeing it "work" before you buy. This is also not the time for "impulse buying." Some species appear very spectacular in the store, but they must also have the appropriate physical attributes to ensure they will thrive in your home, classroom, or laboratory.

When making the wise choice of buying a captive-reared amphibian or reptile instead of catching and trying to maintain a local species, one should keep in mind that it should not be released into the wild when it is no longer needed. One reason is that pet store herps are, by law, not native California species, because the latter cannot be legally sold, and it is illegal to introduce nonnative animals into California. This "no release" law is a sound one, in that it deters the introduction of exotic species that could thrive in one or more of California's many and varied habitats, and eventually compete with native forms. It also negates the possibility of introducing harmful bacteria, fungi, viruses, and parasites into wild native populations. The dangers associated with this situation are discussed in the introductory section of this book.

CHECKLIST OF CALIFORNIA AMPHIBIAN AND REPTILE SPECIES

The following abbreviations denote the assigned protective category:

FPS   Fully Protected Species
FE   Federally Endangered
SE   State Endangered
FT   Federally Threatened
ST   State Threatened
CSSC   California Species of Special Concern

Amphibians (Class Amphibia)
SALAMANDERS (ORDER CAUDATA)

Mole Salamanders (Family Ambystomatidae)
- ☐ California Tiger Salamander (*Ambystoma californiense*), FT, ST
  (A population in Santa Barbara Co. is being considered for subspecies status.)
- ☐ Tiger Salamander (*Ambystoma tigrinum*), Introduced
- ☐ Barred Tiger Salamander (*Ambystoma mavortium*), Introduced

Long-toed Salamander (*Ambystoma macrodactylum*)
- ☐ Santa Cruz Long-toed Salamander (*Ambystoma macrodactylum croceum*), FPS, FE, SE
- ☐ Southern Long-toed Salamander (*Ambystoma macrodactylum sigillatum*)
- ☐ Northwestern Salamander (*Ambystoma gracile*)

Giant Salamanders (Family Dicamptodontidae)
- ☐ California Giant Salamander (*Dicamptodon ensatus*)
- ☐ Pacific Giant Salamander (*Dicamptodon tenebrosus*)

Torrent or Seep Salamanders (Family Rhyacotritonidae)
- ☐ Southern Torrent Salamander (*Rhyacotriton variegatus*), CSSC

Newts (Family Salamandridae)
- [ ] Coast Range Newt (*Taricha torosa*), CSSC
- [ ] Sierra Newt (*Taricha sierrae*)
- [ ] Rough-skinned Newt (*Taricha granulosa*)
- [ ] Red-bellied Newt (*Taricha rivularis*)

Lungless Salamanders (Family Plethodontidae)
- [ ] Dunn's Salamander (*Plethodon dunni*)
- [ ] Del Norte Salamander (*Plethodon elongatus*), CSSC
- [ ] Siskiyou Mountains Salamander (*Plethodon stormi*), ST
- [ ] Scott Bar Salamander (*Plethodon asupak*), ST

Ensatina (*Ensatina eschscholtzii*)
- [ ] Monterey Ensatina (*Ensatina eschscholtzii eschscholtzii*)
- [ ] Yellow-eyed Ensatina (*Ensatina eschscholtzii xanthoptica*)
- [ ] Painted Ensatina (*Ensatina eschscholtzii picta*)
- [ ] Sierra Nevada Ensatina (*Ensatina eschscholtzii platensis*)
- [ ] Yellow-blotched Ensatina (*Ensatina eschscholtzii croceater*), CSSC

- [ ] Large-blotched Ensatina (*Ensatina klauberi*), CSSC
- [ ] Arboreal Salamander (*Aneides lugubris*)

Black Salamander (*Aneides flavipunctatus*)
- [ ] Santa Cruz Black Salamander (*Aneides flavipunctatus niger*)
- [ ] Speckled Black Salamander (*Aneides flavipunctatus flavipunctatus*)

- [ ] Wandering Salamander (*Aneides vagrans*)
- [ ] Clouded Salamander (*Aneides ferreus*)
- [ ] Inyo Mountains Salamander (*Batrachoseps campi*), CSSC
- [ ] Kern Plateau Salamander (*Batrachoseps robustus*)
- [ ] Tehachapi Slender Salamander (*Batrachoseps stebbinsi*), ST
- [ ] Kern Canyon Slender Salamander (*Batrachoseps simatus*), ST
- [ ] Black-bellied Slender Salamander (*Batrachoseps nigriventris*)
- [ ] Gregarious Slender Salamander (*Batrachoseps gregarius*)
- [ ] Channel Islands Slender Salamander (*Batrachoseps pacificus*)

Garden Slender Salamander (*Batrachoseps major*)
- [ ] Garden Slender Salamander (*Batrachoseps major major*)
- [ ] Desert Slender Salamander (*Batrachoseps major aridus*), FE, SE

- [ ] San Gabriel Mountains Slender Salamander (*Batrachoseps gabrieli*)
- [ ] Gabilan Mountains Slender Salamander (*Batrachoseps gavilanensis*)
- [ ] Santa Lucia Mountains Slender Salamander (*Batrachoseps luciae*)
- [ ] Lesser Slender Salamander (*Batrachoseps minor*)
- [ ] San Simeon Slender Salamander (*Batrachoseps incognitus*)
- [ ] Kings River Slender Salamander (*Batrachoseps regius*)
- [ ] Sequoia Slender Salamander (*Batrachoseps kawia*)
- [ ] Relictual Slender Salamander (*Batrachoseps relictus*), CSSC
- [ ] Hell Hollow Slender Salamander (*Batrachoseps diabolicus*)
- [ ] California Slender Salamander (*Batrachoseps attenuatus*)
- [ ] Mount Lyell Salamander (*Hydromantes platycephalus*), CSSC
- [ ] Limestone Salamander (*Hydromantes brunus*), ST
- [ ] Shasta Salamander (*Hydromantes shastae*), ST

FROGS AND TOADS (ORDER ANURA)

Tailed Frogs (Family Ascaphidae)
☐ Coastal Tailed Frog (*Ascaphus truei*), CSSC

Spadefoots (Family Pelobatidae)
☐ Western Spadefoot (*Spea hammondii*), CSSC
☐ Great Basin Spadefoot (*Spea intermontana*)
☐ Couch's Spadefoot (*Scaphiopus couchii*), CSSC

True Toads (Family Bufonidae)
☐ Sonoran Desert Toad (Colorado River Toad) (*Bufo alvarius*), CSSC
☐ Great Plains Toad (*Bufo cognatus*)
☐ Red-spotted Toad (*Bufo punctatus*)
☐ Arroyo Toad (*Bufo californicus*), FE, CSSC
Western Toad (*Bufo boreas*)
    ☐ Boreal Toad (*Bufo boreas boreas*)
    ☐ California Toad (*Bufo boreas halophilus*)
☐ Black Toad (*Bufo exsul*), ST
☐ Yosemite Toad (*Bufo canorus*), CSSC
☐ Woodhouse's Toad (*Bufo woodhousii*)

Chorus Frogs (Family Hylidae)
☐ Pacific Chorus Frog (Pacific Treefrog) (*Pseudacris regilla*)
☐ California Chorus Frog (California Treefrog) (*Pseudacris cadaverina*)

True Frogs (Family Ranidae)
☐ California Red-legged Frog (*Rana draytonii*), FT, CSSC
☐ Northern Red-legged Frog (*Rana aurora*), CSSC
☐ Cascades Frog (*Rana cascadae*), CSSC
☐ Oregon Spotted Frog (*Rana pretiosa*), CSSC
☐ Foothill Yellow-legged Frog (*Rana boylii*), CSSC
☐ Southern Mountain Yellow-legged Frog (*Rana muscosa*), FE
☐ Sierra Nevada Yellow-legged Frog (*Rana sierrae*), ST
☐ Northern Leopard Frog (*Rana pipiens*), Native (CSSC) and introduced
☐ Southern Leopard Frog (*Rana sphenocephala*), Introduced
☐ Rio Grande Leopard Frog (*Rana berlandieri*), Introduced
☐ Lowland Leopard Frog (*Rana yavapaiensis*)
☐ American Bullfrog (*Rana catesbeiana*), Introduced

Tongueless Frogs (Family Pipidae)
☐ African Clawed Frog (*Xenopus laevis*), Introduced

Reptiles (Class Reptilia)
SQUAMATE REPTILES (ORDER SQUAMATA)
LIZARDS (SUBORDER SAURIA)

Eyelid Geckos (Family Eublepharidae)
Western Banded Gecko (*Coleonyx variegatus*)
    ☐ Desert Banded Gecko (*Coleonyx variegatus variegatus*)
    ☐ San Diego Banded Gecko (*Coleonyx variegatus abbotti*)
☐ Barefoot Gecko (Switak's Banded Gecko) (*Coleonyx switaki*), ST

Geckos (Family Gekkonidae)
- [ ] Leaf-toed Gecko (*Phyllodactylus nocticolus*)
- [ ] Mediterranean House Gecko (*Hemidactylus turcicus*), Introduced
- [ ] Moorish Wall Gecko (*Tarantola mauritanica*), Introduced

Iguanids (Family Iguanidae)
- [ ] Desert Iguana (*Dipsosaurus dorsalis*)
- [ ] Common Chuckwalla (*Sauromalus ater*)

Collared and Leopard Lizards (Family Crotaphytidae)
- [ ] Great Basin Collared Lizard (*Crotaphytus bicinctores*)
- [ ] Baja California Collared Lizard (*Crotaphytus vestigium*)
- [ ] Long-nosed Leopard Lizard (*Gambelia wislizenii*)
- [ ] Blunt-nosed Leopard Lizard (*Gambelia sila*), FE, SE
- [ ] Cope's Leopard Lizard (*Gambelia copeii*)

Zebra-tailed, Fringe-toed, Spiny, Tree, Side-blotched, and Horned Lizards (Family Phrynosomatidae)
- [ ] Zebra-tailed Lizard (*Callisaurus draconoides*)
- [ ] Coachella Valley Fringe-toed Lizard (*Uma inornata*), FE, ST
- [ ] Colorado Desert Fringe-toed Lizard (*Uma notata*), CSSC
- [ ] Mojave Fringe-toed Lizard (*Uma scoparia*), CSSC

Western Fence Lizard (*Sceloporus occidentalis*)
   - [ ] Northwestern Fence Lizard (*Sceloporus occidentalis occidentalis*)
   - [ ] Sierra Fence Lizard (*Sceloporus occidentalis taylori*)
- [ ] Island Fence Lizard (*Sceloporus becki*)
- [ ] Sagebrush Lizard (*Sceloporus graciosus*)
- [ ] Southern Sagebrush Lizard (*Sceloporus vandenburgianus*), CSSC
- [ ] Desert Spiny Lizard (*Sceloporus magister*)
- [ ] Granite Spiny Lizard (*Sceloporus orcutti*)
- [ ] Common Side-blotched Lizard (*Uta stansburiana*)
- [ ] Long-tailed Brush Lizard (*Urosaurus graciosus*)
- [ ] Black-tailed Brush Lizard (*Urosaurus nigricaudus*)
- [ ] Ornate Tree Lizard (*Urosaurus ornatus*)
- [ ] Banded Rock Lizard (*Petrosaurus mearnsi*)
- [ ] Coast Horned Lizard (*Phrynosoma blainvillii*), CSSC

Desert Horned Lizard (*Phrynosoma platyrhinos*)
   - [ ] Northern Desert Horned Lizard (*Phryosoma platyrhinos platyrhinos*)
   - [ ] Southern Desert Horned Lizard (*Phryosoma platyrhinos calidiarum*)
- [ ] Flat-tailed Horned Lizard (*Phrynosoma mcallii*), CSSC
- [ ] Pigmy Short-horned Lizard (*Phrynosoma douglasii*)

Night Lizards (Family Xantusiidae)
Desert Night Lizard (*Xantusia vigilis*)
   - [ ] Yucca Night Lizard (*Xantusia vigilis vigilis*)
   - [ ] Sierra Night Lizard (*Xantusia vigilis sierrae*)
- [ ] Wiggins' Night Lizard (*Xantusia wigginsi*)
- [ ] Granite Night Lizard (*Xantusia henshawi*)
- [ ] Sandstone Night Lizard (*Xantusia gracilis*), CSSC
- [ ] Island Night Lizard (*Xantusia riversiana*), FT

Skinks (Family Scincidae)
- ☐ Western Skink (*Plestiodon skiltonianus*)
- ☐ Gilbert's Skink (*Plestiodon gilberti*)

Teiids (Family Teiidae)
Western Whiptail (*Aspidoscelis tigris*)
- ☐ California Whiptail (*Aspidoscelis tigris munda*)
- ☐ Coastal Whiptail (*Aspidoscelis tigris stejnegeri*)
- ☐ Great Basin Whiptail (*Aspidoscelis tigris tigris*)
- ☐ Orange-throated Whiptail (*Aspidoscelis hyperythra*)

Alligator Lizards and Relatives (Family Anguidae)
Southern Alligator Lizard (*Elgaria multicarinata*)
- ☐ California Alligator Lizard (*Elgaria multicarinata multicarinata*)
- ☐ Oregon Alligator Lizard (*Elgaria multicarinata scincicauda*)
- ☐ San Diego Alligator Lizard (*Elgaria multicarinata webbii*)

Northern Alligator Lizard (*Elgaria coerulea*)
- ☐ Northwestern Alligator Lizard (*Elgaria coerulea principis*)
- ☐ San Francisco Alligator Lizard (*Elgaria coerulea coerulea*)
- ☐ Shasta Alligator Lizard (*Elgaria coerulea shastensis*)
- ☐ Sierra Alligator Lizard (*Elgaria coerulea palmeri*)
- ☐ Panamint Alligator Lizard (*Elgaria panamintina*), CSSC

North American Legless Lizards (Family Anniellidae)
- ☐ California Legless Lizard (*Anniella pulchra*), CSSC

Venomous Lizards (Family Helodermatidae)
Gila Monster (*Heloderma suspectum*)
- ☐ Banded Gila Monster (*Heloderma suspectum cinctum*), CSSC

SNAKES (SUBORDER SERPENTES [OPHIDIA])

Blind Snakes (Family Leptotyphlopidae)
- ☐ Western Blind Snake (*Leptotyphlops humilis*)

Boas (Family Boidae)
- ☐ Northern Rubber Boa (*Charina bottae*)
- ☐ Southern Rubber Boa (*Charina umbratica*), ST
- ☐ Rosy Boa (*Lichanura trivirgata*)

Colubrids (Family Colubridae)
- ☐ Ring-necked Snake (*Diadophis punctatus*)
- ☐ Common Sharp-tailed Snake (*Contia tenuis*)
- ☐ Forest Sharp-tailed Sanke (*Contia longicaudae*)
- ☐ Spotted Leaf-nosed Snake (Western Leaf-nosed Snake) (*Phyllorhynchus decurtatus*)

Racer (*Coluber constrictor*)
- ☐ Western Yellow-bellied Racer (*Coluber constrictor mormon*)

Coachwhip (*Masticophis flagellum*)
- ☐ San Joaquin Coachwhip (*Masticophis flagellum ruddocki*), CSSC
- ☐ Red Racer (*Masticophis flagellum piceus*)

- ☐ Baja California Coachwhip (*Masticophis fuliginosus*)

Striped Racer (*Masticophis lateralis*)
- ☐ California Striped Racer (*Masticophis lateralis lateralis*)
- ☐ Alameda Striped Racer (*Masticophis lateralis euryxanthus*), FT, ST
- ☐ Striped Whipsnake (Desert Striped Whipsnake) (*Masticophis taeniatus*)

Western Patch-nosed Snake (*Salvadora hexalepis*)
- ☐ Coast Patch-nosed Snake (*Salvadora hexalepis virgultea*), CSSC
- ☐ Desert Patch-nosed Snake (*Salvadora hexalepis hexalepis*)
- ☐ Mojave Patch-nosed Snake (*Salvadora hexalepis mojavensis*)

Gopher Snake (*Pituophis catenifer*)
- ☐ Great Basin Gopher Snake (*Pituophis catenifer deserticola*)
- ☐ Pacific Gopher Snake (*Pituophis catenifer catenifer*)
- ☐ San Diego Gopher Snake (*Pituophis catenifer annectens*)
- ☐ Santa Cruz Island Gopher Snake (*Pituophis catenifer pumilus*), CSSC
- ☐ Sonoran Gopher Snake (*Pituophis catenifer affinis*)

Glossy Snake (*Arizona elegans*)
- ☐ California Glossy Snake (*Arizona elegans occidentalis*)
- ☐ Desert Glossy Snake (*Arizona elegans eburnata*)
- ☐ Mojave Glossy Snake (*Arizona elegans candida*)

- ☐ Common Kingsnake (*Lampropeltis getula*)
  - ☐ California Kingsnake (*Lampropeltis getula californiae*)

California Mountain Kingsnake (*Lampropeltis zonata*)
- ☐ Coast Mountain Kingsnake (*Lampropeltis zonata multifasciata*)
- ☐ San Bernardino Mountain Kingsnake (*Lampropeltis zonata parvirubra*), CSSC
- ☐ San Diego Mountain Kingsnake (*Lampropeltis zonata pulchra*), CSSC
- ☐ Sierra Mountain Kingsnake (*Lampropeltis zonata multicincta*)
- ☐ St. Helena Mountain Kingsnake (*Lampropeltis zonata zonata*)

- ☐ Western Long-nosed Snake (*Rhinocheilus lecontei*)
- ☐ Baja California Rat Snake (*Bogertophis rosaliae*), CSSC

Common Garter Snake (*Thamnophis sirtalis*)
- ☐ San Francisco Garter Snake (*Thamnophis sirtalis tetrataenia*), FE, SE
- ☐ California Red-sided Garter Snake (*Thamnophis sirtalis infernalis*), CSSC (in southern portion of range)
- ☐ Valley Garter Snake (*Thamnophis sirtalis fitchi*)

Western Terrestrial Garter Snake (*Thamnophis elegans*)
- ☐ Coast Garter Snake (*Thamnophis elegans terrestris*)
- ☐ Mountain Garter Snake (*Thamnophis elegans elegans*)
- ☐ Wandering Garter Snake (*Thamnophis elegans vagrans*)

Aquatic Garter Snake (*Thamnophis atratus*)
- ☐ Diablo Range Garter Snake (*Thamnophis atratus zaxanthus*)
- ☐ Oregon Aquatic Garter Snake (*Thamnophis atratus hydrophilus*)
- ☐ Santa Cruz Aquatic Garter Snake (*Thamnophis atratus atratus*)

- ☐ Sierra Garter Snake (*Thamnophis couchii*)
- ☐ Giant Garter Snake (*Thamnophis gigas*), FT, ST

- ☐ Two-striped Garter Snake (*Thamnophis hammondii*), CSSC
- ☐ Northwestern Garter Snake (*Thamnophis ordinoides*)
- ☐ Checkered Garter Snake (*Thamnophis marcianus*)

Western Shovel-nosed Snake (*Chionactis occipitalis*)
- ☐ Colorado Desert Shovel-nosed Snake (*Chionactis occipitalis annulata*)
- ☐ Mojave Shovel-nosed Snake (*Chionactis occipitalis occipitalis*)
- ☐ Nevada Shovel-nosed Snake (*Chionactis occipitalis talpina*)

- ☐ Western Ground Snake (*Sonora semiannulata*)
- ☐ California Black-headed Snake (*Tantilla planiceps*)
- ☐ Southwestern Black-headed Snake (*Tantilla hobartsmithi*)
- ☐ California Lyre Snake (*Trimorphodon lyrophanes*)
- ☐ Sonoran Lyre Snake (*Trimorphodon lamba*)
- ☐ Coast Night Snake (*Hypsiglena ochrorhynchus*)
- ☐ Desert Night Snake (*Hypsiglena chlorophaea*)

Sea Snakes (Family Hydrophiidae [Elapidae])
- ☐ Yellow-bellied Sea Snake (*Pelamis platurus*)

Vipers and Relatives (Family Viperidae)

Western Rattlesnake (*Crotalus oreganus*)
- ☐ Great Basin Rattlesnake (*Crotalus oreganus lutosus*)
- ☐ Northern Pacific Rattlesnake (*Crotalus oreganus oreganus*)
- ☐ Southern Pacific Rattlesnake (*Crotalus oreganus helleri*)

- ☐ Western Diamond-backed Rattlesnake (*Crotalus atrox*)
- ☐ Mojave Rattlesnake (*Crotalus scutulatus*)
- ☐ Speckled Rattlesnake (*Crotalus mitchellii*)
- ☐ Panamint Rattlesnake (*Crotalus stephensi*)
- ☐ Red Diamond Rattlesnake (*Crotalus ruber*), CSSC
- ☐ Sidewinder (*Crotalus cerastes*)

TORTOISES AND TURTLES (ORDER TESTUDINES)

Land Tortoises (Family Testudinidae)
- ☐ Desert Tortoise (*Gopherus agassizii*), FT, ST

Pond Turtles (Family Emydidae)

Western Pond Turtle (*Actinemys marmorata*)
- ☐ Northwestern Pond Turtle (*Actinemys marmorata marmorata*), CSSC
- ☐ Southwestern Pond Turtle (*Actinemys marmorata pallida*), CSSC

- ☐ Pond Slider Turtle (*Trachemys scripta*), Introduced
- ☐ Western Painted Turtle (*Chrysemys picta*), Introduced

Softshell Turtles (Family Trionychidae)
- ☐ Spiny Softshell Turtle (*Apalone spinifera*), Introduced

Snapping Turtles (Family Chelydridae)
- ☐ Common Snapping Turtle (*Chelydra serpentina*), Introduced

Sea Turtles (Family Cheloniidae)
- ☐ Green Sea Turtle (*Chelonia mydas*), FT
- ☐ Black Sea Turtle (*Chelonia agassizii*), FT

- [ ] Loggerhead Sea Turtle (*Caretta caretta*), FT
- [ ] Olive Ridley Sea Turtle (*Lepidochelys olivacea*), FT
- [ ] Hawksbill Sea Turtle (*Eretmochelys imbricata*), FT

Leatherback Sea Turtle (Family Dermochelyidae)
- [ ] Leatherback Sea Turtle (*Dermochelys coriacea*), FE

ABBREVIATIONS

For Units of Measure

C	Centigrade
F	Fahrenheit
mm	millimeter
cm	centimeter
m	meter
g	gram
kg	kilogram
L	liter
in.	inch
ppm	parts per million
ppt	parts per thousand

Other Abbreviations

Bd	*Batrachochytrium dendrobatidis* (a fungus)
CDFG	California Department of Fish and Game
ESA	Endangered Species Act
TTX	tetrodotoxin
USFWS	United States Fish and Wildlife Service
UV	ultraviolet radiation

GLOSSARY

adult A sexually mature animal that has attained most or all of its growth.

ambient temperature Temperature of the surrounding environment.

amplexus The sexual embrace of a male amphibian. There are two types: *pectoral amplexus*, in which the male clasps the female around the chest region from her dorsal side, and *pelvic amplexus*, in which the male clasps the female about the waist from her dorsal side.

anal scent glands Scent glands in some snakes and lizards that open on each side of the vent.

anal spur A hindlimb vestige in boas and pythons in the shape of a pointed and sometimes hooked horny spur on either side and just in front of the vent.

anterior Toward the front of the body.

anuran A collective term for frogs and toads (tail-less amphibians).

arboreal Dwelling above ground level in trees or shrubs.

auricular scales Enlarged scales at the anterior border of the ear opening of some lizards.

axilla The armpit area.

axolotl The larvae of several species of the mole salamander family (genus *Ambystoma*) that become sexually mature and breed while still in the larval state but that can eventually undergo metamorphosis to the adult form.

basking A behavior by which heliothermic reptiles and some anurans elevate their body temperature to an optimal range by positioning all or portions of their body to receive maximum solar radiation.

boss A bony or glandular protuberance or swelling between the eyes of some members of the true toad and spadefoot families.

capsular cavity The chamber in the amphibian egg surrounded by one or more gelatinous envelopes within which the egg resides.

carapace The upper part of a turtle's shell, containing bony plates and horny shields.

chaparral A plant community composed of large shrub species associated with dry upland areas.

chytrid fungus A generic term for a fungus that causes the thickening of skin epidermal cells and that in amphibians can severely reduce the ability to absorb water in moist terrestrial habitats.

cloaca A chamber that opens to the vent and that receives fecal, urinary, and reproductive products.

clutch The total number of eggs laid by a female in one nesting.

Coast ranges Low mountains and high ridges along the California coast. South of San Francisco Bay these consist of the outer Coast Range (toward the Pacific Ocean) and the inner Coast Range (toward the Great Valley).

colubrid Any snake that belongs to the family Colubridae.

columella The single bone that in most amphibians, lizards, and turtles transmits vibrations from the eardrum to the inner ear.

costal folds Vertical skin folds on the sides of some salamander bodies separated by costal grooves.

costal grooves Vertical furrows on the sides of some salamander bodies separated by costal folds.

countersunk Sunk below the normal margin of an area, used in this book to describe the lower jaw of some burrowing snakes that fits snugly within the margins of the upper jaw.

cranial crests Ridges along the inner edge of the upper eyelids in toads.

crepuscular Describes an animal that is most active at dusk, dawn, or both.

cycloid scales Scales with smoothly rounded free posterior borders.

diploid Having double the number of chromosomes found in the ovarian egg or spermatozoa of a species.

diurnal Active during daytime.

dorsal The upper surface of an animal, usually its back side.

dorsolateral The upper side area of an animal body.

dorsolateral fold A lengthwise glandular ridge between the middle side area and the dorsal midline of certain frog species.

dorsum The upper or dorsal surface of an animal body.

ectotherm An animal that obtains most or all of its body heat from outside sources. (See heliotherm and thigmotherm.)

egg capsule The covering of amphibian eggs, consisting of one or more gelatinous envelopes.

emergence range Refers to the body temperature range through which a heliothermic reptile normally emerges from nightly retreat and begins warming, usually through basking.

endotherm Animals such as birds and mammals that derive their internal body warmth from the production of metabolic heat.

estivation A dormant state assumed by animals in response to hot or arid conditions. Also called *aestivation*.

eurytherm An animal that has a wide range of body temperatures through which it is able to conduct coordinated neuromuscular activities.

femoral pores Pores on the underside of the thighs of certain lizard species that secrete a waxlike substance.

fossorial A burrowing or underground life.

frontal scale A large scale located on the head of lizards and snakes between the dorsal margins of the eyes.

gill arch A bony or cartilaginous support for gill filaments in amphibians and fishes.

gill rakers Bony or cartilaginous projections on the gill arch that prevent food from exiting through the gill slits.

granular scales Small, smooth, round scales that do not overlap.

gravid Laden with eggs that will eventually be laid.

gular fold A fold of skin across the posterior throat area of some lizards and salamanders.

gular horn A long projection of the plastron beneath the throat area of male land tortoises.

heliotherm An animal that obtains its body warmth from direct exposure to solar radiation.

hemipenis One of a pair of copulatory organs (the *hemipenes*) of snakes and lizards.

herps Herpetological "slang" referring to amphibians and reptiles.

hibernaculum A site into which an animal retreats for hibernation.

hibernation A dormant state characterized by very low metabolic activity and entered into in response to cold winter conditions.

high ecological lethal A high body temperature at which an animal's enzymes begin to denature, resulting in a lack of neuromuscular coordination.

high lethal A term defining the body temperature at which an animal's enzyme systems break down and life ceases.

hybrid The offspring resulting from the breeding of one species with another.

intergrade Refers to individuals whose phenotype and/or genotype is intermediate between that of two distinct subspecies.

interorbital The region on the top of the head between the eyes.

Jacobson's organs A pair of small cavities lined with sensory epithelium on the mouth roof of many reptiles that analyzes scent molecules that are usually transported to it by the tongue.

juvenile A young or immature individual, often displaying proportions and coloration that differ from those of the adult.

keeled scale A scale with a lengthwise narrow ridge down its center.

labial scales Scales that line the upper and lower margins of the jaw of reptiles.

labial teeth Small horny teeth arranged like teeth of a comb and attached in crosswise rows to the lips of tadpoles.

larva As used in this book it refers to the aquatic stage of an amphibian before metamorphosis.

loreal region The region that lies between the eye and nostril of a snake.

low ecological lethal A low body temperature at which an animal's enzymes start to become inactive, resulting in a lack of neuromuscular coordination.

low lethal The body temperature at which cellular water freezes, resulting in death.

melanistic The condition in which black pigment is accentuated, sometimes to the point of obscuring all other color.

melanophore Pigment cells containing black or brown pigment that in some amphibians and reptiles can disperse or contract the pigment within the cell wall, causing an animal's overall color to lighten or darken.

mental gland A gland on the chin of the males of some salamander species whose secretion appears to make the females more receptive to mating.

metabolic water Water produced by the oxidation-reduction process within body cells.

metamorph A term describing an amphibian that has recently undergone transformation (metamorphosis) from the larval to the subadult stage.

morph An individual or individuals that differ in form from others of their species.

nasolabial groove A very small groove extending from each nostril to the upper lip in all salamanders of the family Plethodontidae.

neoteny A condition in which some species of salamanders prolong the larval state, become sexually mature, and often breed. These are referred to as *neotenic* larvae.

nuptial pad A patch of roughened and usually dark skin on the underside of the feet of some male amphibians that helps to hold the female during amplexus.

oviparous Describes animals that produce eggs that hatch after they are laid.

ovoviviparous Describes animals that retain eggs with shells or membranous coverings that hatch within the female's oviduct, after which the young are born alive.

ovum The ovulated egg; a female germ cell apart from any enclosing membrane or shell.

paraphyletic Describes a group of organisms with the same distant common ancestor but without all of the decendants of the most recent common ancestor.

parietal eye A "third eye" located on the top of the head of some lizards that functions as a radiation dosimeter.

parthenogenesis Reproduction by the development of an unfertilized egg.

plastron The underside of a turtle's shell.

plethodontid A term referring to members of the lungless salamanders (family Plethodontidae).

postanal scale A scale situated directly behind the vent at the base of the tail of snakes and lizards. In some male lizards these are paired and enlarged.

preformed water Water contained in food and made available during the digestive process.

scutes The large ventral scales of snakes and some lizards.

species A reproductively isolated group of organisms that can produce viable, fertile offspring only through intraspecific breeding.

stenotherm An animal that has only a narrow temperature range within which it is able to conduct coordinated neuromuscular activities.

subspecies A term describing a consistent variation among organisms of a species within a certain segment of its range, the individuals of which are interfertile with other subspecies of their species.

supraocular The area immediately above the eye.

sympatric A term describing two or more distinct populations within a species or subspecies that overlap geographically yet are distinct.

taxon A general term for a taxonomic unit without reference to rank.

thermoregulation The use of behavioral and/or physiological activities to achieve and maintain a constant or near-constant body temperature.

thigmotactile Surface-loving behavior in an animal.

thigmotherm An animal that absorbs heat by contact with warm surfaces.

tonic immobility A hypotonic-like state assumed by some animals when suddenly grabbed or flipped over by a predator, including humans, who sometimes extend the state by rubbing the animal's ventral surface.

uricotelism An animal that excretes its nitrogenous wastes chiefly in the form of uric acid (characteristic of birds and reptiles).

venomous Having the ability to inflict a poisonous bite.

vent The external opening of the cloaca through which fecal, urinary, and reproductive materials are discharged.

venter The ventral or underside of an animal body.

ventral The underside of an animal body.

viviparous A term describing females in which developing young are retained in the oviduct or uterus and are supported by an exchange of materials between the maternal and fetal blood supplies.

vocal sac A sac of loose skin on the throat of anurans that becomes distended and functions as a resonating chamber when vocalizing.

Zzyzx The name of a former mineral spring retreat on the west shore of the dry Soda Lake bed in the Mojave Desert, now the site of the Desert Studies Center of the California State Universities.

SELECTED REFERENCES

The primary reference source for this book was the senior author's large collection of journal articles and books covering all aspects of North American and California herpetology over the past seven decades. However, many of these sources are hard to find in public libraries and on the Internet. The primary goal of this section of the California Natural History Guide series is to present a selection of references that are readily obtainable from college, university, and larger public libraries and that will help to further a reader's interest in and exploration of the subject at hand. We feel the following list serves that purpose well.

Ashton, K. G., and A. de Queiroz. 2001. "Molecular Systematics of the Western Rattlesnake, *Crotalus viridis* (Viperidae), with Comments on the Utility of the D-Loop in Phylogenetic Studies of Snakes." *Molecular Phylogenetics and Evolution* 21: 176–189.

Berry, K. H. 1972. "The Ecology and Social Behavior of the Chuckwalla, *Sauromalus obesus*." Doctoral Thesis, University of California, Berkeley.

Brandley, M., A. Schmitz, and T. Reeder. 2005. "Partitioned Analysis, Partition Choice, and the Phylogenetic Relationships of Scincid Lizards. *Systematic Biology* 54: 373–390.

Brennan, T. C., and A. T. Holycross. 2006. *A Field Guide to Amphibians and Reptiles in Arizona.* Phoenix: Arizona Game and Fish Department.

Brown, P. R. 1997. *A Field Guide to Snakes of California.* Houston: Gulf Publishing Co.

Center for North American Herpetology web site: http.//www.cnah.org

Cochran, D. M. 1967. *Living Amphibians of the World.* Garden City, N.Y.: Doubleday.

Conant, R., and J. T. Collins. 1998. *A Field Guide to Reptiles and Amphibians of Eastern and Central North America.* 3rd ed. (extended). Boston: Houghton-Mifflin Co.

Douglas, M. E., M. R. Douglas, G. W. Schuett, L. W. Pooras, and A. T. Holycross. 2002. "Phylogeography of the Western Rattlesnake (*Crotalus viridis*) Complex, with Emphasis on the Colorado Plateau." In *Biology of the Vipers*, edited by Douglas, M. E., H. W. Green, M. Hoggren, and G. W. Scguett. Eagle Mountain, Utah: Eagle Mountain Publishing, pp. 11–50.

———, H. W. Greene, M. Hoggren, and G. W. Schuett, eds. 2002. *Biology of the Vipers.* Eagle Mountain, Utah: Eagle Mountain Publishing.

Duellman, W. E., and L. Trueb. 1994. *Biology of Amphibians* (first published 1986). Baltimore: John Hopkins University Press.

Ernst, C. H., J. E. Lovich, and R. W. Barbour. 1994. *Turtles of the United States and Canada.* Washington, D.C.: The Smithsonian Institution Press.

Fellers, G. M., and P. M. Kleeman. 2007. "California Red-legged Frog (*Rana draytonii*) Movement and Habitat Use: Implications for Conservation." *Journal of Herpetology* 41: 271–281.

Gans, C., and others, eds. 1969–1998. *Biology of the Reptiles.* Vols. 1-13. Academic Press, London.

Gans, C., and others, eds. 1969–1998. *Biology of the Reptiles.* Vols. 14 and 15. New York: John Wiley & Sons.

Gans, C., and others, eds. 1969–1998. *Biology of the Reptiles.* Vol. 16. Ann Arbor: Banta Books.

Gans, C., and others, eds. 1969–1998. *Biology of the Reptiles.* Vols. 17 and 18. Chicago: University of Chicago Press.

Gans, C., and others, eds. 1969–1998. *Biology of the Reptiles.* Vol. 19. St. Louis, Mo.: Society for the Study of Amphibians and Reptiles.

Goin, C. J., O. B. Goin, and G. R. Zug. 1978. *Introduction to Herpetology.* San Francisco: W. H. Freeman.

Greene, H. W. 1997. *Snakes: The Evolution of Mystery in Nature.* Berkeley: University of California Press.

Grismer, L. 2002. *Amphibians and Reptiles of Baja California, including Its Pacific Islands and the Sea of Cortés.* Berkeley: University of California Press.

Hammerson, G. A. 1979. "Thermal Ecology of the Striped Racer (*Masticophis lateralis*)." *Herpetologica* 35: 267–273.

———. 1999. *Amphibians and Reptiles in Colorado.* Niwot, Colo.: University Press of Colorado.

Harding, J. H. 2006. *Amphibians and Reptiles of the Great Lakes Region.* Ann Arbor: The University of Michigan Press.

Hickman, J. C., ed. 1993. *The Jepson Manual: Higher Plants of California.* Berkeley and Los Angeles: University of California Press.

International Commission on Zoological Nomenclature web site: http://iczn.org/.

Jaeger, E. C. 1967. *The North American Deserts.* Stanford, Calif.: Stanford University Press.

Jennings, M. R., and M. P. Hayes. 1994. *Amphibian and Reptile Species of Special Concern in California.* Rancho Cordova, Calif.: California Department of Fish and Game.

Klauber, L. M. 1956. *Rattlesnakes: Their Habits, Life Histories, and Influence on Mankind.* Berkeley: University of California Press.

Lannoo, M., ed. 2005. *Amphibian Declines.* Berkeley: University of California Press.

Larsen, S. S., K. E. Swaim, and S. M. McGinnis. 1991. "Innate Response of the San Francisco Garter Snake and the Alameda Whipsnake to Specific Prey Items." *Transactions of the Western Section of the Wildlife Society* 27: 37–41.

Latting, J., and P. G. Rowlands, eds. 1995. *Desert: An Introduction to Natural Resources and Man's Impact.* Riverside, Calif.: University of California Press.

Lemm, J. M. 2006. *Amphibians and Reptiles of the San Diego Region*. Berkeley: University of California Press.

Lowe, C. L., C. R. Schwalbe, and T. B. Johnson. 1986. *The Venomous Reptiles of Arizona*. Phoenix: Arizona Game and Fish Department.

Mayer, K. E., and W. F. Laudenslayer. 1988. *A Guide to Wildlife Habitats of California*. Sacramento: California Department of Forestry and Fire Protection.

McGinnis, S. M. 1966. "Preferred Body Temperature of the Western Fence Lizard (*Sceloporus occidentalis*)." *Science* 152: 1090–1991.

———. 1988. "Life History of the San Francisco Garter Snake (*Thamnophis sirtalis tetrataenia*)." California Department of Fish and Game Interagency Agreement C-2045 Report.

———. 2006. *Field Guide to Freshwater Fishes of California*. Berkeley: University of California Press.

———, and L. L. Dickson. 1967. "Thermoregulation in the Desert Iguana (*Dipsosaurus dorsalis*)." *Science* 156: 1757–1759.

———, and W. G. Voigt. 1971. "Thermoregulation in the Desert Tortoise (*Gopherus agassizii*). *Comparative Biochemistry and Physiology* 40: 119–126.

Middendorf, G. A, III, and W. C. Sherbrooke. 1992. "Canid Elicitation of Blood-Squirting in a Horned Lizard (*Phrynosoma cornutum*)." *Copeia* 1992: 519–527.

Murry, J. M., and F. Foote. 1979. "The Origin of Fear of Snakes." *Behaviour Research and Therapy* 3: 45–53.

Norris, K. S., and C. H. Lowe. 1964. "An Analysis of Background Color Matching in Amphibians and Reptiles." *Ecology* 45: 565–580.

Pianka, E. R., and L. J. Vitt. 2003. *Lizards, Windows to the Evolution of Diversity*. Berkeley: University of California Press.

Pough, H. F., and others. 2001. *Herpetology*. 2nd ed. Upper Saddle River, N. J.: Prentice-Hall.

Rababtski, A. M. 2005. "Rattle Loss in Insular Rattlesnake Species." Paper presented at the *Biology of Rattlesnakes Symposium*, Loma Linda University, Loma Linda, Calif.

Reeder, T. W., C. J. Cole, and H. C. Dessauer. 2002. "Phylogenic Relationships of Whiptail Lizards of the Genus *Cnemidophorusi* (Squamata: Teidae). A Test of Monophyly, Reevaluation of Karotypic Evolution, and Review of Hybrid Origins." *American Museum Novitates* 3305: 1–61.

Richmond, J. Q., and G. W. Reeder. 2002. "Evidence for Parallel Ecological Speciation in Scincid Lizards of the *Eumeces skiltonianus* Species Group (Squamata: Scincidae)." *Evolution* 56: 1498–1513.

Rowe, M. P., T. M. Farrell, and P. G. May. 2002. "Rattle Loss in Pigmy Rattlesnakes (*Sistrurus miliarius*): Causes, Consequences, and Implications for Rattle Function and Evolution." In *Biology of the Vipers*, edited by Douglas, M. E., H. W. Green, M. Hoggren, and G. W. Scguett. Eagle Mountain, Utah: Eagle Mountain Publishing, pp. 385–404.

Sherbrooke, W. C. 2003. *Introduction to the Horned Lizards of North America*. Berkeley: University of California Press.

Smith H. 2005. "Plestiodon: A Replacement Name for Most Members of the Genus *Eumeces* in North America." *Journal of Kansas Herpetology* 14: 15–16.

Smith, H. M. 1995. *Handbook of Lizards*. Ithaca, N. Y.: Comstock Publishing Co.

Southwestern Herpetologists Society. 1994. *Herpetology of the North American Deserts*. Special Publication No. 5. Excelsior, Minn.: Serpent's Tale Books (distributor).

St. John, A. 2002. *Reptiles of the Northwest*. Renton, Wash.: Lone Pine Publishing.

Stebbins, R. C. 1951. *Amphibians of Western North America*. Berkeley: University of California Press.

———. 1954. *Amphibians and Reptiles of Western North America*. New York: McGraw-Hill.

———. 1954. "Natural History of the Salamanders of the Plethodontid Genus *Ensatina*." *University of California Publications in Zoology* 54: 47–124.

———. 1960. "Effect of parietalectomy in the Western Fence Lizard." *Copeia* 1960: 276–280.

———. 2003. *A Field Guide to Western Reptiles and Amphibians*. 3rd ed. Boston: Houghton-Mifflin Co.

———, and N. W. Cohen. 1995. *A Natural History of Amphibians*. Princeton, N. J.: Princeton University Press.

Swaim, K. E., and S. M. McGinnis. 1992. "Habitat Requirements of the Alameda Whipsnake (*Masticophis lateralis euryxanthus*)." *Transactions of the Western Section of the Wildlife Society* 28: 107–111.

Wake, D. B., and C. J. Schneider. 1998. "Taxonomy of the Plethodontid Salamander Genus *Ensatina. Herpetologica* 54: 279–298.

Wiens, J. J., and T. W. Reeder. 1997. "Phylogeny of the Spiny Lizards (*Sceloporus*) Based on Molecular and Morphological Evidence." *Herpetological Monographs* 11: 1–101.

INDEX

Page numbers in **bold** indicate the main discussion of the taxon and a color species illustration.

Actinemys
  marmorata, **430–435,** 431 (map), 431 (plate)
    basking, 432
    conservation note, 433–435
    food, 432
    habitat, 431–432
    hibernation, 432
    identification, 430–431, 431 (plate)
    natural history, 431–433
    photography, 464–465
    protective status, 430
    range, 431, 431 (map)
    reproduction, 432–433
    thermoregulation, 432
  marmorata marmorata, **430–435,** 431 (map)
  marmorata pallida, **430–435,** 431 (map)
advertisement call, anurans, 58
African Clawed Frog, **212–215**
  adult, 213, 214 (plate)
  breeding, 214
  chytrid fungus infection, 27–28, 215
  conservation note, 214–215
  eggs, 213
  food, 214
  habitat, 214
  identification, 213, 213 (fig.), 214 (plate)
  as lab frog, 214–215
  larvae, 213, 213 (fig.), 214
  natural history, 214
  range, 213, 213 (map)
  voice, 213
Age of Reptiles, 218
Alameda Striped Racer, **366–369**
  conservation note, 368–369
  habitat, 367–368
  identification, 367, 367 (plate)
  prey, 368, 368 (plate)
  protective status, 367
  range, 367, 367 (map)
  reproduction, 368
  thermoregulation, 228, 367
  trapping, 475
alligator lizards, 325–331
  behavior, 325
  body form, 325
  California Alligator Lizard, **326–328,** 327 (map)
  as eurytherms, 225–226
  food, 242
  habitat, 15
  Northern Alligator Lizard, **329–330,** 329 (map), 330 (plate)
  Northwestern Alligator Lizard, **329–330,** 329 (map)
  Oregon Alligator Lizard, **326–328,** 327 (map)
  Panamint Alligator Lizard, **330–331,** 331 (map)
  prehensile tail, 325, 326 (fig.)
  reproduction, 325
  San Diego Alligator Lizard, **326–328,** 327 (map)
  San Francisco Alligator Lizard, **329–330,** 329 (map)
  Shasta Alligator Lizard, **329–330,** 329 (map)
  Sierra Alligator Lizard, **329–330,** 329 (map)
  Southern Alligator Lizard, **326–328,** 327 (map), 327 (plate)
Alligator Snapping Turtle, 442
Alvarez, Jeff, 189
Amblyrhynchus cristatis, 253
Ambystoma
  californiense, 43, **70–75,** 71 (map), 71 (plate), 72 (plate), 463, 479

Ambystoma (continued)
  gracile, **82–84,** 83 (map), 83 (plate)
  macrodactylum, 62, **78–82,** 79 (map), 79–81 (plates)
  macrodactylum croceum, **78–82,** 79 (map), 80 (plate), 81 (plate)
  macrodactylum sigillatum, **78–82,** 79 (map), 79 (plate)
  mavortium, **75–77,** 76 (map)
  mexicanum, 39
  tigrinum, **75–77,** 76 (map)
Ambystomatidae, 69–84
  California Tiger Salamander, **70–75,** 71 (map), 71 (plate), 72 (plate)
  Long-toed Salamander, **78–82,** 79 (map), 79–81 (plates)
  Northwestern Salamander, **82–84,** 83 (map), 83 (plate)
  Tiger Salamander, **75–77,** 76 (map)
ambystomids, photography, 463
Ameiurus melas, and California Red-legged Frog, 192
American Badger, 428
American Bullfrog, **208–212**
  adult, 208
  aquatic predation pressure by, 26
  breeding, 210
  and California Red-legged Frog, 191–192
  defense, 209–210
  eggs, 208, 210
  food, 210
  habitat, 18, 209–210, 209 (plate)
  hunting of, 210–211
  identification, 208
  larvae, 208, 210
  metamorphosis, 62, 210
  natural history, 209–212, 209 (plate)
  range, 209, 209 (map)
  thermoregulation, 56
  voice, 58, 208
American Eel, 3–4
amniote egg, 219
Amphibia, 36
amphibian(s), 35–215
  anatomy, physiology, and behavior, 36–65
  body form and locomotion, 37–41, 38 (fig.), 40 (fig.)
  body temperature and thermoregulation, 54–56, 54 (fig.)
  distribution throughout California's habitat complex, 5–23, 6 (map), 7–15 (plates)
  frogs, 149–215. *See also* frogs
  hearing, 47–48
  olfaction and chemoreception, 44–46, 46 (fig.)
  osmoregulation, 52–54
  recognition in field of individual, 479–480
  reproduction, 56–65
  respiration, 48–52, 49 (fig.), 50 (fig.)
  salamanders, 67–148. *See also* salamanders
  sensory input, 44–48
  skin, 41–44, 42–45 (figs.)
  taxonomy, 36–37
  toads, 149–215. *See also* toads
  as vertebrate terrestrial pioneers, 2–4, 3 (fig.)
  vision, 46–47
amphibian chytridiomycosis, 27–28
amphibian papilla, 48
Amphisbaenia, 233
amplexing
  anurans, 59, 60 (fig.)
  aquatic-breeding salamanders, 62, 63 (fig.)
  newts, 92 (fig.), 93–94
anatomy
  amphibians, 36–65
  reptiles, 218–232
Anaxyrus, 161
Aneides, 116–124
  ferreus, **122–124,** 123 (map)
  flavipunctatus, **119–121,** 120 (map), 121 (plate)
  flavipunctatus flavipunctatus, **119–121,** 120 (map)
  flavipunctatus niger, **119–121,** 120 (map)
  lugubris, **116–119,** 117 (fig.), 117 (map), 118 (plate)
  vagrans, 15, **122–124,** 123 (map)
Anguidae, 325–331
Anguilla anguilla, 3–4
Anniella pulchra, 233, **332–334,** 332 (plate), 333 (map)
Anniellidae, 331–334
antivenin, 408, 409
Anura, 149–215
anuran(s)
  body form and locomotion, 39–41, 40 (figs.)
  fertilization, 59–60, 60 (fig.)
  metamorphosis, 60–62, 61 (fig.)
  photography, 463
  reproduction, 57–62
  voice, 57–59, 58 (fig.)
anuran larvae, 59–60
  gills, 49–50, 50 (fig.)
Apalone
  spinifera, **440–442,** 441 (map)
  spinifera emoryi, 441

appendage regeneration, salamanders, 39
Aquatic Garter Snake, **387–388,** 388 (map), 388 (plate)
aquatic habitats, chemical contamination, 26–27
aquatic predation pressure, habitat loss due to, 24–26
aquatic-breeding salamanders, 62–63, 63 (fig.)
Arboreal Salamander, **116–119**
  adults, 116–117
  defense, 118–119
  eggs, 117, 117 (fig.), 119
  habitat, 117–118, 118 (plate)
  identification, 116–117
  natural history, 117–119, 118 (plate)
  range, 117, 117 (map)
  reproduction, 119
Arborimus spp., 123
Archoplites interruptus, aquatic predation pressure by, 25
Arizona
  elegans, 18, **373–374,** 374 (map)
  elegans candida, **373–374,** 374 (map)
  elegans eburnata, **373–374,** 374 (map)
  elegans occidentalis, **373–374,** 374 (map)
Arroyo Toad, 20, 21, **168–169,** 169 (map)
Ascaphidae, 150–154
Ascaphus, 59
  truei, 15, **151–154,** 151 (map), 153 (plate)
Asian Mustard, 276, 305
Aspidoscelis, 319–324
  endothermy, 229
  food, 242
  hyperythra, 20, **323–324,** 323 (map), 324 (plate)
  tigris, **320–322,** 321 (map), 321 (plate)
  tigris munda, **320–322,** 321 (map)
  tigris stejnegeri, **320–322,** 321 (map)
  tigris tigris, **320–322,** 321 (map)
atrazine, chemical contamination of aquatic habitat by, 26–27
Australian Lungfish, 2

backswimmers, 72
bacterial infections, 481–482
Baja California Coachwhip, **363–366,** 365 (map)
Baja California Collared Lizard, **263,** 263 (map)
Baja California Rat Snake, **379,** 379 (map)
balancer organs, aquatic-breeding salamanders, 63
banded geckos, 247–251
Banded Gila Monster, **335–337,** 335 (map)

Banded Rock Lizard, **296–297,** 296 (map), 297 (plate)
Barefoot Gecko, **250–251,** 250 (map)
Barred Tiger Salamander, **75–77,** 76 (map)
basilar papilla, amphibians, 48
basking position, reptiles, 228
basking range, reptiles, 227–229
Batrachochytrium dendrobatidis (Bd)
  habitat loss due to, 27–28
  and yellow-legged frogs, 202
Batrachoseps
  attenuatus, 39, **140–142,** 140 (map), 141 (plate)
  campi, 17, 22, **127–128,** 127 (map)
  diabolicus, **139–140,** 140 (map)
  gabrieli, **135–136,** 135 (map)
  gavilanensis, **136,** 136 (map)
  gregarius, **132,** 132 (map)
  incognitus, **137–138,** 138 (map)
  kawia, **138,** 138 (map)
  luciae, **137,** 137 (map)
  major, **134–135,** 134 (map)
  major aridus, 22, **134–135,** 134 (map)
  major major, **134–135,** 134 (map)
  minor, **137,** 137 (map)
  nigriventris, **131,** 131 (map)
  pacificus, **133,** 133 (map)
  regius, **138,** 138 (map)
  relictus, **139,** 139 (map)
  robustus, **128,** 128 (map)
  simatus, **130,** 130 (map)
  stebbinsi, 21, **129,** 129 (map)
  tail, 37–39
Bd *(Batrachochytrium dendrobatidis)* fungus
  habitat loss due to, 27–28
  and yellow-legged frogs, 202
behavior
  amphibians, 36–65
  field notes, 458–460
  reptiles, 218–232
Benes, Elinore, 322
Black Bullhead, and California Red-legged Frog, 192
Black Salamander, **119–121,** 120 (map), 121 (plate)
Black Sea Turtle, **448–451**
  conservation note, 450–451
  food, 449
  habitat, 449
  identification, 448–449
  migration, 449–450
  natural history, 449–450
  protective status, 448
  range, 449

Black Sea Turtle (*continued*)
reproduction, 450
taxonomic note, 448
Black Toad, **174–175,** 174 (map)
Black-bellied Slender Salamander, **131,** 131 (map)
black-headed snakes
California Black-headed Snake, **397–398,** 398 (map)
Southwestern Black-headed Snake, **397–398,** 397 (map)
Black-necked Garter Snake, 393 (plate)
Black-tailed Brush Lizard, **292–294,** 293 (map), 293 (plate)
Black-tailed Rattlesnake, reproduction, 348
blind snakes, 350–352
"blue bellies," 278–279
Bluegill, aquatic predation pressure by, 25
Blunt-nosed Leopard Lizard, 264, **267–269,** 267 (map)
boa(s), 352–355
pit organs, 345
prey, 347
Boa Constrictor, 352
Boa constrictor, 352
body adaptations, snakes, 342–346
body form, amphibians, 37–41, 38 (fig.), 40 (fig.)
body temperature
amphibians, 54–56, 54 (fig.)
reptiles, 225–230, 226 (fig.), 227 (plate), 256 (plate)
Bogertophis rosaliae, **379,** 379 (map)
boid(s), pit organs, 345
Boidae, 352–355
Boreal Toad, **170–173**
adults, 170, 173 (plate)
breeding, 172–173
eggs, 170, 172
habitat, 171
identification, 170
larvae, 170
metamorphosis, 172
natural history, 171–173, 171 (plate), 173 (plate)
range, 171, 171 (map)
thermal preference, 171–172
voice, 170
Botta's Pocket Gopher, burrows, 73
Brassica tournefortii, 276, 305
Brode, John, 81
Bromus madritensis rubens, 430
brooding, terrestrial-breeding salamanders, 64–65, 64 (fig.)
Brook Trout, and yellow-legged frogs, 202

brush lizards, 290–294
Black-tailed Brush Lizard, **292–294,** 293 (map), 293 (plate)
Long-tailed Brush Lizard, **290–292,** 291 (map), 291 (plate)
buccopharyngeal respiration, amphibians, 51
Bufo
alvarius, **161–163,** 162 (map), 162 (plate)
body form and locomotion, 39–41, 40 (fig.)
boreas, 16, **170–173,** 171 (map), 171 (plate), 173 (plate)
boreas boreas, **170–173,** 171 (map)
boreas halophilus, **170–173,** 171 (map)
californicus, 20, 21, **168–169,** 169 (map)
canorus, 19, **176–178,** 177 (map), 177 (plate)
cognatus, **164–165,** 164 (map), 165 (plate)
debilis, 41
exsul, **174–175,** 174 (map)
punctatus, 21, 53, **166–167,** 167 (map), 167 (plate)
woodhousii, **178–179,** 179 (map)
woodhousii fowleri, 179
Bufonidae, 161–179
Bullfrog, American, **208–212.** See also American Bullfrog
Bunchgrass Lizard, eggs, 219, 220 (plate)
burrowing, lizards, 238–239

Caiman crocodilus, osmoregulation, 231
California Alligator Lizard, **326–328**
behavior, 328
food, 328
habitat, 327–328
identification, 326–327
natural history, 327–328
range, 327, 327 (map)
reproduction, 328
California Black-headed Snake, **396–397,** 397 (map)
California Chorus Frog, **183–185**
adults, 184
breeding, 185
cryptic coloration, 43 (fig.)
eggs, 184, 185
food, 185
habitat, 20, 184–185, 185 (plate)
identification, 184
larvae, 184
natural history, 184–185, 185 (plate)
range, 184, 184 (map)
voice, 184
California Giant Salamander, **85–88**
adult, 85, 85 (plate)
breeding pool, 87–88, 88 (plate)
eggs, 85, 87

gills, 48–49
habitat, 17, 86
identification, 85–86, 85 (plate)
larvae, 85–86, 86 (plate), 87–88
metamorphosis, 88
natural history, 86–88, 88 (plate)
range, 86, 86 (map)
reproduction, 87–88
voice, 87
California Glossy Snake, **373–374,** 374 (map)
California Ground Squirrel
burrows, 73, 74, 432
California Kingsnake, **375–376,** 375 (map)
California Legless Lizard, **332–334**
behavior, 333–334
conservation note, 334
food, 334
habitat, 333
identification, 332–333, 332 (plate)
natural history, 333–334
protective status, 332
range, 333, 333 (map)
reproduction, 334
California Lyre Snake, 20, 344, **398,** 398 (map)
California Mountain Kingsnake, 19, **377–378,** 377 (map)
California Red-legged Frog, **186–192**
adults, 187, 190 (plate)
aquatic predation pressure on, 25–26
breeding, 60 (fig.), 189
chemical contamination of aquatic habitat, 26
conservation note, 190–192
eggs, 187, 188 (plate), 189
food, 190
habitat, 188–189
habitat loss, 23
identification, 187–188, 190 (plate)
larvae, 187–188, 189
metamorphosis, 189–190
natural history, 188–190, 190 (plate)
olfaction and chemoreception, 46
protective status, 187
range, 188, 188 (map)
thermoregulation, 56
voice, 58, 187
California Red-sided Garter Snake, **381–384**
habitat, 382–383
identification, 381
natural history, 382–383
prey, 383
protective status, 381
range, 382, 382 (map)
reproduction, 383

California Slender Salamander, **140–142**
adults, 140, 141 (plate)
eggs, 141 (plate), 142
habitat, 141–142
identification, 140
natural history, 141–142
range, 140, 140 (map)
reproduction, 142
tail, 39
California species of special concern (CSSC), 30
California Striped Racer, **366–369,** 367 (map)
California Tiger Salamander, **70–75**
adults, 70
breeding ponds, 71–72, 72 (plate)
conservation note, 74–75
disruptive coloration, 43
eggs, 70
identification, 70
larvae, 70
larval cannibalism, 72
longevity, 74
metamorphosis, 72–73
migration, 71, 71 (plate)
natural history, 71–74, 71 (plate), 72 (plate)
photography, 463
protective status, 70
range, 71, 71 (map)
recognition of individual animals, 479
retreats, 73–74
California Toad, **170–173**
adults, 170, 173 (plate)
breeding, 172–173
eggs, 170, 172
habitat, 171
identification, 170
larvae, 170
metamorphosis, 172
natural history, 171–173, 171 (plate), 173 (plate)
range, 171, 171 (map)
thermal preference, 171–172
voice, 170
California Tree Frog, **183–185.** *See also* California Chorus Frog
California Whiptail, **320–322**
behavior, 321–322
food, 322
habitat, 321
identification, 320–321
natural history, 321–322
range, 321, 321 (map)
reproduction, 322

call notes, anurans, 57–59
Callisaurus draconoides, **270–272**
    breeding, 272
    coloration, 272
    food, 272
    habitat, 22, 271–272
    identification, 270–271, 271 (plate)
    limbs, 238
    natural history, 271–272
    range, 271, 271 (map)
camouflage, amphibians, 42–44, 43 (fig.)
Canis latrans, trap robbing, 476–477
captive animals
    husbandry, 481–483
    photography, 466–468
capture, 469–480
    cover boards, 477–478, 477 (fig.)
    hand, 470–472
    hunting strategy, 470, 471 (fig.)
    net and noose, 472–474, 473 (plate)
    rules and regulations, 469
    signs of presence and, 470, 470 (plate)
    site selection, 469–470
    snakes, 474
    techniques, 470–474
    trapping, 474–477, 475 (plate), 476 (plate)
Carboniferous Period, vertebrate terrestrial pioneers in, 4
Caretta caretta, **451–452**
Carpenter, Charles, 243
Cascade Range, 6 (map), 8 (plate), 16
Cascades Frog, **193–194**
    adult, 193
    aquatic predation pressure on, 26
    breeding, 194
    conservation note, 194
    eggs, 193
    habitat, 16, 194
    identification, 193
    larvae, 193
    natural history, 194
    protective status, 193
    range, 194, 194 (map)
    voice, 193
caterpillar action, snakes, 342–343
Caudata, 67–148
Channel Islands Slender Salamander, **133,** 133 (map)
Charina
    bottae, 16, **353–354,** 353 (map), 466
    umbratica, **353–354,** 353 (map)
Checkered Garter Snake, **392–393,** 392 (map), 393 (plate)

Chelonia
    agassizii, **448–451**
    mydas, 229, 446, **448–451,** 450 (plate)
Cheloniidae, 445–455
Chelydra serpentina, 442, **443–445,** 443 (map)
Chelydridae, 442–445
chemical contamination, habitat loss due to, 26–27
chemoreception, amphibians, 44–46, 46 (fig.)
chilling, of captive specimen, for photography, 468
Chionactis
    occipitalis, 21, **393–394,** 394 (map), 394 (plate)
    occipitalis annulata, **393–394,** 394 (map)
    occipitalis occipitalis, **393–394,** 394 (map)
    occipitalis talpina, **393–394,** 394 (map)
chorus frogs, 179–185
    body form and locomotion, 39–40
    California Chorus Frog, **183–185,** 184 (map), 185 (plate)
    Pacific Chorus Frog, **180–183,** 181 (map), 181 (plate)
Chrysemys picta, **437–439**
    food, 438
    habitat, 437–438
    hibernation, 438
    identification, 437, 438 (plate)
    natural history, 437–439
    range, 437, 437 (map)
    reproduction, 438–439
    and Western Pond Turtle, 433
Chuckwalla. *See* Common Chuckwalla
chytrid fungus infection
    habitat loss due to, 27–28
    yellow-legged frogs, 202
chytridiomycosis, amphibian, 27–28
Clarias batrachus, 3–4
Clemens, Samuel, 191
Clemmys marmorata, **430–435**
    basking, 432
    conservation note, 433–435
    food, 432
    habitat, 431–432
    hibernation, 432
    identification, 430–431, 431 (plate)
    natural history, 431–433
    photography, 464–465
    protective status, 430
    range, 431, 431 (map)
    reproduction, 432–433
    thermoregulation, 432
climatic conditions, in field notes, 460–461

climbing salamanders, 116–124
　Arboreal Salamander, **116–119,** 117 (fig.), 117 (map), 118 (plate)
　Black Salamander, **119–121,** 120 (map), 121 (plate)
　Clouded Salamander, **122–124,** 123 (map)
　Santa Cruz Black Salamander, **119–121,** 120 (map)
　Speckled Black Salamander, **119–121,** 120 (map)
　Wandering Salamander, **122–124,** 123 (map)
Clouded Salamander, **122–124,** 123 (map)
Cnemidophorus, 319–324
　hyperythra, **323–324,** 323 (map), 324 (plate)
　tigris, **320–322,** 321 (map), 321 (plate)
Coach Patch-nosed Snake, **370–371,** 370 (map)
Coachella Valley Fringe-toed Lizard, 23, **274–276,** 275 (map), 305
Coachwhip, 20, **363–366,** 365 (map), 365 (plate)
Coast Garter Snake, **385–387**
　habitat, 16, 19, 386
　identification, 385, 386 (plate)
　natural history, 386–387
　prey, 386–387
　range, 386, 386 (map)
　reproduction, 387
Coast Horned Lizard, **300–302**
　conservation note, 301–302
　display behavior, 301
　food, 301
　habitat, 18, 300–301
　identification, 300, 301 (plate)
　natural history, 300–301
　protective status, 300
　range, 300, 300 (map)
　reproduction, 301
Coast Mountain Kingsnake, **377–378,** 377 (map)
Coast Night Snake, **399–400,** 399 (map)
Coast Range Newt, **96–100**
　adults, 96
　conservation, 99–100
　eggs, 94 (fig.), 95 (plate), 96, 97
　food, 99
　habitat, 97
　identification, 96–97
　larvae, 96, 97
　natural history, 97–99
　protective status, 96
　range, 97, 97 (map)
　reproduction, 98
　taxonomic note on, 96
　ventral skin coloration, 43–44, 44 (fig.)
Coastal Southern California, 6 (map), 11 (plate), 19–20
Coastal Tailed Frog, **151–154,** 153 (plate)
　adults, 151, 152
　conservation note, 153–154
　eggs, 151, 152
　food, 153
　habitat, 15, 151–152
　identification, 151
　larvae, 151, 152–153
　natural history, 151–153
　protective status, 151
　range, 151, 151 (map)
　reproduction, 152
Coastal Whiptail, **320–322**
　behavior, 321–322
　food, 322
　habitat, 321
　identification, 320
　natural history, 321–322
　range, 321, 321 (map)
　reproduction, 322
Coelacanth, as vertebrate terrestrial pioneer, 2
Coenus hypiglena, prey, 347
Cohen, Nate, 24, 470
Coleonyx
　switaki, **250–251,** 250 (map)
　variegatus, 21, **247–249,** 248 (map), 248 (plate)
　variegatus abbotti, **247–249,** 248 (map), 248 (plate)
　variegatus variegatus, **247–249,** 248 (map), 248 (plate)
collared lizards, 260–263
　Baja California Collared Lizard, **263,** 263 (map)
　coloration, 244
　Great Basin Collared Lizard, **261–263,** 261 (map), 262 (plate)
　habitat, 22
color vision
　amphibians, 47
　reptiles, 223
Colorado Desert, 6 (map), 15 (plate), 22–23
Colorado Desert Fringe-toed Lizard, **276–277,** 276 (map), 305
Colorado Desert Shovel-nosed Snake, **393–394,** 394 (map)
Colorado River Toad, **161–163,** 162 (map), 162 (plate). *See also* Sonoran Desert Toad

coloration
    amphibians, 42–44, 43 (fig.), 44 (fig.)
    lizards, 244
Coluber, 361–374
    constrictor, 16, **362–363,** 362 (map)
    constrictor mormon, 362–363, 363 (plate), 466
    vision, 223
colubrid(s), 355–400, 356 (fig.)
    Baja California Rat Snake, **379,** 379 (map)
    California Black-headed Snake, **396–397,** 397 (map)
    California Lyre Snake, **398,** 398 (map)
    Coast Night Snake, **399–400,** 399 (map)
    Common Sharp-tailed Snake, **358–360,** 359 (map), 359 (plate), 360 (fig.)
    Desert Night Snake, **399–400,** 399 (map)
    Forest Sharp-tailed Snake, **358–360,** 359 (map)
    garter snakes, 379–400
    Glossy Snake, **373–374,** 374 (map)
    Gopher Snake, **371–373,** 372 (map), 372 (plate)
    head scalation, 356, 356 (fig.)
    kingsnakes, 374–379
    racers, 361–374
    Ring-necked Snake, **356–358,** 357 (map), 357 (plate), 360 (fig.)
    Sonoran Lyre Snake, **398,** 398 (map)
    Southwestern Black-headed Snake, **397–398,** 397 (map)
    Spotted Leaf-nosed Snake, **360–361,** 361 (map), 361 (plate)
    Western Ground Snake, **395,** 395 (map), 396 (plate)
    Western Long-nosed Snake, **378,** 378 (map), 379 (plate)
    Western Shovel-nosed Snake, **393–394,** 394 (map), 394 (plate)
    whipsnakes, 361–374
Colubridae, 355–400, 356 (fig.)
columella, reptiles, 224
Common Carp
    aquatic predation pressure by, 25
    and California Red-legged Frog, 192
Common Chuckwalla, **257–260**
    activity cycle, 258
    conservation note, 260
    defense, 258–259
    food, 242, 259–260
    habitat, 21, 22, 258
    identification, 257, 258 (plate)
    juveniles, 257
    natural history, 258–260
    osmoregulation, 231
    photography, 465
    range, 258, 258 (map)
    reproduction, 245, 259
    territoriality, 259
Common Garter Snake, **381–384**
    conservation note, 383–384
    habitat, 382–383, 384
    identification, 381, 382 (plate)
    natural history, 382–383
    prey, 383, 383 (plate), 384
    protective status, 381
    range, 380, 382, 382 (map)
    reproduction, 383
Common Kingsnake, **375–376**
    defensive behavior, 376
    habitat, 16, 375
    as habitat generalist, 5
    identification, 375
    natural history, 375–376
    prey, 375–376
    range, 375, 375 (map)
    reproduction, 376
common name changes, 31
Common Sharp-tailed Snake, **358–360**
    behavior, 359–360
    habitat, 16, 359–360
    head and jaw, 359–360, 360 (fig.)
    identification, 358–359, 359 (plate)
    natural history, 359–360
    range, 359, 359 (map)
    taxonomic note, 358
Common Side-blotched Lizard, **288–290**
    breeding, 290
    habitat, 17, 18, 22, 289
    hearing, 225
    identification, 288, 289 (plate)
    natural history, 289–290
    range, 288–289, 288 (map)
    territoriality, 289–290
Common Snapping Turtle, 442, **443–445,** 443 (map)
concertina locomotion, snakes, 343
Conolophus spp, 253
constrictors, prey, 347
Contia
    longicaudae, **358–360**
    tenuis, 16, **358–360,** 359 (plate), 360 (fig.)
Cope's Leopard Lizard, **269–270,** 269 (map)
Corixidae, 72
costal folds, 37, 38 (fig.)
costal grooves, 37, 38 (fig.)
Couch's Spadefoot, **159–161**
    adults, 159
    conservation note, 160–161
    eggs, 160

food, 160
habitat, 22, 160
identification, 159–160
larvae, 160
natural history, 160
protective status, 159
range, 160, 160 (map)
reproduction, 160
voice, 159–160
courtship, terrestrial-breeding salamanders, 63, 64 (fig.)
cover boards, 477–478, 477 (fig.)
Coyote, trap robbing by, 476–477
Cretaceous Period, vertebrate terrestrial pioneers in, 4
critical minimum temperature, amphibians, 55
Crocodilia, 218–219
Crocodilians, respiration, 221
Crotalinae, 401–421, 402 (fig.), 403 (fig.)
Crotalus
  atrox, 406, **413–415,** 414 (map), 414 (plate), 415 (plate)
  cerastes, 21, 343, 343 (fig.), **419–421,** 420 (map), 420 (plate), 421 (plates), 466
  mitchellii, **417–418,** 417 (map), 418 (plate)
  molossus, 348
  oreganus, 16, **410–413,** 411 (map), 412 (plate)
  oreganus helleri, **410–413,** 411 (map), 412 (plate)
  oreganus lutosus, **410–413,** 411 (map)
  oreganus oreganus, **410–413,** 411 (map)
  ruber, 20, **418–419,** 418 (map)
  ruber ruber, **418–419,** 418 (map)
  scutulatus, 21, **415–417,** 416 (map)
  stephensi, **417–418,** 417 (map)
  viridis, 411, 413
  viridis helleri, 411
Crotaphytidae, 260–270
Crotaphytus, 260–263
  bicinctores, 17, **261–263,** 261 (map), 262 (plate)
  coloration, 244
  habitat, 260
  vestigium, **263,** 263 (map)
cryptic coloration, amphibians, 42–44, 43 (fig.)
CSSC (California species of special concern), 30
Cyprinus carpio
  aquatic predation pressure by, 25
  and California Red-legged Frog, 192

DAP (display action pattern), lizards, 243
Declining Amphibian Population Task Force (DAPTF), 24
defense mechanisms
  skin coloration, 42–44, 43 (fig.), 44 (fig.)
  skin toxins, 41–42, 42 (fig.)
Del Norte Salamander, 15, **108–109,** 108 (map)
Dermochelyidae, 454–455
Dermochelys coriacea, **454–455**
  body form, 446–447
  endothermy, 229
Desert Banded Gecko, **247–249,** 248 (map)
Desert Glossy Snake, **373–374,** 374 (map)
Desert Horned Lizard, **302–303,** 302 (map), 303 (plate)
Desert Iguana, **254–256**
  breeding, 256
  conservation note, 260
  food, 242, 256
  habitat, 21, 22, 254
  identification, 254, 255 (plate)
  natural history, 255–256
  range, 254, 254 (map)
  territoriality, 256
  thermoregulation, 255, 256 (plate)
Desert Night Lizard, **307–309**
  habitat, 21, 22, 308–309
  identification, 307–308, 309 (plate)
  natural history, 308–309
  range, 308, 308 (map)
  reproduction, 309
  taxonomic note, 307
Desert Night Snake, **399–400,** 399 (map)
Desert Patch-nosed Snake, **370–371,** 370 (map)
Desert Salamander, **134–135,** 134 (map)
Desert Slender Salamander, habitat, 22
Desert Spiny Lizard, **284–286**
  breeding, 286
  coloration, 244
  habitat, 21, 22, 285
  identification, 284, 285 (plate)
  natural history, 285–286
  range, 285, 285 (map)
Desert Striped Whipsnake, 17, **369,** 369 (map). See also Striped Whipsnake
Desert Tortoise, **426–430**
  basking, 229, 428
  conservation note, 430
  drinking, 429
  habitat, 23, 427–428
  identification, 427
  natural history, 427–430

Desert Tortoise (*continued*)
  photography, 464
  protective status, 427
  range, 427, 427 (map)
  reproduction, 429–430
  sex, 429
  thermoregulation, 428–429
Devonian Period, vertebrate terrestrial pioneers in, 2–4, 3 (fig.)
Diablo Range Garter Snake, **387–388**, 388 (map)
Diadophis punctatus, **356–358**
  behavior, 357–358
  food, 358
  habitat, 16, 357
  identification, 356–357, 357 (plate)
  natural history, 357–358
  photography, 466
  range, 357, 357 (map)
  reproduction, 358
  tail, 344
diamond-backed rattlesnake, 406, **413–415**, 414 (map), 414 (plate), 415 (plate)
Dicamptodon ensatus, **85–88**
  adult, 85, 85 (plate)
  breeding pool, 87–88, 88 (plate)
  eggs, 85, 87
  gills, 48–49
  habitat, 17, 86
  identification, 85–86, 85 (plate)
  larvae, 85–86, 86 (plate), 87–88
  metamorphosis, 88
  natural history, 86–88, 88 (plate)
  range, 86, 86 (map)
  reproduction, 87–88
  voice, 87
Dicamptodontidae, 84–89
dinosaurs, 218
Dipsochelys gigantea, 426
Dipsosaurus dorsalis, **254–256**
  breeding, 256
  conservation note, 260
  food, 242, 256
  habitat, 21, 22, 254
  identification, 254, 255 (plate)
  natural history, 255–256
  range, 254, 254 (map)
  territoriality, 256
  thermoregulation, 255, 256 (plate)
display action pattern (DAP), lizards, 243
disruptive coloration, amphibians, 43
documentation, 457–461
  field notes, 457–458, 459 (fig.)
  journal, 458, 460–461
  special accounts, 458–460, 459 (fig.)

drift fence, 474–475, 475 (plate)
Dunn's Salamander, 15, **107**, 107 (map)

Eakin, Richard, 472
eardrum
  amphibians, 47–48
  reptiles, 224–225
ectothermy, 5
  amphibians, 54–56, 54 (fig.)
  reptiles, 225–230
Eel, American, 3–4
eggs
  amniote, 219
  anurans, 59, 60 (fig.)
  aquatic-breeding salamanders, 62
  lizards, 244–245
  reptiles, 219–221, 220 (plates)
  salamanders, 69, 69 (fig.)
  snakes, 347–348, 348 (plate)
  terrestrial-breeding salamanders, 64–65, 64 (fig.), 65 (plate)
Elapidae, 400–401
Elgaria, 325–331
  coerulea, 15, **329–330**, 329 (map), 330 (plate)
  coerulea coerulea, **329–330**, 329 (map)
  coerulea palmeri, **329–330**, 329 (map)
  coerulea principis, **329–330**, 329 (map)
  coerulea shastensis, **329–330**, 329 (map)
  eurythermy, 225–226
  food, 242
  multicarinata, 16, 226, **326–328**, 327 (map), 327 (plate)
  multicarinata multicarinata, **326–328**, 327 (map)
  multicarinata scincicauda, **326–328**, 327 (map)
  multicarinata webbii, **326–328**, 327 (map)
  panamintina, **330–331**, 331 (map)
Ellicott Pond, 81
emergence range, reptiles, 227
Emydidae, 430–439
Endangered Species Act, 29
Endangered Species Conservation Act, 29
Endangered Species Preservation Act, 29
endothermy, reptiles, 229–230
Ensatina, 110–115
  adults, 112–113
  conservation note, 115
  courtship, 64 (fig.)
  disruptive coloration, 43
  eggs, 113, 115
  eschscholtzii, 16, 42, 42 (fig.), 54–55, 54 (fig.), **110–115**

eschscholtzii croceater, **110–115,** 115 (plate)
eschscholtzii eschscholtzii, **110–115**
eschscholtzii oregonensis, 112
eschscholtzii picta, **110–115**
eschscholtzii platensis, **110–115**
eschscholtzii xanthoptica, **110–115**
habitats, 16, 114
identification, 112–113
intergrade, 112, 113 (plate), 114
klauberi, 21, **110–115**
nasolabial grooves, 45–46, 46 (fig.)
natural history, 114–115
protective status, 112
range, 113–114, 113 (map)
reproduction, 115
tail, 37
taxonomic note, 112
toxic mucus, 114–115, 115 (plate)
Eretmochelys imbricata, **453–454**
estivation, amphibians, 53
Eublepharidae, 245–253
Eumeces (now *Plestiodon*), 314–318
gilberti, **316–318,** 317 (plates), 318 (map)
skiltonianus, **314–316,** 315 (map), 315 (plate)
eurythermy, reptiles, 225–226
eye(s), amphibians, 46–47
eyelid geckos, 247–253
eyesight
amphibians, 46–47
reptiles, 222–223
snakes, 344
turtles, 425

families, new, 31–32
fangs, snakes, 345–346
fear, of snakes, 348–350, 349 (plate)
federally threatened species (FT), 29
femoral glands, lizards, 239
fertilization
anurans, 59–60, 60 (fig.)
aquatic-breeding salamanders, 62, 63 (fig.)
field identification keys, 32–33
field notes, 457–458, 459 (fig.)
field observations, 457–461
field notes, 457–458, 459 (fig.)
journal, 458, 460–461
special accounts, 458–460, 459 (fig.)
fighting, lizards, 243–244
Fitch, Henry, 313–314, 474
flash photography, 461–462
Flat-tailed Horned Lizard, **304–305,** 304 (map)
Florida Worm Lizards, osmoregulation, 230

Flowerpot Snake, 350
food
captive animals, 482–483
lizards, 242
Foothill Yellow-legged Frog, **196–199**
adult, 196–197, 198 (plate)
breeding, 198
chemical contamination of aquatic habitat, 26
conservation note, 198–199
eggs, 197, 198
food, 198
habitat, 17, 197–198
identification, 196–197, 198 (plate)
larvae, 197, 198
metamorphosis, 198
natural history, 197–198, 198 (plate)
protective status, 196
range, 197, 197 (map)
voice, 58, 197
Forest Sharp-tailed Snake, **358–360**
behavior, 359–360
habitat, 359–360
head and jaw, 359–360
identification, 358–359
natural history, 359–360
range, 359, 359 (map)
taxonomic note, 358
forked tongue, snakes, 223–224
FPS (fully protected species), 30
fringe-toed lizards, 272–278
body form, 272–273
Coachella Valley Fringe-toed Lizard, **274–276,** 275 (map)
Colorado Desert Fringe-toed Lizard, **276–277,** 276 (map)
habitat, 22, 273
as habitat specialists, 6
Mojave Fringe-toed Lizard, 272 (plate), **277–278,** 278 (map)
reproduction, 274
territoriality, 273–274
frogs, 149–215
body form and locomotion, 39–41, 40 (fig.)
chorus, 179–185
classification, 149
key, 150
photography, 463–464
tailed, 150–154
tongueless, 212–215
true, 186–212
frogs legs, as human food, 210–211
frothy saliva, reptiles, 227
FT (federally threatened species), 29

fully protected species (FPS), 30
fungal infections, 481–482
funnel trap, 474–475, 476 (plate)

Gabilan Mountains Slender Salamander, **136,** 136 (map)
Gambelia, 263–270
  coloration, 244, 260
  copeii, **269–270,** 269 (map)
  habitat, 260
  sila, 264, **267–269,** 267 (map)
  wislizenii, 17, **264–266,** 264 (map), 265 (plate)
Gambusia affinis, aquatic predation pressure by, 25, 75
Garden Slender Salamander, **134–135,** 134 (map)
garter snakes, 379–400
  Aquatic Garter Snake, **387–388,** 388 (map), 388 (plate)
  Black-necked Garter Snake, 393 (plate)
  body form, 380
  California Lyre Snake, **398,** 398 (map)
  California Red-sided Garter Snake, **381–384,** 382 (map)
  Checkered Garter Snake, **392–393,** 392 (map), 393 (plate)
  classification, 380
  Coast Garter Snake, **385–387,** 386 (map), 386 (plate)
  Common Garter Snake, **381–384,** 381 (map), 382 (plate), 383 (plate)
  Giant Garter Snake, **389–390,** 390 (map)
  habitat, 380
  Mountain Garter Snake, **385–387,** 386 (map)
  Northwestern Garter Snake, **391,** 391 (map)
  photography, 466
  prey, 347
  reproduction, 347, 380
  San Francisco Garter Snake, **381–384,** 382 (map), 382 (plate), 383 (plate)
  Sierra Garter Snake, **388–389,** 389 (map)
  Sonoran Lyre Snake, **398,** 398 (map)
  Two-striped Garter Snake, **391,** 391 (map)
  Valley Garter Snake, **381–384,** 382 (map)
  Wandering Garter Snake, **385–387,** 386 (map)
  Western Terrestrial Garter Snake, **385–387,** 386 (map)
geckos, 245–246
  banded, 247–251
  Barefoot Gecko, **250–251,** 250 (map)
  body form, 245–246
  Desert Banded Gecko, **247–249,** 248 (map), 248 (plate)
  eyelid, 247–253
  habitat, 246
  Leaf-toed Gecko, **251–252,** 251 (map)
  Mediterranean House Gecko, **252–253,** 252 (map)
  Moorish Wall Gecko, **253**
  reproduction, 246
  San Diego Banded Gecko, **247–249,** 248 (map), 248 (plate)
  voice, 246
  Western Banded Gecko, **247–249,** 248 (map), 248 (plate)
Gekkonidae, 245–246
genera, new, 31–32
Geochelone, 426
Geococcyx californianus, 298
Giant Garter Snake, 18, **389–390,** 390 (map)
giant salamanders, 84–89
  California Giant Salamander, **85–88,** 85 (plate), 86 (map), 86 (plate), 88 (plate)
  Pacific Giant Salamander, **88–89,** 89 (map)
Gila Monster, 242, **335–337,** 335 (map)
Gilbert's Skink, **316–318,** 317 (plates)
gill(s), amphibians, 48–52, 49 (fig.), 50 (fig.)
gill rakers, amphibians, 48
Glossy Snake, 18, 20, **373–374,** 374 (map)
Golden Trout, and yellow-legged frogs, 202
Gopher Snake, **371–373**
  conservation note, 373
  eggs, 348 (plate)
  habitat, 16, 18, 372–373, 372 (plate)
  hissing, 344–345
  identification, 371–372, 372 (plate)
  natural history, 372–373
  osmoregulation, 231
  photography, 466
  prey, 347
  protective status, 371
  range, 372, 372 (map)
Gopherus agassizii, **426–430**
  basking, 229, 428
  conservation note, 430
  drinking, 429
  habitat, 23, 427–428
  identification, 427, 427 (plate)
  natural history, 427–430
  photography, 464
  protective status, 427
  range, 427
  reproduction, 429–430
  sex, 429
  thermoregulation, 428–429

Granite Night Lizard, 20, **310–311,** 311 (map)
Granite Spiny Lizard, **286–288,** 286 (map), 287 (plate)
Great Basin Collared Lizard, **261–263**
  breeding, 262–263
  food, 263
  habitat, 17, 262
  identification, 261, 262 (plate)
  natural history, 262–263
  range, 261, 261 (map)
Great Basin Desert, 6 (map), 16–17
Great Basin Gopher Snake, **371–373,** 372 (map)
Great Basin Rattlesnake, **410–413,** 411 (map)
Great Basin Spadefoot, 17, **158–159,** 159 (map)
Great Basin Whiptail, **320–322**
  behavior, 321–322
  food, 322
  habitat, 321
  identification, 320–321
  natural history, 321–322
  range, 321, 321 (map)
  reproduction, 322
Great Plains Toad, **164–165,** 164 (map), 165 (plate)
Great Valley, 6 (map), 10 (plate), 18
Greater Roadrunner, 298
Green Iguana, 253
Green Sea Turtle, **448–451**
  captured for meat, 450–451, 450 (plate)
  conservation note, 450–451
  food, 449
  habitat, 449
  identification, 448–449
  migration, 449–450
  natural history, 449–450
  protective status, 448
  range, 449
  reproduction, 450
  taxonomic note, 448
  thermoregulation, 229, 446
Green Toad, skin toxins, 41
Gregarious Slender Salamander, **132,** 132 (map)
ground snakes, **395,** 395 (map), 396 (plate)
gular fold, 38 (fig.)

habitat complex, amphibian and reptile distribution throughout, 5–23, 6 (map), 7–15 (plates)
habitat generalists, amphibians and reptiles as, 5
habitat loss, 23–29
  due to aquatic predation pressure, 24–26
  due to chemical contamination, 26–27
  due to chytrid fungus infection, 27–28
  due to ultraviolet radiation, 28–29
habitat specialists, amphibians and reptiles as, 5–6
Hammerson, Geoff, 228
hand capture, 470–472
Hawksbill Sea Turtle, **453–454**
Hayes, Marc, 194
head, lizards, 238 (fig.)
head scalation, colubrids, 356, 356 (fig.)
head-neck basking, reptiles, 228
hearing
  amphibians, 47–48
  reptiles, 224–225
  snakes, 345
  turtles, 425
heliotherms, reptiles as, 225–229
Hell Hollow Slender Salamander, **139–140,** 140 (map)
Heloderma
  suspectum, 242, **335–337,** 335 (map)
  suspectum cinctum, **335–337,** 335 (map)
Helodermatidae, 334–337
Hemidactylus turcicus, **252–253,** 252 (map)
hemipenis(es)
  lizards and snakes, 233–234
  snakes, 347
Herman, Andrea, 147–148
herpetology, 36
hindlimbs, frogs and toads, 39–41, 40 (figs.)
hissing, snakes, 344–345
Hodge, Hillary, 189
horned lizards, 298–307
  body form, 298
  Coast Horned Lizard, **300–302,** 300 (map), 301 (plate)
  conservation note, 299–300
  defensive behavior, 299
  Desert Horned Lizard, **302–303,** 302 (map), 303 (plate)
  Flat-tailed Horned Lizard, **304–305,** 304 (map)
  habitat, 298
  osmoregulation, 230
  Pygmy Short-horned Lizard, **305–307,** 306 (map), 306 (plate)
  reproduction, 244, 299
"horned toads," 298
husbandry, 481–483
Hydromantes, 142–148
  brunus, 142 (plate), **144–146,** 145 (map)
  as habitat specialists, 6
  locomotion, use of tail, 37, 38 (fig.)

Hydromantes (*continued*)
   platycephalus, 19, **143–144,** 143 (map)
   shastae, 16, **146–148,** 146 (map)
Hydrophiidae, 400–401
Hylidae, 179–185
   body form and locomotion, 39–40
hyperthermia, amphibians, 55–56
hypothermia, amphibians, 55
Hypsiglena
   chlorophaea, **399–400,** 399 (map)
   ochrorhynchus, **399–400,** 399 (map)
   prey, 347

ICZN (International Committee for Zoological Nomenclature), 31
Iguana iguana, 254
Iguanidae, 253–260
   Common Chuckwalla, **257–260,** 258 (map), 258 (plate)
   Desert Iguana, **254–256,** 254 (map), 255 (plate), 256 (plate)
   parietal eye, 241 (fig.)
Incilius, 161
individual animals, recognition in field, 479–480
International Committee for Zoological Nomenclature (ICZN), 31
Inyo Mountain Salamander, 17, 22, **127–128,** 127 (map)
Island Fence Lizard, **279–282**
   basking behavior, 281
   breeding, 282
   displacement behavior, 282
   display behavior, 281–282
   food, 282
   as habitat generalist, 280
   identification, 279–280
   natural history, 280–282
   parietal eye, 241–242
   range, 280, 280 (map)
   taxonomic note on, 279
Island Night Lizard, **312–313,** 312 (map)

Jacobson's organ
   amphibians, 45
   lizards, 242
   reptiles, 223–224
   snakes, 344
jaw, snakes, 345–346
Jennings, Mark, 194
journal, 458, 460–461
Jurassic Period, vertebrate terrestrial pioneers in, 4

Kern Canyon Slender Salamander, **130,** 130 (map)

Kern Plateau Salamander, **128,** 128 (map)
key(s)
   field identification, 32–33
   frogs and toads, 150
   lizards, 236–237, 238 (fig.)
   salamanders, 68
   snakes, 340–341
   tortoises and turtles, 423–424
kidney function, amphibians, 53–54
Kings River Slender Salamander, **138,** 138 (map)
kingsnakes, 374–379
   California Kingsnake, **375–376,** 375 (map)
   California Mountain Kingsnake, **378,** 378 (map)
   Coast Mountain Kingsnake, **377–378,** 377 (map)
   coloration, 376 (plate)
   Common Kingsnake, **375–376,** 375 (map)
   eggs, 220 (plate)
   habitat, 374
   prey, 347
   San Bernardino Mountain Kingsnake, **377–378,** 377 (map)
   San Diego Mountain Kingsnake, **377–378,** 377 (map)
   Sierra Mountain Kingsnake, **377–378,** 377 (map)
   St. Helena Mountain Kingsnake, **377–378,** 377 (map)
Klauber, Laurence, 412–413
Knapp, Roland, 202

labial teeth, anurans, 59, 60 (fig.)
Lampropeltis
   getula, 5, 16, **375–376,** 375 (map)
   getula californiae, **375–376,** 375 (map)
   zonata, 19, **377–378,** 377 (map)
   zonata multicincta, **377–378,** 377 (map)
   zonata multifasciata, **377–378,** 377 (map)
   zonata parvirubra, **377–378,** 377 (map)
   zonata pulchra, **377–378,** 377 (map)
   zonata zonata, **377–378,** 377 (map)
Land Iguana, 253
land tortoises, 426–430
Lanius ludovicianus, 298
Large-blotched Ensatina, **110–115**
   adults, 112–113
   conservation note, 115
   defense, 114–115
   eggs, 113, 115
   habitats, 114
   identification, 112–113
   natural history, 114–115
   protective status, 112
   range, 113–114, 113 (map)

reproduction, 115
taxonomic note, 112
Largemouth Bass
aquatic predation pressure by, 25
and California Red-legged Frog, 192
Larsen, Sheila, 126, 384
lateral undulatory locomotion, snakes, 342
Latimeria chalumnae, 2
Latin name changes, 31
Leaf-toed Gecko, **251–252,** 251 (map)
skin, 246
Leatherback Sea Turtle, **454–455**
body form, 446–447
endothermy, 229
legless lizards, 233, 331–334
lens, reptiles, 222–223
"lens-squeezing" mechanism, reptiles, 222
Leopard Frog
Lowland, **207,** 207 (map)
Northern, **203–205,** 204 (map)
Rio Grande, **206,** 206 (map)
Southern, **203–205,** 204 (map)
leopard lizards, 260, 263–270
Blunt-nosed Leopard Lizard, 264, **267–269,** 267 (map)
coloration, 244, 260
Cope's Leopard Lizard, **269–270,** 269 (map)
Long-nosed Leopard Lizard, **264–266,** 264 (map), 265 (plate)
Lepidochelys olivacea, **452–453**
Lepomis macrochirus, aquatic predation pressure by, 25
Leptotyphlopidae, 350–352
Leptotyphlops
dulcis, 352
humilis, 350, **351–352,** 351 (map), 351 (plate)
Lesser Slender Salamander, **137,** 137 (map)
Lichanura trivirgata, 20, **354–355,** 355 (map)
Licht, Paul, 231
life cycle
anurans, 57–62
aquatic-breeding salamanders, 62–63, 63 (fig.)
terrestrial-breeding salamanders, 63–65, 64 (figs.), 65 (plates)
limbs, turtle, 425
Limestone Salamander, 142 (plate), **144–146,** 146 (map)
Lithobates, 186
lizards, 235–337
alligator, 325–331
basking position, 228
breeding, 244–245
brush, 290–294
burrowing, 238–239
classification, 235–236
collared, 260–263
coloration, 244
eyelid geckos, 247–253
feeding habits, 242
geckos, 245–246
head anatomy, 238 (fig.)
horned, 298–307
iguanids, 253–260
key, 236–237, 238 (fig.)
legless, 233, 331–334
leopard, 260, 263–270
limbs, 238–239
night, 307–313
noose capture, 472–474, 473 (plate)
olfaction and taste, 223–224
parietal eye ("third eye"), 240–242, 241 (fig.)
photography, 465
Phrynosomatidae, 270–307
preanal and femoral glands, 239
reproduction, 233–234
rock, 295–297
scale types, 238 (fig.)
sex determination, 234
side-blotched, 288–290
skinks, 313–318
spiny, 278–288
tail breakage, 239–240, 239 (fig.)
teiids, 318–324
territoriality, 243–244
tree, 294–295
venomous, 334–337
vision, 222–223
whiptails, 319–324
zebra-tailed, 270–272
lobefin fishes, as vertebrate terrestrial pioneers, 2–4
locomotion
amphibians, 37–41, 38 (fig.), 40 (fig.)
snakes, 342–343, 343 (fig.)
Loggerhead Sea Turtle, **451–452**
Loggerhead Shrike, 298
Long-nosed Leopard Lizard, **264–266**
breeding, 265–266
coloration, 265
conservation note, 266
food, 266
habitat, 17, 22, 264–265
identification, 264, 265 (plate)
natural history, 264–266
range, 264, 264 (map)
territoriality, 265

Long-tailed Brush Lizard, **290–292,** 291 (map), 291 (plate)
Long-toed Salamander, **78–82**
  adult, 78, 80 (plate)
  conservation note, 81–82
  eggs, 78
  gills, 49
  habitat, 16, 17, 79
  identification, 78, 80 (plate)
  larvae, 78, 79 (plate)
  migration, 79
  natural history, 79–81, 79–81 (plates)
  protective status, 78
  range, 79, 79 (map)
  reproduction, 62, 79–80
  spawning pools, 80
loreal pit, 402, 402 (fig.)
Lowland Leopard Frog, **207,** 207 (map)
lungless salamanders, 104–148
  climbing salamanders, 116–124
  Ensatina, 110–115
  habitat, 104
  nasolabial grooves, 105
  reproduction, 63–65, 104, 105–106
  respiration, 50, 104–105
  slender salamanders, 124–142
  tongue protrusion, 105
  web-toed salamanders, 142–148
  woodland salamanders, 106–110
lymphatic tourniquet, 409
lyre snakes, 20, 344, **398,** 398 (map)

Macrochelys temminckii, 442
Marantelli, Gerry, 215
marginal endothermy, reptiles, 229
Marine Iguana, 253
Masticophis, 361–374
  flagellum, 20, **363–366,** 365 (map), 365 (plate)
  flagellum piceus, **364–366,** 365 (map), 365 (plate)
  flagellum ruddocki, **363–366,** 365 (map)
  fuliginosus, **363–366,** 365 (map)
  lateralis, **366–369,** 367 (map), 367 (plate), 368 (plate)
  lateralis euryxanthus, 228, **366–369,** 367 (map), 367 (plate), 368 (plate), 475
  lateralis lateralis, **366–369,** 367 (map)
  taeniatus, **369,** 369 (map)
  taeniatus taeniatus, 17
  vision, 223
mechanical snake stick, 474
Mediterranean House Gecko, **252–253,** 252 (map)

melanophore-directed color change, amphibians, 42–43
metabolism, reptiles, 225–230
metamorphosis
  anurans, 60–62, 61 (fig.)
  aquatic-breeding salamanders, 63
Mexican Salamander, appendage regeneration, 39
Micropterus
  coosae, 199
  salmoides, 25, 192
microtransponders, for recognition of individual animals, 479
mite infestations, snakes, 481
Mojave Desert, 6 (map), 12–14 (plates), 22–23
Mojave Desert Shovel-nosed Snake, **393–394,** 394 (map)
Mojave Fringe-toed Lizard, 272 (plate), **277–278,** 278 (map)
Mojave Glossy Snake, **373–374,** 374 (map)
Mojave Patch-nosed Snake, **370–371,** 370 (map)
Mojave Rattlesnake, **415–417**
  habitat, 21, 416
  identification, 415, 416
  natural history, 416–417
  range, 416, 416 (map)
  venom, 416–417
mole salamanders, 69–84
  California Tiger Salamander, **70–75,** 71 (map), 71 (plate), 72 (plate)
  Long-toed Salamander, **78–82,** 79 (map), 79–81 (plates)
  Northwestern Salamander, **82–84,** 83 (map), 83 (plate)
  Tiger Salamander, **75–77,** 76 (map)
Moloch, osmoregulation, 230
Moloch horridus, osmoregulation, 230
Monterey Ensatina, **110–115**
  adults, 112–113
  conservation note, 115
  defense, 114–115
  eggs, 113, 115
  habitats, 114
  identification, 112–113
  natural history, 114–115
  range, 113–114, 113 (map)
  reproduction, 115
  taxonomic note, 112
Moore, Beatrice, 92
Moorish Wall Gecko, **253**
Mosquitofish, aquatic predation pressure by, 25, 75

Mount Lyell Salamander, 19, **143–144,** 143 (map)
Mountain Garter Snake, **385–387,** 386 (map)
Mountain Yellow-legged Frog, 194
mucus glands, amphibians, 41
mudskippers, 3, 3 (fig.)

name changes, 31
nasolabial grooves, amphibians, 45–46, 46 (fig.)
natural field photography, 462–466
    frogs and toads, 463–464
    lizards, 465
    salamanders, 462–463
    snakes, 465–466
    turtles and tortoises, 464–465
"nature-faking," 467–468
Neoceratodus forsteri, 2
nest site, lizards, 244–245
net capture, 472
Nevada Desert Shovel-nosed Snake, **393–394,** 394 (map)
newt(s), 92–103
    breeding pond, 98 (plate)
    Coast Range Newt, **96–100,** 97 (map)
    defensive posture, 44 (fig.), 95
    eggs, 94, 94 (fig.), 95 (plate)
    feeding, 95
    longevity, 95–96
    metamorphs, 99 (plate)
    migration, 92, 94
    Red-bellied Newt, **102–103,** 103 (map)
    reproduction, 92–94, 92 (fig.), 93 (plate), 94 (fig.), 95 (plate)
    Rough-skinned Newt, **100–102,** 101 (map), 101 (plate)
    Sierra, **96–100,** 97 (map), 97 (plate)
    tetrodotoxin (TTX), 94–95
newt ball, 93–94, 93 (plate)
nictitating membrane, amphibians, 46–47
night lizards, 307–313
    body form, 307
    Desert Night Lizard, **307–309,** 308 (map), 309 (plate)
    Granite Night Lizard, **310–311,** 311 (map)
    Island Night Lizard, **312–313,** 312 (map)
    reproduction, 245
    Sandstone Night Lizard, **310–311,** 311 (map)
    Wiggins' Night Lizard, **307–309,** 308 (map)
night snakes
    Coast Night Snake, **399–400,** 399 (map)
    Desert Night Snake, **399–400,** 399 (map)
    prey, 347

"no release" law, 483
noose capture, 472–474, 473 (plate)
North Coast Range, 6 (map), 7 (plate), 15–16
North Desert Horned Lizard, **302–303,** 302 (map)
Northern Alligator Lizard, 19, **329–330,** 329 (map), 330 (plate)
Northern Leopard Frog, **203–205,** 204 (map)
Northern Pacific Rattlesnake, **410–413,** 411 (map)
Northern Red Rattlesnake, **418–419,** 418 (map)
Northern Red-legged Frog, **186–192**
    adults, 187
    aquatic predation pressure on, 26
    eggs, 187, 189
    food, 190
    habitat, 188–189
    identification, 187–188
    larvae, 187–188, 189
    natural history, 188–190
    protective status, 187
    range, 188, 188 (map)
    voice, 187
Northern Rubber Boa, **353–354**
    food, 354
    habitat, 16, 19, 353–354
    identification, 353
    natural history, 353–354
    photography, 466
    protective status, 353
    range, 353, 353 (map)
    reproduction, 354
Northwestern Alligator Lizard, **329–330,** 329 (map)
Northwestern Fence Lizard, **279–282,** 280 (map)
Northwestern Garter Snake, **391–392,** 392 (map)
Northwestern Pond Turtle, **430–435**
    basking, 432
    conservation note, 433–435
    food, 432
    habitat, 431–432
    hibernation, 432
    identification, 430–431
    natural history, 431–433
    photography, 464–465
    protective status, 430
    range, 431, 431 (map)
    reproduction, 432–433
    thermoregulation, 432
Northwestern Salamander, **82–84,** 83 (map), 83 (plate)
Notonectidae, 72
nuptial pads, newts, 93

observations, 457–461
  field notes, 457–458, 459 (fig.)
  journal, 458, 460–461
  special accounts, 458–460, 459 (fig.)
olfaction
  amphibians, 44–46, 46 (fig.)
  reptiles, 223–224
  turtles, 425
olfactory epithelium, amphibians, 45
Olive Ridley Sea Turtle, **452–453**
Oncorhynchus
  mykiss, 202
  mykiss aguabonita, 202
operculum, amphibians, 48
Ophidia, 339–421
ophidiophobia, 348–350, 349 (plate)
oral papillae, anurans, 59, 60 (fig.)
Orange-throated Whiptail, 20, **323–324,** 323 (map), 324 (plate)
Oregon Alligator Lizard, **326–328**
  behavior, 328
  food, 328
  habitat, 327–328
  identification, 326–327
  natural history, 327–328
  range, 327, 327 (map)
  reproduction, 328
Oregon Aquatic Garter Snake, **387–388,** 388 (map)
Oregon Ensatina, 112
Oregon Spotted Frog, **195–196,** 195 (map)
Ornate Tree Lizard, **294–295,** 294 (map), 295 (plate)
osmoregulation
  amphibians, 52–54
  reptiles, 230–232
ovoviviparous reproductive mode, reptiles, 219

Pacific Chorus Frog, **180–183**
  adults, 180, 181 (plate)
  body form and locomotion, 40
  breeding, 182–183
  camouflage, 42
  eggs, 180, 182–183
  food, 183
  habitat, 17, 18, 181
  as habitat generalist, 5, 181
  identification, 180, 181 (plate)
  larvae, 180, 183
  natural history, 181–183, 181 (plate)
  photography, 463
  range, 181, 181 (map)
  spawning site, 57
  voice, 58–59, 180, 181–182

Pacific Giant Salamander, **88–89,** 89 (map)
Pacific Gopher Snake, **371–373,** 372 (map)
Pacific Hawksbill Sea Turtle, **453–454**
Pacific Tree Frog, **180–183.** *See also* Pacific Chorus Frog
Painted Ensatina, **110–115**
  adults, 112–113
  conservation note, 115
  defense, 114–115
  eggs, 113, 115
  habitats, 114
  identification, 112–113
  natural history, 114–115
  range, 113–114, 113 (map)
  reproduction, 115
  taxonomic note, 112
Panamint Alligator Lizard, **330–331,** 331 (map)
Panamint Rattlesnake, **417–418,** 417 (map)
"pancake turtles," 439–442
panting, reptiles, 227, 227 (plate)
parietal eye, lizards, 240–242, 241 (fig.)
parthenogenesis, snakes, 350
pectoral amplexus, anurans, 59
Pelamis platurus, 400, **401**
Pelobatidae, 154–161
  body form and locomotion, 40–41, 40 (fig.)
  voice, 58
pelvic amplexus, anurans, 59
Peninsular Range, 6 (map), 12 (plate), 21–22
Periophthalmus, 3, 3 (fig.)
pet stores, 481–483
Petrosaurus, 295–297
photo terrariums, 467
photography, 461–468, 461 (plate)
  captive or temporarily restrained specimens, 466–468
  flash, 461–462
  frogs and toads, 463–464
  lizards, 465
  natural, 462–466
  for recognition of individual animals, 479–480
  salamanders, 462–463
  snakes, 465–466
  telephoto lenses, 462
  turtles and tortoises, 464–465
Phrynosoma, 298–307
  blainvillii, 18, **300–302,** 300 (map), 301 (plate)
  douglasii, **305–307,** 306 (map), 306 (plate)
  mcallii, **304–305,** 304 (map)
  osmoregulation, 230

platyrhinos, **302–303,** 302 (map), 303 (plate)
platyrhinos calidiarum, **302–303,** 302 (map)
platyrhinos platyrhinos, **302–303,** 302 (map)
reproduction, 244
Phrynosomatidae, 270–307
Phyllodactylus nocticolus, 246, **251–252,** 251 (map)
Phyllorhynchus decurtatus, **360–361,** 361 (map), 361 (plate)
physiology
  amphibians, 36–65
  reptiles, 218–232
Pipidae, 212–215
pit organs, snakes, 345
"pit tags," 479
pit trap, 475 (plate)
pit viper(s), 401–421
  body form, 401–402, 402 (fig.), 403 (fig.)
  conservation note, 409–410
  defensive strategies, 404–405
  Great Basin Rattlesnake, **410–413,** 411 (map)
  head anatomy, 402, 402 (fig.)
  and humans, 405–407
  Mojave Rattlesnake, **415–417,** 416 (map)
  Northern Pacific Rattlesnake, **410–413,** 411 (map), 412 (plate)
  Panamint Rattlesnake, **417–418,** 417 (map)
  pit organs, 345
  prey, 403–404
  Red Diamond Rattlesnake, **418–419,** 418 (map)
  Sidewinder, **419–421,** 420 (map), 420 (plate), 421 (plates)
  Southern Pacific Rattlesnake, **410–413,** 411 (map)
  Speckled Rattlesnake, **417–418,** 417 (map), 418 (plate)
  tail rattles, 402–403, 403 (fig.)
  treatment of bites, 407–409
  Western Diamond-backed Rattlesnake, 406, **413–415,** 414 (map), 414 (plate), 415 (plate)
  Western Rattlesnake, **410–413,** 411 (map), 412 (plate)
Pituophis
  catenifer, 16, 231, 344–345, **371–373,** 372 (map), 372 (plate), 466
  catenifer affinis, **371–373,** 372 (map)
  catenifer annectens, **371–373,** 372 (map)
  catenifer catenifer, **371–373,** 372 (map)

  catenifer deserticola, **371–373,** 372 (map)
  catenifer pumilis, **371–373,** 372 (map)
Plestiodon (formerly *Eumeces*), 314–318
  gilberti, **316–318,** 317 (plates), 318 (map)
  skiltonianus, 15–16, **314–316,** 315 (map), 315 (plate)
Plethodon, 106–110
  asupak, **109–110,** 109 (map)
  dunni, 15, **107,** 107 (map)
  elongatus, 15, **108–109,** 108 (map)
  stormi, **109–110,** 109 (map)
plethodontid salamanders
  photography, 463
  respiration, 50
  tail, breakage and regeneration, 37–39
Plethodontidae, 104–148
  reproduction, 63–65
polliwogs, 59–60
Pond Slider, **435–436,** 435 (map)
pond turtles, 430–439
  Northwestern Pond Turtle, **430–435,** 431 (map)
  Pond Slider, **435–436,** 435 (map)
  Southwestern Pond Turtle, **430–435,** 431 (map)
  Western Painted Turtle, **437–439,** 437 (map), 438 (plate)
  Western Pond Turtle, **430–435,** 431 (map), 431 (plate)
Prairie Rattlesnake, 411
preanal glands, lizards, 239
preferred thermal range, reptiles, 227
prey, snakes, 346–347
Procyon lotor, trap robbing, 476–477
protective status, 29–30
Pseudacris
  cadaverina, 20, **183–185,** 184 (map), 185 (plate)
  regilla, 5, 40, **180–183,** 181 (map), 181 (plate), 463
"Psychedelic Toad," 163
Ptychocheilus grandis, 25
Pygmy Short-horned Lizard, **305–307,** 306 (map), 306 (plate)
Python
  molurus, 229
  reticulatus, 352
pythons, 352–355
  pit organs, 345

Raccoon, trap robbing by, 476–477
Racer, **362–363,** 362 (map), 363 (plate)
racer(s), 361–374
  Coast Patch-nosed Snake, **370–371,** 370 (map)

racer(s) (*continued*)
  Desert Patch-nosed Snake, **370–371,** 370 (map)
  habitat, 16
  Mojave Patch-nosed Snake, **370–371,** 370 (map)
  photography, 466
  prey, 347
  Racer, **362–363,** 362 (map), 363 (plate)
  Red Racer, **363–366,** 365 (map), 365 (plate)
  Striped Racer, **366–369,** 367 (map), 367 (plate), 368 (plate)
  vision, 223
Rainbow Trout, and yellow-legged frogs, 202
Ramphotyphlops braminus, 350
Rana
  aurora, 26, 56, **186–192,** 188 (map), 188 (plate), 190 (plate)
  berlandieri, **206,** 206 (map)
  body form and locomotion, 39–41, 40 (fig.)
  boylii, 17, 58, **196–199,** 197 (map), 198 (plate)
  cascadae, 16, **193–194,** 194 (map)
  catesbeiana, 18, 56, **208–212,** 209 (map), 209 (plate)
  draytonii, 23, 25, 46, **186–192,** 188 (map), 188 (plate), 190 (plate)
  muscosa, 19, 56, **199–202,** 200 (map)
  pipiens, **203–205,** 204 (map)
  pretiosa, **195–196,** 195 (map)
  sierrae, 194, **199–202,** 200 (map), 201 (plate)
  sphenocephala, **203–205,** 204 (map)
  yavapaiensis, **207,** 207 (map)
ranid frogs, photography, 463–464
Ranidae, 186–212
rattlesnakes, 401–421, 402 (fig.), 403 (fig.)
  body form, 401–402, 402 (fig.), 403 (fig.)
  conservation note, 409–410
  defensive strategies, 404–405
  eyesight, 344, 404
  Great Basin Rattlesnake, **410–413,** 411 (map)
  head anatomy, 402, 402 (fig.)
  and humans, 405–407
  Mojave Rattlesnake, **415–417,** 416 (map)
  Northern Pacific Rattlesnake, **410–413,** 411 (map), 412 (plate)
  Panamint Rattlesnake, **417–418,** 417 (map)
  photography, 465–466
  Prairie Rattlesnake, 411
  prey, 347, 403–404
  Red Diamond Rattlesnake, **418–419,** 418 (map)
  Sidewinder, **419–421,** 420 (map), 420 (plate), 421 (plates)
  skin, 342
  Southern Pacific Rattlesnake, **410–413,** 411 (map)
  Speckled Rattlesnake, **417–418,** 417 (map), 418 (plate)
  tail rattles, 402–403, 403 (fig.)
  treatment of bites, 407–409
  Western Diamond-backed Rattlesnake, 406, **413–415,** 414 (map), 414 (plate), 415 (plate)
  Western Rattlesnake, **410–413,** 411 (map), 412 (plate)
rear-fanged venomous snakes, prey, 347
recognition, of individual amphibians and reptiles in field, 479–480
recording observations, 457–461
  field notes, 457–458, 459 (fig.)
  journal, 458, 460–461
  special accounts, 458–460, 459 (fig.)
rectilinear locomotion, snakes, 342–343
Red Brome, 430
Red Diamond Rattlesnake, 20, **418–419,** 418 (map)
Red Racer, **363–366,** 365 (map), 365 (plate)
Red-bellied Newt, 93 (plate), **102–103,** 103 (map)
Red-eared Slider, 433, 435, 436
Redeye Bass, and Foothill Yellow-legged Frog, 199
Red-legged Frog
  California, **186–192,** 188 (map), 188 (plate), 190 (plate)
  Northern, **186–192,** 188 (map)
Red-spotted Toad, **166–167**
  adults, 166, 167 (plate)
  breeding, 167
  eggs, 166, 167
  habitat, 21, 22, 167
  identification, 166
  larvae, 166
  osmoregulation, 53
  range, 167, 167 (range)
  voice, 58, 166, 167 (plate)
Relictual Slender Salamander, **139,** 139 (map)
reproduction
  anurans, 57–62
  aquatic-breeding salamanders, 62–63, 63 (fig.)
  lizards, 233–234, 244–245
  reptiles, 219–221, 220 (plates)

snakes, 233–234, 347–348, 348 (plate)
terrestrial-breeding salamanders, 63–65, 64 (figs.), 65 (plates)
turtles, 425
reptile(s), 217–455
anatomy, physiology, and behavior, 218–232
distribution throughout California's habitat complex, 5–23, 6 (map), 7–15 (plates)
hearing, 224–225
olfaction and taste, 223–224
osmoregulation, 230–232
recognition in field of individual, 479–480
reproduction, 219–221, 220 (plates)
respiration, 221
sensory input, 222–225
skin, 221–222
taxonomy, 218
thermoregulation and metabolism, 225–230, 226 (fig.), 227 (plate)
as vertebrate terrestrial pioneers, 2–4, 3 (fig.)
vision, 222–223
Reptilia, 218–219
respiration
amphibians, 48–52, 49 (fig.), 50 (fig.)
reptiles, 221
respiratory organ, skin as, 41
restrained specimens, photography, 466–468
Reticulate Python, 352
Rhineura floridana, osmoregulation, 230
Rhinocheilus lecontei, **378**, 378 (map), 379 (plate)
habitat, 18, 20
Rhyacotriton variegatus, 15, **90–92**, 90 (map), 91 (plate)
Rhyacotritonidae, 89–92
Rhynchocephalia, 218
rib cage, reptiles, 221
Ring-necked Snake, **356–358**
behavior, 357–358
food, 358
habitat, 16, 357
identification, 356–357, 357 (plate)
natural history, 357–358
photography, 466
range, 357, 357 (map)
reproduction, 358
tail, 344
Rio Grande Leopard Frog, **206,** 206 (map)
rock lizards, 295–297
Rock Pythons, endothermy, 229
Rosy Boa, 20, 21, **354–355,** 355 (map)

Rough-skinned Newt, **100–102,** 101 (map), 101 (plate)
Rubber Boa, **353–354**
food, 354
habitat, 16, 19, 353–354
identification, 353
natural history, 353–354
photography, 466
protective status, 353
range, 353, 353 (map)
reproduction, 354

Sacramento Perch, aquatic predation pressure by, 25
Sacramento Pikeminnow, aquatic predation pressure by, 25
Sagebrush Lizard, **282–284**
habitat, 15, 17, 19, 283
identification, 283
natural history, 283–284
range, 283, 283 (map)
reproduction, 284
salamander(s), 67–148
appendage regeneration, 39
aquatic-breeding, 62–63, 63 (fig.)
body form and locomotion, 37–39, 38 (figs.)
classification, 67–68
climbing, 116–124
eggs, 69, 69 (fig.)
Ensatina, 110–115
giant, 84–89
hand capture, 470–471
key, 68
lungless, 104–148
mole, 69–84
newts, 92–103
photography, 462–463
skin shedding, 44, 45 (fig.)
slender, 124–142
tail, 37–39, 38 (fig.)
terrestrial-breeding, 63–65, 64 (figs.), 65 (plates)
torrent or seep, 89–92
trapping, 475 (plate)
web-toed, 142–148
woodland, 106–110
salamander larvae
aquatic-breeding, 62–63
gills, 48–49, 49 (figs.)
Salamandridae, 92–103
Salmo trutta, and yellow-legged frogs, 202
Salmonella, in pet turtles, 435
salmonid spp., 87
salt excretion, reptiles, 231–232

salt-secreting glands, reptiles, 232
Salvadora
    hexalepis, 20, **370–371,** 370 (map)
    hexalepis hexalepis, 20, **370–371,** 370 (map)
    hexalepis mojavensis, 20, **370–371,** 370 (map)
    hexalepis virgultea, **370–371,** 370 (map)
Salvelinus fontinalis, and yellow-legged frogs, 202
San Bernardino Mountain Kingsnake, **377–378,** 377 (map)
San Diego Alligator Lizard, **326–328**
    behavior, 328
    food, 328
    habitat, 327–328
    identification, 326–327
    natural history, 327–328
    range, 327, 327 (map)
    reproduction, 328
San Diego Banded Gecko, **247–249,** 248 (map)
San Diego Gopher Snake, **371–373,** 372 (map)
San Diego Mountain Kingsnake, **377–378,** 377 (map)
San Francisco Alligator Lizard, **329–330,** 329 (map)
San Francisco Garter Snake, **381–384**
    conservation note, 383–384
    habitat, 382–383, 384
    habitat loss, 23
    identification, 381, 382 (plate)
    natural history, 382–383
    prey, 383, 383 (plate), 384
    protective status, 381
    range, 380, 382, 382 (map)
    reproduction, 383
San Gabriel Mountains Slender Salamander, **135–136,** 135 (map)
San Joaquin Coachwhip, **363–366,** 365 (map)
San Simeon Slender Salamander, **137–138,** 138 (map)
Sandstone Night Lizard, **310–311,** 311 (map)
Santa Cruz Aquatic Garter Snake, **387–388,** 388 (map), 388 (plate)
Santa Cruz Black Salamander, **119–121,** 120 (map)
Santa Cruz Island Gopher Snake, **371–373,** 372 (map)
Santa Cruz Long-toed Salamander, **78–82**
    adult, 78, 80 (plate)
    conservation note, 81–82
    eggs, 78
    gills, 49
    habitat, 17, 79
    identification, 78, 80 (plate)
    larvae, 78
    migration, 79
    natural history, 80–81, 80 (plate), 81 (plate)
    protective status, 78
    range, 79, 79 (map)
    reproduction, 62, 79–80
    spawning pools, 80
Santa Lucia Mountains Slender Salamander, **137,** 137 (map)
Sarcopterygii, as vertebrate terrestrial pioneers, 2–4
Sauria, 235–337
Sauromalus
    ater, 21, 231, 242, 245, **257–260,** 258 (map), 258 (plate), 465
    obesus, **257–260,** 258 (map), 258 (plate)
scales
    colubrids, 356, 356 (fig.)
    lizards, 238 (fig.)
Scaphiopus couchii, habitat, 22
Sceloporus, 278–288
    becki, **279–282,** 280 (map)
    graciosus, 15, **282–284,** 283 (map)
    magister, 21, 244, **284–286,** 285 (map), 285 (plate)
    occidentalis, 5, 226, 241–242, **279–282,** 280 (map), 281 (plate), 465
    occidentalis occidentalis, **279–282,** 280 (map)
    occidentalis taylori, **279–282,** 280 (map)
    orcutti, **286–288,** 286 (map), 287 (plate)
    scalaris, 219, 220 (plate)
    vandenburgianus, **282–284,** 283 (map)
scent glands, snakes, 344
Scinidae, 313–318
Scott Bar Salamander, **109–110,** 109 (map)
sea snakes, 400–401
sea turtles, 445–455
    Black Sea Turtle, **448–451**
    body form, 446–447
    conservation note, 447–448
    Green Sea Turtle, 446, **448–451,** 450 (plate)
    Hawksbill Sea Turtle, **453–454**
    Leatherback Sea Turtle, 446–447, **454–455**
    Loggerhead Sea Turtle, **451–452**
    Olive Ridley Sea Turtle, **452–453**
    reproduction, 447
seat patch, amphibians, 52–53
seep salamanders, 89–92
    Southern Torrent Salamander, **90–92,** 91 (map), 92 (plate)

sensory input
  amphibians, 44–48
  reptiles, 222–225
  snakes, 344–345
Sequoia Slender Salamander, **138,** 138 (map)
Serpentes, 339–421
sex
  lizards, 234
  snakes, 234, 347
Shadle, Sid, 192
Shasta Alligator Lizard, **329–330,** 329 (map)
Shasta Salamander, 16, **146–148,** 146 (map)
shell, turtle, 424
shovel-nosed snakes, **393–394,** 394 (map), 394 (plate)
Sidewinder, **419–421**
  "cratering," 420
  in desert brush, 421 (plate)
  habitat, 21, 420
  identification, 407–408, 419
  locomotion, 343, 343 (fig.)
  natural history, 420–421
  on open ground, 420 (plate)
  photography, 466
  prey, 421
  range, 420, 420 (map)
  reverse "J"-shaped track, 420, 421 (plate)
sidewinding locomotion, snakes, 343, 343 (fig.)
Sierra Alligator Lizard, **329–330,** 329 (map)
Sierra Fence Lizard, **279–282**
  basking behavior, 281
  breeding, 282
  displacement behavior, 282
  display behavior, 281–282
  food, 282
  as habitat generalist, 280
  identification, 279–280
  natural history, 280–282
  parietal eye, 241–242
  range, 280, 280 (map)
Sierra Garter Snake, **388–389,** 389 (map)
Sierra Mountain Kingsnake, **377–378,** 377 (map)
Sierra Nevada, 6 (map), 10 (plate), 19
Sierra Nevada Ensatina, **110–115**
  adults, 112–113
  conservation note, 115
  defense, 114–115
  eggs, 113, 115
  habitats, 114
  identification, 112–113
  natural history, 114–115

range, 113–114, 113 (map)
reproduction, 115
taxonomic note, 112
Sierra Nevada Yellow-legged Frog, **199–202**
  adult, 199–200, 201 (plate)
  breeding, 201
  conservation note, 201–202
  eggs, 200
  food, 201
  habitat, 200–201
  identification, 199–200, 201 (plate)
  larvae, 200, 201
  natural history, 194, 200–201, 201 (plate)
  protective status, 199
  range, 200, 200 (map)
  voice, 200
Sierra Newt, **96–100**
  adults, 96, 97 (plate)
  conservation, 99–100
  eggs, 96, 97
  food, 99
  habitat, 97
  identification, 96–97
  larvae, 96, 97
  natural history, 97–99
  range, 97, 97 (map)
  reproduction, 98
  taxonomic note on, 96
Sierra Night Lizard, **307–309,** 308 (map)
Siskiyou Mountains Salamander, **109–110,** 109 (map)
site description, in field notes, 460
skin
  amphibians, 41–44, 42–45 (figs.)
  reptiles, 221–222
  as respiratory organ, 41, 50–51
  snakes, 342
skin coloration, amphibians, 42–44, 43 (fig.), 44 (fig.)
skin hydration, amphibians, 52–54
skin shedding
  amphibians, 44, 45 (fig.)
  reptiles, 222
skin toxins, amphibians, 41–42, 42 (fig.)
skinks, 313–318
  body form, 313
  chemoreception, 313
  Gilbert's Skink, **316–318,** 317 (plates), 318 (map)
  juvenile, 315 (plate)
  northern, 314–318
  reproduction, 313–314
  tail breakage, 313
  Western Skink, **314–316,** 315 (map)

slender salamanders, 124–142
  Black-bellied Slender Salamander, **131,** 131 (map)
  California Slender Salamander, **140–142,** 140 (map), 141 (plate)
  Channel Islands Slender Salamander, **133,** 133 (map)
  defense, 126
  description, 125
  Desert Salamander, **134–135,** 134 (map)
  food, 127
  Gabilan Mountains Slender Salamander, **136,** 136 (map)
  Garden Slender Salamander, **134–135,** 134 (map)
  Gregarious Slender Salamander, **132,** 132 (map)
  habitat, 125–126
  Hell Hollow Slender Salamander, **139–140,** 140 (map)
  Inyo Mountain Salamander, **127–128,** 127 (map)
  Kern Canyon Slender Salamander, **130,** 130 (map)
  Kern Plateau Salamander, **128,** 128 (map)
  Kings River Slender Salamander, **138,** 138 (map)
  Lesser Slender Salamander, **137,** 137 (map)
  natural history, 125–127
  Relictual Slender Salamander, **139,** 139 (map)
  reproduction, 126–127
  San Gabriel Mountains Slender Salamander, **135–136,** 135 (map)
  San Simeon Slender Salamander, **137–138,** 138 (map)
  Santa Lucia Mountains Slender Salamander, **137,** 137 (map)
  Sequoia Slender Salamander, **138,** 138 (map)
  Tehachapi Slender Salamander, **129,** 129 (map)
smell
  amphibians, 44–46, 46 (fig.)
  reptiles, 223–224
  turtles, 425
Smiley, Leonidas W., 191
Smits, Alan, 171
snake(s), 339–421
  basking position, 228
  blind, 350–352
  boas and pythons, 352–355
  body adaptations, 342–346
  captive, 481–483
  capturing, 474
  classification, 339–340
  colubrids, 355–400, 356 (fig.)
  fangs, 345–346
  fear of, 348–350, 349 (plate)
  garter snakes, 379–400
  hearing, 224, 345
  hissing, 344–345
  Jacobson's organ, 344
  key, 340–341
  kingsnakes, 374–379
  locomotion, 342–343, 343 (fig.)
  mite infestations, 481
  olfaction and taste, 223–224
  photography, 465–466
  pit organs, 345
  pit vipers, 401–421, 402 (fig.), 403 (fig.)
  prey, 346–347
  racers, 361–374
  rattlesnakes, 401–421, 402 (fig.), 403 (fig.)
  reproduction, 233–234, 347–348, 348 (plate)
  scent glands, 344
  sea snakes, 400–401
  sensory input, 344–345
  sex determination, 234
  skin, 342
  swallowing by, 346
  tail, 344
  teeth and jaw, 345–346
  trapping, 475–476, 476 (plate)
  vipers, 401–421, 402 (fig.), 403 (fig.)
  vision, 222–223
  whipsnakes, 361–374
  worm snakes, 350–352
snake stick, mechanical, 474
snapping turtle(s), 442–445
softshell turtles, **439–442,**
Sonora semiannulata, **395,** 395 (map), 396 (plate)
Sonoran Desert Toad, **161–163,** 162 (map), 162 (plate)
Sonoran Gopher Snake, **371–373,** 372 (map)
Sonoran Lyre Snake, **398,** 398 (map)
South Coast Range, 3–4 (plate), 6 (map), 17–18
South Desert Horned Lizard, **302–303,** 302 (map)
Southern Alligator Lizard, **326–328**
  behavior, 328
  food, 328
  habitat, 16, 327–328
  identification, 326–327, 327 (plate)
  natural history, 327–328
  range, 327, 327 (map)

reproduction, 328
thermoregulation, 226
Southern Leopard Frog, **203–205,** 204 (map)
Southern Long-toed Salamander, **78–82**
  adult, 78
  aquatic predation pressure on, 25
  eggs, 78
  gills, 49
  habitat, 16, 17, 79
  identification, 78
  larvae, 78, 79 (plate)
  migration, 79
  natural history, 79–80, 79 (plate)
  range, 79, 79 (map)
  reproduction, 62, 79–80
  spawning pools, 80
Southern Mountain Yellow-legged Frog, **199–202**
  adult, 199–200
  aquatic predation pressure on, 25
  breeding, 201
  conservation note, 201–202
  eggs, 200
  food, 201
  habitat, 19, 200–201
  identification, 199–200
  larvae, 200, 201
  natural history, 200–201, 201 (plate)
  protective status, 199
  range, 200, 200 (map)
  thermoregulation, 56
  ultraviolet radiation effect on, 28
  voice, 200
Southern Pacific Rattlesnake, **410–413,** 411 (map)
Southern Rubber Boa, **353–354,** 353 (map), 466
Southern Sagebrush Lizard, **282–284,** 283 (map)
Southern Torrent Salamander, **90–92**
  adult, 90
  conservation note, 92
  eggs, 90, 91
  habitat, 15, 90–91
  identification, 90
  larvae, 90, 91
  natural history, 90–91, 91 (plate)
  protective status, 90
  range, 90, 90 (map)
  reproduction, 91, 91 (plate)
  respiration, 50
Southwestern Black-headed Snake, **397–398,** 397 (map)
Southwestern Pond Turtle, **430–435**
  basking, 432
  conservation note, 433–435
  food, 432
  habitat, 431–432
  hibernation, 432
  identification, 430–431
  natural history, 431–433
  photography, 464–465
  protective status, 430
  range, 431, 431 (map)
  reproduction, 432–433
  thermoregulation, 432
spadefoot toads, 154–161
  body form and locomotion, 40–41, 40 (fig.)
  Couch's Spadefoot, **159–161,** 160 (map)
  desiccation, 155, 155 (plate)
  dormancy, 154
  Great Basin Spadefoot, **158–159,** 159 (map)
  larvae, 155, 155 (plate)
  reproduction, 154–155
  voice, 58
  Western Spadefoot, **156–158,** 157 (map), 157 (plate)
spawning site
  anurans, 57
  aquatic-breeding salamanders, 62
Spea
  couchii, **159–161,** 160 (map)
  hammondii, 18, 57, **156–158,** 157 (map), 157 (plate)
  intermontana, 17, **158–159,** 159 (map)
special accounts, 458–460, 459 (fig.)
species, new, 31–32
Speckled Black Salamander, **119–121,** 120 (map)
Speckled Rattlesnake, **417–418,** 417 (map), 418 (plate)
spectacle, reptiles, 223
Spectacled Caiman, osmoregulation, 231
spermatophore
  aquatic-breeding salamanders, 62
  terrestrial-breeding salamanders, 63
spermatozoa, anurans, 59
Spermophilus beecheyi, burrows, 73, 74, 432
Sphenodon, 218
spiny lizards, 278–288
  body form, 278
  coloration, 278–279
  Desert Spiny Lizard, **284–286,** 285 (map), 285 (plate)
  Granite Spiny Lizard, **286–288,** 286 (map), 287 (plate)
  habitat, 278
  Island Fence Lizard, **279–282,** 280 (map)

spiny lizards (*continued*)
    reproduction, 279
    Sagebrush Lizard, **282–284,** 283 (map)
    Sierra Fence Lizard, **279–282,** 280 (map)
    Southern Sagebrush Lizard, **282–284,** 283 (map)
    Western Fence Lizard, **279–282,** 280 (map), 281 (plate)
Spiny Softshell Turtle, **440–442,** 441 (map)
spiracle, anuran larvae, 50, 50 (fig.)
Spotted Frog, Oregon, **195–196,** 195 (map)
Spotted Leaf-nosed Snake, **360–361,** 361 (map), 361 (plate)
Squamata, 219, 233–234
St. Helena Mountain Kingsnake, **377–378,** 377 (map)
ST (state-threatened species), 30
stapes, amphibians, 48
state-threatened species (ST), 30
stenothermy, reptiles, 225–226, 228
Striped Racer, **366–369,** 367 (map), 367 (plate), 368 (plate)
Striped Whipsnake, 17, **369,** 369 (map)
subspecies, endangered, 30
Swaim, Karen, 228, 367
swallowing, by snakes, 346
"swifts," 278–279
Switak's Banded Gecko, **250–251,** 250 (map)

tadpoles, 59–60, 60 (fig.)
tail
    salamanders, 37–39, 38 (fig.)
    snakes, 344
    turtles, 425
tail breakage, lizards, 239–240, 239 (fig.)
tail rattles, 402–403, 403 (fig.)
tailed frog(s), 150–154
    reproduction of, 59
tail-nudging walk, aquatic-breeding salamanders, 62
tail-straddling walk, terrestrial breeding salamanders, 63, 64 (fig.)
Tantilla
    hobartsmithi, **397–398,** 397 (map)
    planiceps, **396–397,** 397 (map)
Tarentola mauritanica, **253**
Taricha
    granulosa, **100–102,** 101 (map), 101 (plate)
    rivularis, 93 (plate), **102–103,** 103 (map)
    sierrae, **96–100,** 97 (map), 97 (plate)
    skin toxins, 41–42
    torosa, 94 (fig.), 95 (plate), **96–100,** 97 (map)
tarichatoxin, 41
taste, reptiles, 223–224

Taxidea taxus, 428
taxonomy, amphibians, 36–37
Taylor, Mike, 172
teeth
    lizards, 242
    snakes, 345–346
Tehachapi Range, 6 (map), 11 (plate), 20–21
Tehachapi Slender Salamander, 21, **129,** 129 (map)
teiid(s), 318–324
Teiidae, 318–324
telephoto lenses, 462
temperature measurement, in field notes, 460
terrariums, 482
    photo, 467
terrestrial-breeding salamanders, 63–65, 64 (figs.), 65 (plates)
territoriality, lizards, 243–244
Testudines, 219
Testudinidae, 426–430
tetrodotoxin (TTX), newts, 94–95
Texas Blind Snake, 352
Texas Spiny Softshell, 441
Thamnophis, 379–400
    atratus, **387–388,** 388 (map), 388 (plate)
    atratus atratus, **387–388,** 388 (map), 388 (plate)
    atratus hydrophilus, **387–388,** 388 (map)
    atratus zaxanthus, **387–388,** 388 (map)
    couchii, **388–389,** 389 (map)
    cyrtopsis, 393 (plate)
    elegans, 16, **385–387,** 386 (map)
    elegans elegans, **385–387,** 386 (map)
    elegans terrestris, **385–387,** 386 (map), 386 (plate)
    elegans vagrans, **385–387,** 386 (map)
    gigas, 18, **389–390,** 390 (map)
    hammondii, **391,** 391 (map)
    marcianus, **392–393,** 392 (map), 393 (plate)
    ordinoides, **391–392,** 392 (map)
    reproduction, 347
    sirtalis, 380, **381–384,** 382 (map), 382 (plate), 383 (plate)
    sirtalis fitchi, **381–384,** 382 (map)
    sirtalis infernalis, **381–384,** 382 (map)
    sirtalis tetrataenia, 23, **381–384,** 382 (map), 382 (plate), 383 (plate)
thermal acclimatization, amphibians, 55
thermoregulation
    amphibians, 54–56, 54 (fig.)
    reptiles, 225–230, 226 (fig.), 227 (plate)
thigmothermy, reptiles, 225
"third eye," lizards, 240–242, 241 (fig.)

Thomomys bottae, burrows, 73
Tiger Salamander, **75–77**
   adult, 75–76
   disruptive coloration, 43
   eggs, 76
   identification, 75–76
   larvae, 76
   larval cannibalism, 77
   longevity, 77
   natural history, 76–77
   range, 76, 76 (map)
   reproduction, 62
toads, 149–215
   body form and locomotion, 39–41, 40 (fig.)
   classification, 149
   hand capture, 470–471
   key, 150
   photography, 463–464
   "Psychedelic Toad," 163
   spadefoot, 154–161
   true, 161–179
toe disks, Hylidae, 40
toe pads, Hylidae, 40
tongue, snakes, 223–224
tongueless frogs, 212–215
torrent salamanders, 89–92
   Southern Torrent Salamander, **90–92,** 91 (map), 92 (plate)
tortoises, 426–430
   basking position, 228
   key, 423–424
   photography, 464–465
toxins, skin, 41–42, 42 (fig.)
Trachemys
   scripta, **435–436,** 435 (map)
   scripta elegans, 433, 435, 436
transponders, for recognition of individual animals, 479
Transverse Range, 6 (map), 20–21
trap robbers, 476–477
trapping, 474–477, 475 (plate), 476 (plate)
tree lizards, 294–295
tree voles, 123
treefrog, body form and locomotion, 39–40
Trimorphodon
   eyesight, 344
   lamba, **398,** 398 (map)
   lyrophanes, 20, **398,** 398 (map)
Trionychidae, 439–442
trout
   aquatic predation pressure by, 25
   and yellow-legged frogs, 202
true frogs, 186–212
   American Bullfrog, **208–212,** 209 (map), 209 (plate)
   California Red-legged Frog, **186–192,** 188 (map), 188 (plate), 190 (plate)
   Cascades Frog, **193–194,** 194 (map)
   eggs, 186
   Foothill Yellow-legged Frog, **196–199,** 197 (map), 198 (plate)
   identification, 186
   Lowland Leopard Frog, **207,** 207 (map)
   Northern Leopard Frog, **203–205,** 204 (map)
   Northern Red-legged Frog, **186–192,** 188 (map), 188 (plate), 190 (plate)
   Oregon Spotted Frog, **195–196,** 195 (map)
   reproduction, 186
   Rio Grande Leopard Frog, **206,** 206 (map)
   Sierra Nevada Yellow-legged Frog, **199–202,** 200 (map), 201 (plate)
   Southern Leopard Frog, **203–205,** 204 (map)
   Southern Mountain Yellow-legged Frog, **199–202,** 200 (map)
true toads, 161–179
   Arroyo Toad, **168–169,** 169 (map)
   Black Toad, **174–175,** 174 (map)
   Boreal Toad, **170–173,** 171 (map)
   breeding, 161
   California Toad, **170–173,** 171 (map)
   defense, 161
   Great Plains Toad, **164–165,** 164 (map), 165 (plate)
   Red-spotted Toad, **166–167,** 167 (map), 167 (plate)
   Sonoran Desert Toad, **161–163,** 162 (map), 162 (plate)
   tadpoles, 161
   Western Toad, **170–173,** 171 (map), 171 (plate), 173 (plate)
   Woodhouse's Toad, **178–179,** 179 (map)
   Yosemite Toad, **176–170,** 177 (map), 177 (plate)
TTX (tetrodotoxin), newts, 94–95
tuataras, 218
turtles, 423–455
   basking position, 228
   classification, 423
   hearing, 224–225, 425
   key, 423–424
   leatherback sea turtle, 454–455
   limbs, 425
   olfaction, 425
   photography, 464–465
   pond turtles, 430–439
   reproduction, 425

turtles (*continued*)
   respiration, 221
   Salmonella in, 435
   sea turtles, 445–455
   shell, 424
   snapping turtles, 442–445
   softshell turtles, 439–442
   tail, 425
   vision, 425
Twain, Mark, 191
Twitty, Victor, 93
Two-striped Garter Snake, **391,** 391 (map)
tympanic membrane
   amphibians, 47–48
   reptiles, 224–225

ultraviolet (UV) radiation, habitat loss due to, 28–29
Uma, 272–278
   as habitat specialists, 6
   inornata, 23, **274–276,** 275 (map)
   notata, **276–277,** 276 (map)
   scoparia, 272 (plate), **277–278,** 278 (map)
urea, in osmoregulation, 53
Urosaurus, 290–295
   graciosus, **290–292,** 291 (map), 291 (plate)
   nigricaudus, **292–294,** 293 (map), 293 (plate)
Uta stansburiana, **288–290**
   breeding, 290
   habitat, 17, 18, 22, 289
   hearing, 225
   identification, 288, 289 (plate)
   natural history, 289–290
   range, 288–289, 288 (map)
   territoriality, 289–290
UV (ultraviolet) radiation, habitat loss due to, 28–29

Valencia Lagoon, 80
Valley Garter Snake, **381–384,** 382 (map)
venomous lizards, 334–337
ventral skin coloration, amphibians, 43–44, 44 (fig.)
vertebrate terrestrial pioneers, amphibians and reptiles as, 2–4, 3 (fig.)
videography, 461–468
   captive or temporarily restrained specimens, 466–468
   frogs and toads, 463–464
   lizards, 465
   natural, 462–466
   salamanders, 462–463
   snakes, 465–466
   turtles and tortoises, 464–465

viper(s), 401–421
   body form, 401–402, 402 (fig.), 403 (fig.)
   conservation note, 409–410
   defensive strategies, 404–405
   Great Basin Rattlesnake, **410–413,** 411 (map)
   head anatomy, 402, 402 (fig.)
   and humans, 405–407
   Mojave Rattlesnake, **415–417,** 416 (map)
   Northern Pacific Rattlesnake, **410–413,** 411 (map), 412 (plate)
   Panamint Rattlesnake, **417–418,** 417 (map)
   prey, 403–404
   Red Diamond Rattlesnake, **418–419,** 418 (map)
   Sidewinder, **419–421,** 420 (map), 420 (plate), 421 (plates)
   Southern Pacific Rattlesnake, **410–413,** 411 (map)
   Speckled Rattlesnake, **417–418,** 417 (map), 418 (plate)
   tail rattles, 402–403, 403 (fig.)
   treatment of bites, 407–409
   Western Diamond-backed Rattlesnake, 406, **413–415,** 414 (map), 414 (plate), 415 (plate)
   Western Rattlesnake, **410–413,** 411 (map), 412 (plate)
Viperidae, 401–421, 402 (fig.), 403 (fig.)
vision
   amphibians, 46–47
   reptiles, 222–223
   snakes, 344
   turtles, 425
viviparous reproductive mode, reptiles, 219–221
vocal sac, anurans, 57, 58 (fig.)
voice, anurans, 57–59, 58 (fig.)
Voight, Bill, 428
vomeronasal organs, amphibians, 45
Vonarb, Richard, 190, 192

Wake, David, 24, 116
walking catfish, as vertebrate terrestrial pioneer, 3–4
Wandering Garter Snake, **385–387,** 386 (map)
Wandering Salamander, **122–124,** 123 (map)
   habitat, 15
water boatmen, 72
web-toed salamanders, 142–148
   as habitat specialists, 6
   Limestone Salamander, 142 (plate), **144–146,** 145 (map)

Mount Lyell Salamander, **143–144,** 143 (map)
Shasta Salamander, **146–148,** 146 (map)
Western Banded Gecko, **247–249**
  habitat, 21, 248–249
  identification, 247, 248 (plate)
  natural history, 248–249
  range, 248, 248 (map)
  reproduction, 249
Western Blind Snake, 350, **351–352,** 351 (map), 351 (plate)
Western Diamond-backed Rattlesnake, 406, **413–415,** 414 (map), 414 (plate), 415 (plate)
Western Fence Lizard, **279–282**
  basking behavior, 281
  breeding, 282
  displacement behavior, 282
  display behavior, 281–282
  food, 282
  habitat, 15, 17, 18, 19
  as habitat generalist, 5, 280
  identification, 279–280, 281 (plate)
  natural history, 280–282, 281 (plate)
  parietal eye, 241–242
  photography, 465
  range, 280, 280 (map)
  thermoregulation, 226, 226 (fig.), 227, 227 (plate)
Western Ground Snake, **395,** 395 (map), 396 (plate)
Western Leaf-nosed Snake, **360–361,** 361 (map), 361 (plate)
Western Long-nosed Snake, **378,** 378 (map), 379 (plate)
  habitat, 18, 20
Western Painted Turtle, **437–439**
  food, 438
  habitat, 437–438
  hibernation, 438
  identification, 437, 438 (plate)
  natural history, 437–439
  range, 437, 437 (map)
  reproduction, 438–439
  and Western Pond Turtle, 433
Western Patch-nosed Snake, 20, **370–371,** 370 (map)
Western Pond Turtle, **430–435**
  basking, 432
  conservation note, 433–435
  food, 432
  habitat, 431–432
  hibernation, 432
  identification, 430–431, 431 (plate)
  natural history, 431–433
  photography, 464–465
  protective status, 430
  range, 431, 431 (map)
  reproduction, 432–433
  thermoregulation, 432
Western Rattlesnake, **410–413**
  conservation note, 412–413
  habitat, 16, 411–412
  identification, 411, 412 (plate)
  natural history, 411–412
  prey, 412
  range, 411, 411 (map)
  taxonomic note, 410–411
Western Shovel-nosed Snake, **393–394,** 394 (map), 394 (plate)
  habitat, 21
Western Skink, **314–316**
  breeding, 316
  habitat, 15, 316
  identification, 315
  natural history, 316
  range, 315, 315 (map)
Western Spadefoot, **156–158**
  adults, 156, 157 (plate)
  conservation note, 157–158
  eggs, 156
  habitat, 18, 20, 22, 157
  identification, 156, 157 (plate)
  larvae, 156
  metamorphosis, 61, 61 (fig.)
  natural history, 157
  protective status, 156
  range, 157, 157 (map)
  spawning site, 57
  voice, 156
Western Terrestrial Garter Snake, **385–387**
  habitat, 16, 19, 386
  identification, 385, 386 (plate)
  natural history, 386–387
  prey, 386–387
  range, 386, 386 (map)
  reproduction, 387
Western Toad, **170–173**
  adults, 170, 173 (plate)
  breeding, 172–173
  chemical contamination of aquatic habitat, 26
  chytrid fungus infection, 27
  eggs, 170, 172
  habitat, 16, 17, 171
  identification, 170
  larvae, 170
  metamorphosis, 172
  natural history, 171–173, 171 (plate), 173 (plate)

Western Toad (*continued*)
  range, 171, 171 (map)
  thermal preference, 171–172
  voice, 170
Western Whiptail, **320–322**
  behavior, 321–322
  food, 322
  habitat, 18, 22, 321
  identification, 320–321, 321 (plate)
  natural history, 321–322
  range, 321, 321 (map)
  reproduction, 322
Western Yellow-bellied Racer, 362–363, 363 (plate), 466
whipsnakes, 362–374
  Coachwhip, **363–366**, 365 (map), 365 (plate)
  juvenile, 369 (plate)
  Striped Whipsnake, **369**, 369 (map)
whiptails, 319–324
  body form, 319
  breeding, 320
  California Whiptail, **320–322**, 321 (map)
  Coastal Whiptail, **320–322**, 321 (map)
  food, 242
  foraging, 319–320
  Great Basin Whiptail, **320–322**, 321 (map)
  hearing, 225
  Orange-throated Whiptail, **323–324**, 323 (map), 324 (plate)
  thermoregulation, 229, 319
  tracks and tail drag, 470 (plate)
  Western Whiptail, **320–322**, 321 (map), 321 (plate)
Wiggins' Night Lizard, **307–309**, 308 (map)
Wilhoft, Dan, 240
Woodhouse's Toad, **178–179**, 179 (map)
woodland salamanders, 106–110
  Del Norte Salamander, **108–109**, 108 (map)
  Dunn's salamander, **107**, 107 (map)
  Scott Bar Salamander, **109–110**, 109 (map)
  Siskiyou Mountains Salamander, **109–110**, 109 (map)
worm lizards, 233
worm snakes, 350–352

Xantusia
  gracilis, **310–311,** 311 (map)
  henshawi, 20, **310–311,** 311 (map)
  reproduction, 245
  riversiana, **312–313,** 312 (map)
  vigilis, **307–309,** 308 (map), 309 (plate)
  vigilis sierrae, **307–309,** 308 (map)
  vigilis vigilis, **307–309,** 308 (map)
  wigginsi, **307–309,** 308 (map)
Xenopus laevis, **212–215**
  adult, 213, 214 (plate)
  breeding, 214
  chytrid fungus infection, 27–28, 215
  conservation note, 214–215
  eggs, 213
  food, 214
  habitat, 214
  identification, 213, 213 (fig.), 214 (plate)
  as lab frog, 214–215
  larvae, 213, 213 (fig.), 214
  natural history, 214
  range, 213, 213 (map)
  voice, 2.3

Yellow-bellied Sea Snake, 400, **401**
Yellow-blotched Ensatina, **110–115**
  adults, 112–113
  conservation note, 115
  defense, 114–115
  eggs, 113, 115
  habitats, 114
  identification, 112–113
  natural history, 114–115
  protective status, 112
  range, 113–114, 113 (map)
  reproduction, 115
  taxonomic note, 112
Yellow-eyed Ensatina, **110–115**
  adults, 112–113
  conservation note, 115
  defense, 114–115, 115 (plate)
  eggs, 113, 115
  habitats, 114
  identification, 112–113
  natural history, 114–115
  range, 113–114, 113 (map)
  reproduction, 115
  taxonomic note, 112
Yellow-legged Frog
  Foothill, **196–199**, 197 (map), 198 (plate)
  Sierra Nevada, **199–202**, 200 (map), 201 (plate)
  Southern Mountain, **199–202**, 200 (map)
Yosemite Toad, **176–178**
  adults, 176
  breeding, 178
  chemical contamination of aquatic habitat, 26
  conservation note, 178
  eggs, 176
  habitat, 19, 177–178
  identification, 176–177

larvae, 176–177
natural history, 177–178, 177 (plate)
protective status, 176
range, 177, 177 (map)
ultraviolet radiation effect on, 28
voice, 176
Yucca Night Lizard, **307–309,** 308 (map)

Zebra-tailed Lizard, **270–272**
breeding, 272
coloration, 272
food, 272
habitat, 22, 271–272
identification, 270–271, 271 (plate)
limbs, 238
natural history, 271–272
range, 271, 271 (map)
zoom telephoto lenses, 462

ABOUT THE AUTHORS

R. C. Stebbins and S. M. McGinnis attempting to identify the former owner of a shed snake skin.

Robert C. Stebbins is professor emeritus of zoology at the University of California, Berkeley and a curator emeritus of the University's Museum of Vertebrate Zoology. He is the author of over a dozen books plus numerous scientific papers and the producer of two documentaries. He is also the foremost illustrator of North American amphibians and reptiles, and his paintings and prints are prized by collectors. In addition to his work in herpetology and vertebrate natural history, he has devoted much of his life to promoting "ecological literacy" in both young people and adults through seminars, workshops, and publications. His latest work, *Connecting with Nature: A Naturalist's Perspective* (2009) summarizes his work in this important area.

Samuel M. McGinnis is professor emeritus of biology at California State University, East Bay. After receiving a degree in zoology at the University of Wisconsin, Madison, he migrated to California to work with Robert C. Stebbins on his Ph.D. degree at the University of California, Berkeley. While there he pioneered the use of miniature radio transmitters to study temperature regulation and behavior in a variety of California lizard, snake, and turtle species in their natural habitats. In addition to his university teaching, he has served as a consultant on endangered and threatened amphibian and reptile species for a variety of state and federal agencies. Besides numerous scientific papers and technical reports, he recently published another book in the University of California Press field guide series, *Freshwater Fishes of California* (2006).

Each author brings to this book over five decades of enlightening university students about California's amphibians and reptiles, and much of the information they have imparted during their teaching tenures is now made available to the readers of this book, their newest group of students.